Original Title:
LOGICAL MATHEMATICAL REASONING
Unlocking Critical Thinking and Problem-Solving Skills Across Disciplines

Author:
Luque Zevallos Helbert Justo

Independently published
Available at https://www.amazon.com

Total or partial reproduction of this work is prohibited.

ALL RIGHTS RESERVED

Language: English

2024

First Edition

ISBN: 9798303552661

9 798303 552661

1. Presentation

The present book, titled **Logical Mathematical Reasoning**, has been conceived with the objective of providing a solid and practical tool for developing logical and analytical thinking skills. Logic and mathematics are fundamental pillars in the learning process and in problem-solving, both in academic contexts and in everyday life. This material aims to accompany students in their education, offering them a clear and organized structure of concepts, examples, exercises, and applications that facilitate their understanding and application.

The content has been carefully structured to cover topics ranging from the basic concepts of logical reasoning to advanced applications that develop inference and deduction abilities. Each chapter is designed to introduce concepts progressively, allowing students to strengthen their critical analysis skills and solve problems with rigor and precision. The thematic organization enables each reader to approach the topics autonomously, with a practical and progressive focus that fosters active and reflective learning.

This book offers solved and proposed exercises to practice mathematical reasoning. Each section includes explanations, examples, and self-assessment activities to facilitate the understanding and application of concepts. The selected exercises challenge the reader and promote the development of problem-solving skills, strengthening their logical and mathematical abilities.

Additionally, sections are dedicated to reflecting on the importance of logical reasoning in various fields of knowledge and the professional domain. This interdisciplinary approach allows the reader to understand how mathematical skills are applicable and essential in diverse areas, from the exact sciences and engineering to social sciences and business decision-making.

This book invites readers to discover the power of logical mathematical thinking, to recognize its utility in problem-solving, and to strengthen key skills to face the challenges of a constantly changing world. We hope this work serves as a useful guide for students, educators, and professionals interested in delving deeper into logical and mathematical reasoning.

MBA. Helbert Justo Luque Zevallos

Índice general

2. Summary

The course **Logical Mathematical Reasoning** is designed to strengthen and develop essential skills in analytical and logical thinking, crucial in a world where the ability to solve complex problems is increasingly valued. Through its four units, this course covers the fundamentals of logic, arithmetic, algebra, and geometry, providing students with a solid foundation in mathematics and reasoning. Below is a detailed description of each unit, highlighting its relevance and objectives within the course context.

Unit 1: Logic and Sets

Logic is the foundation of mathematical reasoning, enabling students to structure thoughts coherently and precisely. In this first unit, essential concepts of propositional logic are addressed, including propositions, logical connectives, inference, and the principles of inductive and deductive reasoning. Through the study of propositional functions and set algebra, students learn to organize and analyze information in a structured way.

This unit also introduces practical applications, such as truth and lie problems, logical sequences, orderings, and analogies. These exercises not only reinforce theoretical concepts but also foster the development of skills for analyzing and solving complex problems. Logic and sets are, ultimately, the basis upon which mathematical reasoning is built, proving useful in scientific disciplines and everyday life.

Unit 2: Arithmetic Reasoning

Arithmetic is a fundamental tool in developing mathematical skills. This unit delves into topics ranging from numeral systems and basic operations to more complex concepts such as divisibility, fractional numbers, ratios, proportions, and percentage calculations. The rule of three, arithmetic and geometric sequences and series are also covered.

Understanding and applying these arithmetic concepts are essential for solving everyday and professional problems. The unit includes examples and problems that demonstrate the applicability of arithmetic in contexts such as finance, management, and other fields requiring precise numerical handling. In this way, arithmetic reasoning becomes a key skill for making informed decisions.

Unit 3: Algebraic Reasoning

Algebra enables the abstraction and simplification of problems, facilitating their resolution through equations and relationships between variables. In this unit, students learn to model situations using linear and quadratic equations, solve inequalities, and manage intervals and mathematical operators. The ability to express problems algebraically is fundamental in many areas, from physics to economics, where equations help model phenomena and predict outcomes. The exercises proposed in this unit encourage the practice of mathematical modeling skills, helping students understand how to use algebra to formulate and solve problems efficiently.

Unit 4: Plane and Spatial Geometric Reasoning

Geometric reasoning is essential for understanding and analyzing the physical world around us. This final unit explores concepts of plane and spatial geometry, including proportionality and similarity in geometric figures, metric relationships in triangles, and the calculation of perimeters, areas, and volumes of solids. Additionally, the spatial positioning of points, lines, and planes is addressed, skills that are useful in disciplines such as engineering and architecture.
Through practical problems and exercises, this unit helps students develop spatial visualization skills and apply geometric principles in analyzing shapes and structures. Geometry, due to its direct relationship with physical space, provides an intuitive understanding that reinforces visualization and planning skills, facilitating problem-solving in practical and technical contexts.

Course Integration and Application

The course **Logical Mathematical Reasoning** not only provides mathematical tools but also promotes the development of an analytical and structured mindset. Each unit connects with others and related disciplines, allowing students to apply their knowledge in diverse situations. This integrated approach prepares students to face academic and professional challenges, providing them with a solid foundation for continuing their learning in fields related to logic, mathematics, and data analysis.
In conclusion, this course offers a fundamental foundation in mathematics and reasoning, building competencies that will be highly valuable in any field requiring effective problem-solving and deep analytical skills. We encourage students to make the most of each unit, aiming to strengthen their ability to think logically, critically, and creatively.

MBA. Helbert Justo Luque Zevallos

3. Introduction

The content of this course on **Logical Mathematical Reasoning** is organized into four units, each focused on developing specific and general skills in logic, arithmetic, algebra, and geometry, integrating both theory and practical applications. Below is a summary of each unit, along with comments on their importance in the development of critical thinking and problem-solving skills.

Unit 1: Logic and Sets

Main Topics:

- Logical propositions and logical connectives
- Propositional algebra and logical inference
- Inductive and deductive reasoning
- Propositional functions, sets, and set algebra

Applications:

- Solving truth and lie problems
- Information ordering and temporal relationships
- Series, graphical analogies, and set problems

This unit introduces the fundamental concepts of logical thinking, essential for structuring reasoning and analyzing situations coherently. Understanding logic and sets is key to developing valid arguments and solving complex problems.

Unit 2: Arithmetic Reasoning

Main Topics:

- Numeral systems and arithmetic operations
- Divisibility, GCD, and LCM
- Fractional numbers, ratios, proportions, and the rule of three
- Percentage calculations, arithmetic, and geometric sequences and series

This unit reinforces arithmetic concepts and their applications in everyday and professional problems. Arithmetic skills are the foundation for numerical decision-making, useful in contexts such as economics, administration, and engineering.

Unit 3: Algebraic Reasoning

Main Topics:

- Mathematical modeling using linear and quadratic equations
- Solving linear and quadratic inequalities
- Intervals and mathematical operators

This unit delves into the use of algebra as a tool for representing and solving problems in various areas. Algebraic reasoning simplifies and manipulates numerical relationships and is fundamental for any discipline requiring mathematical modeling.

Unit 4: Plane and Spatial Geometric Reasoning

Main Topics:

- Proportionality and similarity in geometric figures
- Metric relationships in triangles
- Calculation of perimeters, areas, and volumes of geometric solids
- Spatial positioning of points, lines, and planes

This unit explores geometric concepts and their application in visualizing and analyzing spatial structures. Geometric reasoning is essential in disciplines such as architecture, physics, and engineering, where understanding shapes and spaces is fundamental.

This course reinforces general and specific competencies while establishing connections with subjects such as **University Work Methodology** and topics on **National Reality**, promoting an integral approach to the development of logical and mathematical thinking.

Integration with other academic areas allows students to develop a holistic vision, applying logical and mathematical reasoning in different contexts and preparing to face interdisciplinary situations.

MBA. Helbert Justo Luque Zevallos

Logic and Sets

1. Logical Propositions and Logical Connectives

1.1 Types of Propositions: Simple and Compound

1.1.1 Declarative Propositions

Definition 1.1.1 A **declarative proposition** is a sentence or statement that expresses an affirmation or negation which can be classified as true or false, but not both at the same time. Declarative propositions are fundamental in logic as they provide the basic structure for analyzing arguments and logical reasoning. It is important for declarative propositions to be precise and clear to avoid ambiguities, thus enabling their truth value to be determined.

■ **Example 1.1** Let us consider some examples of declarative propositions:

1. **"Water boils at 100 degrees Celsius at sea level."**

This statement can be evaluated as true under standard conditions, making it a declarative proposition. It is a verifiable fact in specific scientific contexts.

2. **"The Amazon is the longest river in the world."**

This proposition can also be evaluated as true or false. Although there is debate over whether the Amazon or the Nile is the longest river, criteria can be investigated and established to determine its truth value.

3. **"2 + 2 = 4"**

In mathematics, this statement is true and universally accepted within the system of real numbers, making it a declarative proposition.

4. **.All swans are white."**

While this proposition was considered true for centuries in some regions, its falsity was demonstrated upon discovering black swans in Australia. This example illustrates how a declarative proposition's truth value can change with new discoveries or additional information. ■

R It is important to note that not all statements in natural language are declarative propositions. For a statement to be a declarative proposition, it must be evaluable in terms of truth or falsity. Below are some examples of statements that are not declarative propositions:

■ **Example 1.2** 1. **"What a beautiful day!"**
This statement is an exclamation and cannot be classified as true or false, as it expresses an opinion or feeling.
2. **.^re you ready to leave?"**
This is a question and cannot be evaluated in terms of truth or falsity. Questions, by their nature, seek information and do not assert anything.
3. **Çlose the door."**
This statement is a command and is not a declarative proposition either, as its purpose is to induce an action rather than express something that can be classified as true or false. ■

R In logic, precision in language is crucial. Some sentences may appear to be declarative propositions, but their ambiguity prevents determining their truth value. For example:

- **"The player is the best."**

This statement may depend on the context, specific criteria, or personal interpretations, making it difficult to establish whether it is true or false without additional information.

- **Ït is a good day."**

Although it may seem like a proposition, its truth value is subjective as it depends on the interpretation of "good day."This illustrates the need for precision for a proposition to be logically evaluable.

R Another important characteristic of declarative propositions is that they can be compound, meaning they are formed by several simple propositions joined by logical connectives. An example of a compound proposition:

- **"The sun is a star and the Earth orbits the Sun."**

This proposition includes two statements that can be individually evaluated as true or false, connected by the conjunction .ªnd.Çompound propositions are useful for forming more complex logical arguments.

■ **Example 1.3** A final example to illustrate the importance of declarative propositions:
- **Ïf it rains, then the street will be wet."**
This is a conditional proposition that can be classified as true or false, and is useful for establishing causality or logical implication in more advanced reasoning. ■

Exercise 1.1 Determine whether the following statements are declarative propositions. Justify your answer in each case.
1. **"The Pacific Ocean is the largest ocean in the world."**
2. **Çould you close the window?"**
3. **"5 is a prime number."**
4. **.^ll mammals can fly."**
5. **Ïncredible what is happening!"**
For each statement, explain whether it can be classified as true or false, and if not, describe why it does not meet the criteria of a declarative proposition. ■

R Declarative propositions, in summary, are the foundation of logic, as they allow for the establishment of a formal and evaluable structure of arguments. Propositions must be clear, precise, and capable of being classified as true or false for their analysis in propositional logic.

1.1.2 Conditional Propositions

Definition 1.1.2 A **conditional proposition** is a type of compound proposition that has the form **Ïf P, then Q"**, where **P** is called the **hypothesis** (or antecedent) and **Q** is called the **conclusion** (or consequent). This proposition is denoted as $P \to Q$ and is true in all cases except when **P** is true and **Q** is false. Conditional propositions are used to express relationships of dependence or implication between two propositions.

▪ **Example 1.4** Let us consider some examples of conditional propositions to illustrate this concept:
1. **Ïf you study, then you will pass the exam."**
In this case, **P** is zou study.ªnd **Q** is zou will pass the exam."The conditional proposition suggests that passing the exam depends on studying.
2. **Ïf it rains, then the street will be wet."**
Here, the hypothesis **P** is ït rains.ªnd the conclusion **Q** is "the street will be wet."This proposition implies a causal relationship where the state of the street depends on the occurrence of rain.
3. **Ïf 2 is an even number, then 2 + 2 = 4."**
In this case, **P** is "2 is an even number.ªnd **Q** is "2 + 2 = 4."The conditional proposition is true because when **P** is true, **Q** is also true.

■

(R) It is important to note that a conditional proposition does not assert that the hypothesis is true; it only states that if the hypothesis is true, then the conclusion must also be true. Therefore, a conditional proposition $P \to Q$ is true in all cases except when P is true and Q is false. In this context, conditional propositions are useful in logical arguments and deductive reasoning.

Exercise 1.2 Verify the truth of the following conditional propositions:
1. **Ïf a number is divisible by 4, then it is divisible by 2."**
Analyze whether this proposition is true or false when applied to different numbers.
2. **Ïf the number 7 is even, then 7 + 1 = 8."**
Determine whether this conditional proposition is true or false. Consider the fact that the hypothesis may be false, and yet the conditional proposition can still be true. ■

(R) A conditional proposition can be rewritten in several equivalent forms without changing its logical meaning. For example, the proposition **Ïf P, then Q"** can be expressed as:

- "P implies Q. "Q is a consequence of P. "Q if P."

These alternative forms can help express the same concept in different ways and allow for a better understanding of the logical relationship between the hypothesis and the conclusion.

▪ **Example 1.5** Another example of a conditional proposition and its evaluation in terms of truth is as follows:
- **Ïf 3 is an even number, then 3 + 1 = 4."**
In this proposition, the hypothesis **P** is "3 is an even number"(which is false), and the conclusion **Q** is "3 + 1 = 4"(which is true). Even though **P** is false, the entire conditional proposition is considered true, as it would only be false if **P** were true and **Q** were false. ■

(R) In logic, the **contrapositive** of a conditional proposition can also be analyzed. Given a conditional proposition $P \to Q$, its contrapositive is $\neg Q \to \neg P$. It is interesting to note that a conditional proposition and its contrapositive always have the same truth value. For example:

- Conditional proposition: **Ïf a number is divisible by 4, then it is divisible by 2."** - Contrapositive: **Ïf a number is not divisible by 2, then it is not divisible by 4."**

Both propositions are true and share the same truth value.

1.2 Logical Connectives: Truth Tables for Connectives (AND, OR, NOT, IF-THEN)

1.2.1 Binary Operators

Definition 1.2.1 A **binary operator** is a logical connective applied to two propositions to form a new compound proposition. The main binary operators in propositional logic are **conjunction** (AND), **disjunction** (OR), and the **conditional** (IF-THEN). Each binary operator has a truth table that shows all possible truth values of the propositions and the resulting truth value of the compound proposition.

■ **Example 1.6** Below are examples of the most common binary operators, along with their truth tables to illustrate their functioning:
1. **Conjunction (AND)**: The conjunction of two propositions **P** and **Q**, denoted as $P \wedge Q$, is true only when both propositions are true. Otherwise, it is false.

P	Q	P ∧ Q
T	T	T
T	F	F
F	T	F
F	F	F

Cuadro 1.2.1: *Truth Table for Conjunction (AND)*

2. **Disjunction (OR)**: The disjunction of two propositions **P** and **Q**, denoted as $P \vee Q$, is true when at least one of the propositions is true. It is false only when both propositions are false.

P	Q	P ∨ Q
T	T	T
T	F	T
F	T	T
F	F	F

Cuadro 1.2.2: *Truth Table for Disjunction (OR)*

■

(R) It is important to remember that the truth value of a compound proposition formed with binary operators depends on the truth values of the individual propositions. Conjunction requires both propositions to be true, while disjunction is satisfied if at least one is true. The conditional fails only when the antecedent is true and the consequent is false.

Exercise 1.3 Evaluate the following compound propositions using the binary operators and their truth tables:
1. Given P : "The Earth is round"(T) and Q : "The Moon is a planet"(F), evaluate $P \wedge Q$ and $P \vee Q$.
2. Given R : "5 is an even number"(F) and S : "10 is divisible by 2"(T), evaluate $R \rightarrow S$ and $S \rightarrow R$.

For each exercise, refer to the corresponding truth tables and determine whether the compound propositions are true or false. ■

R Binary operators are essential in propositional logic for constructing compound propositions and evaluating arguments. By understanding their truth tables, it is possible to analyze the truth of complex propositions based on combinations of simpler propositions.

1.2.2 Conditionals and Biconditionals

Definition 1.2.2 A **conditional** is a compound proposition denoted as $P \to Q$ and read as Ïf **P**, then **Q**."The proposition **P** is called the **hypothesis** or **antecedent**, and the proposition **Q** is called the **conclusion** or **consequent**. The conditional is false only when **P** is true and **Q** is false; in all other cases, it is true.
A **biconditional**, denoted as $P \leftrightarrow Q$, is a compound proposition read as "**P** if and only if **Q**."The biconditional is true only when both propositions, **P** and **Q**, have the same truth value (both true or both false).

■ **Example 1.7** Below are examples of conditional and biconditional propositions, along with their truth tables to understand their functioning.
1. **Conditional (IF-THEN)**: The conditional of two propositions **P** and **Q**, denoted as $P \to Q$, is false only when **P** is true and **Q** is false.

P	**Q**	**P → Q**
T	T	T
T	F	F
F	T	T
F	F	T

Cuadro 1.2.3: *Truth Table for Conditional (IF-THEN)*

2. **Biconditional (IF AND ONLY IF)**: The biconditional of two propositions **P** and **Q**, denoted as $P \leftrightarrow Q$, is true only when both propositions have the same truth value.

P	**Q**	**P ↔ Q**
T	T	T
T	F	F
F	T	F
F	F	T

Cuadro 1.2.4: *Truth Table for Biconditional (IF AND ONLY IF)*

■

R It is important to note that in logic, the conditional and biconditional are distinct operators: while the conditional only requires the conclusion to logically follow the hypothesis, the biconditional establishes an equivalence relationship. In other words, the biconditional $P \leftrightarrow Q$ indicates that **P** and **Q** are true or false together, while the conditional $P \to Q$ focuses on implication.

Exercise 1.4 Evaluate the following propositions using conditionals and biconditionals:
1. Given P : "The number 4 is even"(T) and Q : "The number 4 is divisible by 2"(T), evaluate $P \to Q$ and $P \leftrightarrow Q$.
2. Given R : "5 is an even number"(F) and S : "7 is a prime number"(T), evaluate $R \to S$ and $R \leftrightarrow S$.
For each proposition, refer to the corresponding truth tables and determine whether they are true or false. ■

(R) The use of conditionals and biconditionals is fundamental for constructing logical arguments. The conditional is useful in inferences where one proposition depends on another, while the biconditional represents equivalence relationships, indicating that both propositions are true or false together.

1.3 Sixteen Fundamental Cases in Logic

- **Case 1: Tautology**

A	B	c.1
T	T	T
T	F	T
F	T	T
F	F	T

Cuadro 1.3.1: *Truth Table of Tautology*

In this case, the function is always true, regardless of the values of A and B. This function represents a fixed connection.

- **Case 2: Logical Disjunction**

$A \vee B$

A	B	c.2
T	T	T
T	F	T
F	T	T
F	F	F

Cuadro 1.3.2: *Truth Table of Logical Disjunction*

In this case, the function is true if at least one of the variables A or B is true.

- **Case 3: Opposite Implication**

$A \vee \neg B$

This case is true when A is true or when both variables are false.

- **Case 4: Logical Affirmation (Dependent only on A)**

A	B	c.3
T	T	T
T	F	T
F	T	F
F	F	T

Cuadro 1.3.3: *Truth Table of Opposite Implication*

A	B	c.4
T	T	T
T	F	T
F	T	F
F	F	F

Cuadro 1.3.4: *Truth Table of Logical Affirmation*

$$A$$

The function is true if A is true; the value of B does not affect the result.

- **Case 5: Material Conditional**

$$\neg A \vee B = A \Rightarrow B$$

A	B	c.5
T	T	T
T	F	F
F	T	T
F	F	T

Cuadro 1.3.5: *Truth Table of Material Conditional*

It is true when A is false or B is true.

- **Case 6: Logical Affirmation (Dependent only on** B**)**

$$B$$

A	B	c.6
T	T	T
T	F	F
F	T	T
F	F	F

Cuadro 1.3.6: *Truth Table of Logical Affirmation (Dependent only on B)*

The function is true if B is true; the value of A does not affect the result.

- **Case 7: Biconditional**

A	B	c.7
T	T	T
T	F	F
F	T	F
F	F	T

Cuadro 1.3.7: *Truth Table of Biconditional*

$$(A \land B) \lor (\neg A \land \neg B) = A \Leftrightarrow B$$

It is true only when A and B are both true or both false.

- **Case 8: Logical Conjunction**

$$A \land B$$

A	B	c.8
T	T	T
T	F	F
F	T	F
F	F	F

Cuadro 1.3.8: *Truth Table of Logical Conjunction*

The function is true only when A and B are both true.

- **Case 9: Opposite Conjunction**

$$\neg A \lor \neg B$$

A	B	c.9
T	T	F
T	F	T
F	T	T
F	F	T

Cuadro 1.3.9: *Truth Table of Opposite Conjunction*

In this case, the result is false only if A and B are both true.

- **Case 10: Exclusive Disjunction (XOR)**

$$(A \land \neg B) \lor (\neg A \land B)$$

The function is true only when A and B are different.

- **Case 11: Logical Negation of B**

$$\neg B$$

In this case, the value is the opposite of B, regardless of A.

A	B	c.10
T	T	F
T	F	T
F	T	T
F	F	F

Cuadro 1.3.10: *Truth Table of Exclusive Disjunction (XOR)*

A	B	c.11
T	T	F
T	F	T
F	T	F
F	F	T

Cuadro 1.3.11: *Truth Table of Logical Negation of B*

- **Case 12: Logical Adjunction**

$A \land \neg B$

A	B	c.12
T	T	F
T	F	T
F	T	F
F	F	F

Cuadro 1.3.12: *Truth Table of Logical Adjunction*

The function is true only if A is true and B is false.

- **Case 13: Logical Negation of** A

$\neg A$

A	B	c.13
T	T	F
T	F	F
F	T	T
F	F	T

Cuadro 1.3.13: *Truth Table of Logical Negation of A*

In this case, the value is the opposite of A, regardless of B.

- **Case 14: Opposite Adjunction**

$\neg A \land B$

The function is true only if A is false and B is true.

A	B	c.14
T	T	F
T	F	F
F	T	T
F	F	F

Cuadro 1.3.14: *Truth Table of Opposite Adjunction*

- **Case 15: Opposite Disjunction**

$$\neg A \wedge \neg B$$

A	B	c.15
T	T	F
T	F	F
F	T	F
F	F	T

Cuadro 1.3.15: *Truth Table of Opposite Disjunction*

In this case, the function is true only if A and B are both false.

- **Case 16: Contradiction**

$$F$$

A	B	c.16
T	T	F
T	F	F
F	T	F
F	F	F

Cuadro 1.3.16: *Truth Table of Contradiction*

In the final case, the result is always false, regardless of the values of A and B.

1.4 Negation of Propositions: Application of De Morgan's Laws

1.4.1 De Morgan's Law for the Negation of Conjunctions.

Definition 1.4.1 De Morgan's Law for the Negation of Conjunctions states that the negation of a conjunction of two propositions is equivalent to the disjunction of the negations of those propositions. Formally, for two propositions P and Q, this law is expressed as:

$$\neg(P \wedge Q) \equiv (\neg P) \vee (\neg Q)$$

This means that to negate the compound proposition $P \wedge Q$, it suffices to negate each individual proposition and change the conjunction operator (AND) to disjunction (OR). This law is a fundamental tool in propositional logic and Boolean algebra, as it simplifies and transforms logical expressions.

■ **Example 1.8** To illustrate this law, consider two propositions:
- P: Ït is summer. Q: Ït is sunny."

The conjunction $P \wedge Q$ represents the proposition Ït is summer and it is sunny."The negation of this conjunction, $\neg(P \wedge Q)$, would be Ït is not true that it is summer and it is sunny."
Applying De Morgan's Law, we can rewrite this negation as:

$$\neg(P \wedge Q) \equiv (\neg P) \vee (\neg Q)$$

In words, the result is Ït is not summer or it is not sunny."
We observe that the negation of the conjunction is true whenever at least one of the propositions is false. This translates to the fact that for ït is not summer and it is sunny"to be true, it suffices that it is not summer, or even if it is summer, it is not sunny. ■

■ **Example 1.9** Another practical example:
- P: Ï have free time. Q: Ï am at home."

The proposition $P \wedge Q$ represents Ï have free time and I am at home."The negation of this proposition, $\neg(P \wedge Q)$, would be Ït is not true that I have free time and I am at home."
Using De Morgan's Law, we can equivalently express this negation as:

$$\neg(P \wedge Q) \equiv (\neg P) \vee (\neg Q)$$

Which translates to: Ï do not have free time or I am not at home."
This expression tells us that it suffices for one of the two conditions to be false for the negation of $P \wedge Q$ to be true. That is, it is sufficient for me not to have free time or not to be at home for the proposition to be true. ■

P	Q	P ∧ Q	$\neg(P \wedge Q) \equiv \neg P \vee \neg Q$
T	T	T	F
T	F	F	T
F	T	F	T
F	F	F	T

Cuadro 1.4.1: *Truth Table of De Morgan's Law for the Negation of Conjunctions*

The truth table illustrates how the equivalence $\neg(P \wedge Q) \equiv (\neg P) \vee (\neg Q)$ holds in all possible cases of truth or falsity for propositions P and Q. We see that: - When P and Q are both true, the conjunction $P \wedge Q$ is true, so its negation is false. - In the other three cases (when P or Q is false), the negation of the conjunction is true, which aligns with the disjunction of the negations $\neg P \vee \neg Q$.

De Morgan's Law for the negation of conjunctions is very useful in simplifying logical expressions, particularly when working with Boolean algebra or designing logical circuits. In a digital circuit, this law allows replacing a negated AND gate with an OR gate of the individual negations, thereby optimizing circuit design and reducing complexity.

■ **Example 1.10** Consider an example applied to circuit design:
- P: "Switch 1 is closed. Q: "Switch 2 is closed."

The conjunction $P \wedge Q$ represents a series circuit where both switches must be closed for the current to flow. The negation of $P \wedge Q$, i.e., $\neg(P \wedge Q)$, means that the circuit does not allow the current to flow.
Applying De Morgan's Law, we rewrite this negation as:

$$\neg(P \wedge Q) \equiv (\neg P) \vee (\neg Q)$$

This corresponds to a parallel circuit where it suffices for one of the switches to be open for the current not to flow. In this way, we can simplify a series circuit with an AND gate and a negation into a parallel circuit with an OR gate of the individual negations. ■

Exercise 1.5 Given the following propositions:
1. P: "The motor is running."2. Q: "The control system is active."
Write the negation of the proposition "The motor is running and the control system is activeüsing De Morgan's Law, and express in words the meaning of the resulting proposition.
3. R: "The alarm is activated."4. S: "The motion sensor detects presence."
Apply De Morgan's Law to express the negation of "The alarm is activated and the motion sensor detects presence."Then, explain the logical meaning of this negation. ■

(R) The application of De Morgan's laws not only simplifies expressions but also facilitates logical manipulation in verification processes and system testing, such as in security systems, access control, and decision circuits. For example, in alarm systems, the negation of a condition dependent on multiple sensors can be expressed more efficiently using De Morgan's Law, optimizing logical evaluation.

(R) It is important to remember that De Morgan's Law for the negation of conjunctions has a counterpart for disjunctions, which covers both types of operations in logic. Familiarity with both laws is essential for handling propositional logic and its applications in computer science, electronics, and set theory.

1.4.2 De Morgan's Law for the Negation of Disjunctions.

Definition 1.4.2 De Morgan's Law for the Negation of Disjunctions states that the negation of a disjunction of two propositions is equivalent to the conjunction of the negations of those propositions. Formally, if P and Q are two propositions, this law is expressed as:

$$\neg(P \vee Q) \equiv (\neg P) \wedge (\neg Q)$$

This law allows expressing the negation of a disjunction as a conjunction of negations, which is highly useful for simplifying and rewriting logical expressions in Boolean algebra and propositional logic.

■ **Example 1.11** Consider the following propositions:
- P: "The device is on. Q: "The system is in safe mode."
The disjunction $P \vee Q$ represents the proposition "The device is on or the system is in safe mode."The negation of this disjunction, $\neg(P \vee Q)$, would be Ït is not true that the device is on or the system is in safe mode."
Applying De Morgan's Law, we can equivalently express this negation as:

$$\neg(P \vee Q) \equiv (\neg P) \wedge (\neg Q)$$

In words, this means: "The device is not on and the system is not in safe mode."
We observe that for the negation of the disjunction to be true, both individual propositions must be false. This implies that only when both events fail, the negation of $P \vee Q$ is true. ■

■ **Example 1.12** Let's look at another everyday example to understand the application of this law:
- P: Ï have a gym membership. Q: Ï have access to online classes."

The disjunction $P \vee Q$ represents the statement Ï have a gym membership or I have access to online classes.Ïts negation, $\neg(P \vee Q)$, is Ït is not true that I have a gym membership or access to online classes."
Using De Morgan's Law, we can rewrite this negation as:

$$\neg(P \vee Q) \equiv (\neg P) \wedge (\neg Q)$$

That is, Ï do not have a gym membership and I do not have access to online classes."For this proposition to be true, both conditions must be false. ■

P	Q	P $\vee$ Q	$\neg(P \vee Q) \equiv \neg P \wedge \neg Q$
T	T	T	F
T	F	T	F
F	T	T	F
F	F	F	T

Cuadro 1.4.2: *Truth Table of De Morgan's Law for the Negation of Disjunctions*

The truth table shows how the equivalence $\neg(P \vee Q) \equiv (\neg P) \wedge (\neg Q)$ is verified in all possible cases of truth and falsity for P and Q. We observe that: - The disjunction $P \vee Q$ is false only when both P and Q are false, which makes the negation of the disjunction true only in this case. - In all other cases, the disjunction $P \vee Q$ is true, so its negation is false.

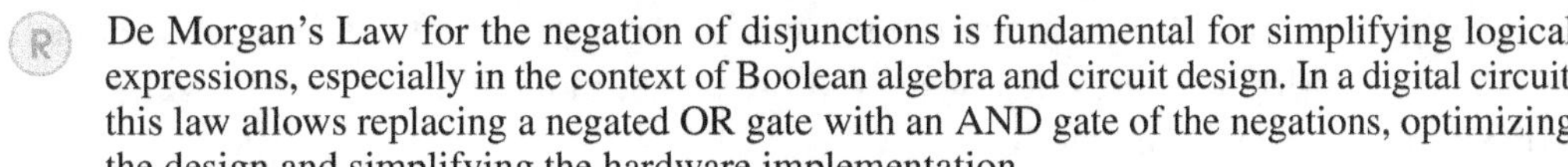

De Morgan's Law for the negation of disjunctions is fundamental for simplifying logical expressions, especially in the context of Boolean algebra and circuit design. In a digital circuit, this law allows replacing a negated OR gate with an AND gate of the negations, optimizing the design and simplifying the hardware implementation.

■ **Example 1.13** Consider an example applied to a security system:
- P: "The intrusion alarm is active. Q: "The motion sensor detects presence."
The disjunction $P \vee Q$ represents a security system that states "The intrusion alarm is active or the motion sensor detects presence."The negation $\neg(P \vee Q)$ means the system denies both conditions: Ït is not true that the intrusion alarm is active or that the motion sensor detects presence."
Using De Morgan's Law, we rewrite this negation as:

$$\neg(P \vee Q) \equiv (\neg P) \wedge (\neg Q)$$

This means: "The intrusion alarm is not active and the motion sensor does not detect presence.Ïn this case, for the system to be inactive, both events must be false. ■

Exercise 1.6 Given the following propositions:
1. P: Ï have the supervisor's permission."2. Q: Ï have completed the form."
Write the negation of the proposition Ï have the supervisor's permission or I have completed the formüsing De Morgan's Law, and express in words the meaning of the resulting proposition.
3. R: "The client approved the contract."4. S: "The payment was received."
Apply De Morgan's Law to express the negation of "The client approved the contract or the payment was received."Then, explain the logical meaning of this negation. ■

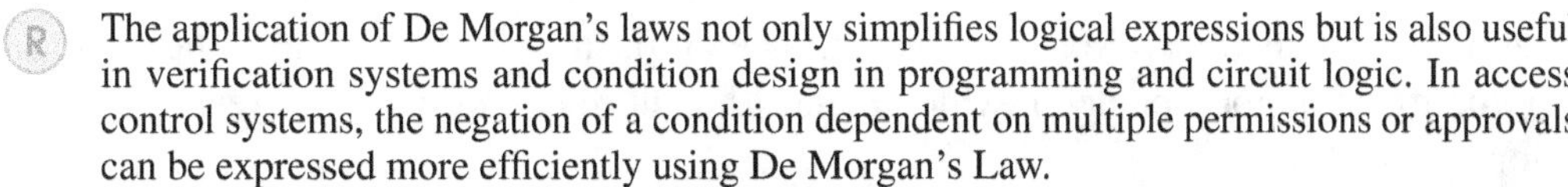

The application of De Morgan's laws not only simplifies logical expressions but is also useful in verification systems and condition design in programming and circuit logic. In access control systems, the negation of a condition dependent on multiple permissions or approvals can be expressed more efficiently using De Morgan's Law.

It is essential to understand both De Morgan's Law for the negation of disjunctions and its counterpart for the negation of conjunctions, as together they provide a complete toolset for handling negations of logical expressions in propositional logic, set theory, and Boolean algebra. This understanding is crucial for optimizing logical processes in computer science and electronics.

1.5 Solved Exercises

Exercise 1.7 Determine whether the following statements are declarative propositions, and explain why or why not in each case:
1. "The speed of light is approximately 300,000 km/s."
2. Çan you help me solve this problem?"
3. "Mount Everest is the tallest mountain in the world."
4. "What an interesting idea!"
5. .All prime numbers are odd."
For each one, indicate whether it can be classified as true or false, and justify your answer. ■

Demostración. 1. "The speed of light is approximately 300,000 km/s."This statement can be classified as true or false because the speed of light in a vacuum is approximately 299,792 km/s. Therefore, it is a declarative proposition and it is true.
2. Çan you help me solve this problem?"This is a question, and it cannot be classified as true or false. Thus, it is not a declarative proposition.
3. "Mount Everest is the tallest mountain in the world."This statement can be classified as true or false. Currently, Mount Everest is considered the tallest mountain above sea level, so it is a declarative proposition and it is true.
4. "What an interesting idea!"This is an exclamatory statement and depends on opinion, so it cannot be classified as true or false. It is not a declarative proposition.
5. .All prime numbers are odd."This statement is false because the number 2 is prime and even. It is a declarative proposition and it is false. ■

Exercise 1.8 Using the definition of conditional proposition, evaluate the truth of the following propositions:
1. Ïf a figure is a square, then it has four sides."
2. Ïf the number 5 is divisible by 2, then 5 + 1 = 6."
3. Ïf today is Monday, then tomorrow is Tuesday."
4. Ïf the moon is a planet, then Mars is a planet."
5. Ïf a number is even, then it is divisible by 4."
Explain in each case why the conditional proposition is true or false. ■

Demostración. 1. Ïf a figure is a square, then it has four sides."The proposition is true because if the hypothesis (.ª figure is a square") is true, then the conclusion (ït has four sides") is also true.
2. Ïf the number 5 is divisible by 2, then 5 + 1 = 6."The proposition is true because the hypothesis is false ("the number 5 is not divisible by 2"), which makes the entire conditional true regardless of the conclusion.
3. Ïf today is Monday, then tomorrow is Tuesday."The proposition is true because if the hypothesis ("today is Monday") is true, the conclusion ("tomorrow is Tuesday") is also true.
4. Ïf the moon is a planet, then Mars is a planet."The proposition is true because the hypothesis is false ("the moon is not a planet"), which makes the entire conditional true regardless of the conclusion.

5. Ïf a number is even, then it is divisible by 4."This proposition is false because if the hypothesis is true (for example, for the number 2), the conclusion is false (2 is not divisible by 4). ■

Exercise 1.9 Use truth tables to evaluate the following compound propositions with binary operators:
1. Given P: "The number 8 is even.ªnd Q: "The number 8 is prime,.ᵉᵛaluate $P \wedge Q$ and $P \vee Q$.
2. Given R: "The number 3 is odd.ªnd S: "The number 6 is even,.ᵉᵛaluate $R \to S$ and $S \to R$.
3. Given A: "2 + 2 = 4.ªnd B: "2 + 3 = 6,.ᵉᵛaluate $A \wedge \neg B$ and $A \vee B$.
4. Given X: "The sun is a star.ªnd Y: "The moon is a planet,.ᵉᵛaluate $X \to Y$ and $X \wedge Y$.
5. Given M: "The sky is blue.ªnd N: "Water is transparent,.ᵉᵛaluate $M \vee \neg N$ and $M \wedge N$. ■

Demostración. 1. P: true, Q: false - $P \wedge Q$: false - $P \vee Q$: true
2. R: true, S: true - $R \to S$: true - $S \to R$: true
3. A: true, B: false - $A \wedge \neg B$: true - $A \vee B$: true
4. X: true, Y: false - $X \to Y$: false - $X \wedge Y$: false
5. M: true, N: true - $M \vee \neg N$: true - $M \wedge N$: true ■

Exercise 1.10 Use De Morgan's Law to simplify the following negations and express the meaning of each resulting proposition:
1. The negation of Ï am on vacation and I am at home."
2. The negation of Ït is summer or it is hot."
3. The negation of "The exam is easy and I am well-prepared."
4. The negation of "We will travel in December or we will go hiking."
5. The negation of "The computer is on and the internet is working." ■

Demostración. 1. The negation of Ï am on vacation and I am at homeïs Ï am not on vacation or I am not at home."
2. The negation of Ït is summer or it is hotïs Ït is not summer and it is not hot."
3. The negation of "The exam is easy and I am well-preparedïs "The exam is not easy or I am not well-prepared."
4. The negation of "We will travel in December or we will go hikingïs "We will not travel in December and we will not go hiking."
5. The negation of "The computer is on and the internet is workingïs "The computer is not on or the internet is not working." ■

Exercise 1.11 Consider the following propositions and construct the corresponding truth tables to evaluate the validity of the following expressions:
1. Given A and B, construct the truth table for $A \to (B \vee \neg A)$.
2. Given P and Q, construct the truth table for $(P \wedge Q) \vee \neg P$.
3. Given X and Y, construct the truth table for $X \leftrightarrow (Y \wedge \neg X)$.
4. Given M and N, construct the truth table for $\neg(M \vee N) \leftrightarrow (\neg M \wedge \neg N)$ and verify whether this expression is always true (tautology).
5. Given R and S, construct the truth table for $\neg(R \to S) \to (R \wedge \neg S)$. ■

Demostración. Construct truth tables for each expression and verify the results.
1. $A \to (B \vee \neg A)$ is always true, thus it is a tautology.
2. $(P \wedge Q) \vee \neg P$ is always true, indicating a tautology.
3. $X \leftrightarrow (Y \wedge \neg X)$ is false for all combinations of truth values, indicating a contradiction.
4. $\neg(M \vee N) \leftrightarrow (\neg M \wedge \neg N)$ is always true, showing that it is a tautology.

5. $\neg(R \to S) \to (R \wedge \neg S)$ is true only in some cases, making it a contingency. ■

1.6 Proposed Exercises

1.6.1 Types of Propositions: Simple and Compound

Exercise 1.12 Determine whether each of the following statements is a simple or compound proposition. Explain your answer. ■

Exercise 1.13 Identify whether the following propositions are true or false and classify them as simple or compound: 1. .ᴬll planets orbit the Sun."
2. "The sky is blue and the clouds are white."
3. Ïf you study, then you will pass the exam." ■

Exercise 1.14 Convert the following statements into compound propositions using the connectives .ªnd,or,.ºr ïf... then": 1. Çats are animals. Dogs are animals."
2. Ïf you exercise, you have good health."
3. "The sun is a star. The moon is a satellite." ■

Exercise 1.15 Given the proposition Ïf a figure is a square, then it has four sides and all its angles are right angles,ïdentify the simple propositions that compose it. ■

Exercise 1.16 Write three examples of simple propositions and three of compound propositions. Indicate which simple propositions form the compound ones. ■

1.6.2 Logical Connectives: Truth Tables for Connectives (AND, OR, NOT, IF-THEN)

Exercise 1.17 Construct the truth table for the compound proposition $P \wedge Q$. ■

Exercise 1.18 Construct the truth table for $P \vee Q$ and explain in which situations it is true. ■

Exercise 1.19 Given the propositions P: Ït is raining today.ªnd Q: Ït is cold,.ᵉᵛaluate the truth of $P \to Q$ in all four possible cases for P and Q. ■

Exercise 1.20 Construct the truth table for the compound proposition $\neg P \vee Q$. ■

Exercise 1.21 Evaluate the truth of the proposition $(P \wedge Q) \to \neg R$ for all possible truth values of P, Q, and R. ■

1.6.3 Sixteen Fundamental Cases in Logic

Exercise 1.22 Identify whether the following cases are tautologies, contradictions, or contingencies: 1. $P \vee \neg P$
2. $P \wedge \neg P$
3. $P \to (Q \vee \neg Q)$ ■

Exercise 1.23 Construct the truth table for $\neg(P \vee Q)$ and determine whether it is equivalent to any of the sixteen fundamental logical functions. ■

Exercise 1.24 Given the propositions P: "The dog barks.ªnd Q: "The cat meows,çonstruct and evaluate the proposition $\neg(P \wedge Q)$. ■

Exercise 1.25 Given the fundamental case of biconditional, construct the truth table for $P \leftrightarrow Q$ and determine when it is true. ■

Exercise 1.26 Use the fundamental cases to identify whether the proposition $(P \vee Q) \wedge (\neg P \vee \neg Q)$ is a tautology, contradiction, or contingency. ■

1.6.4 Negation of Propositions: Application of De Morgan's Laws

Exercise 1.27 Use De Morgan's Law to simplify the negation of Ï am at home and I have free time." ■

Exercise 1.28 Write the negation of Ït is winter or it is coldüsing De Morgan's Law. ■

Exercise 1.29 Simplify the negation of "The exam is easy and the test is short." ■

Exercise 1.30 Express the negation of Ï study or I work.ªnd use De Morgan's Law to simplify. ■

Exercise 1.31 Given the propositions P: "The store is open.ªnd Q: "There is a discount,.ªpply De Morgan's Law to simplify $\neg(P \vee Q)$. ■

2. Propositional Algebra and Logical Inference

2.1 Truth Tables: Construction and Analysis.

2.1.1 Truth Tables of Simple Connectives.

Definition 2.1.1 In propositional logic, **simple connectives** are logical operators that act on one or two propositions and produce a new truth value. The main simple connectives are:
- **Negation** (NOT), denoted by $\neg$ - **Conjunction** (AND), denoted by $\wedge$ - **Disjunction** (OR), denoted by $\vee$ - **Conditional** (IF-THEN), denoted by $\rightarrow$ - **Biconditional** (IF AND ONLY IF), denoted by $\leftrightarrow$
Each of these connectives has a truth table that describes the resulting truth value of the compound proposition based on the truth values of the original propositions.

■ **Example 2.1** Below are the truth tables of the most common simple connectives:
1. **Negation (NOT)**: The negation of a proposition P, denoted by $\neg P$, reverses its truth value. If P is true, $\neg P$ is false; and if P is false, $\neg P$ is true.

P	**¬ P**
T	F
F	T

Cuadro 2.1.1: *Truth Table for Negation (NOT)*

2. **Conjunction (AND)**: The conjunction of two propositions $P \wedge Q$ is true only if both propositions are true. If either proposition is false, the conjunction is false.
3. **Disjunction (OR)**: The disjunction of two propositions $P \vee Q$ is true if at least one of the propositions is true. It is false only if both propositions are false.
4. **Conditional (IF-THEN)**: The conditional $P \rightarrow Q$ is false only when P is true and Q is false. In all other cases, the conditional is true.
5. **Biconditional (IF AND ONLY IF)**: The biconditional $P \leftrightarrow Q$ is true when P and Q have the same truth value, i.e., both are true or both are false.

■

P	Q	P ∧ Q
T	T	T
T	F	F
F	T	F
F	F	F

Cuadro 2.1.2: *Truth Table for Conjunction (AND)*

P	Q	P ∨ Q
T	T	T
T	F	T
F	T	T
F	F	F

Cuadro 2.1.3: *Truth Table for Disjunction (OR)*

P	Q	P → Q
T	T	T
T	F	F
F	T	T
F	F	T

Cuadro 2.1.4: *Truth Table for Conditional (IF-THEN)*

P	Q	P ↔ Q
T	T	T
T	F	F
F	T	F
F	F	T

Cuadro 2.1.5: *Truth Table for Biconditional (IF AND ONLY IF)*

(R) The truth tables of simple connectives provide the foundation for evaluating more complex expressions in propositional logic. Combining these connectives allows for constructing compound propositions that can be analyzed using extended truth tables.

Exercise 2.1 Given the following propositions:
1. P: "Today is Monday."2. Q: Ït is raining."
Construct the truth table for the compound propositions $\neg P$, $P \wedge Q$, $P \vee Q$, $P \rightarrow Q$, and $P \leftrightarrow Q$. Evaluate each row considering all combinations of truth and falsity for P and Q. ■

(R) Understanding truth tables for simple connectives is essential in logic as they allow systematic analysis and verification of the validity of propositions and arguments. Moreover, truth tables are a key tool in Boolean algebra, used in the design and simplification of digital logic circuits.

2.1.2 Analysis of Tautologies and Contradictions.

Definition 2.1.2 In propositional logic, a **tautology** is a compound proposition that is always true, regardless of the truth values of the propositions that compose it. On the other hand, a

contradiction is a compound proposition that is always false, regardless of the truth values of its components. These concepts are fundamental in logic because they help identify propositions that are valid in any situation (tautologies) and propositions that are impossible or inconsistent (contradictions).

■ **Example 2.2** Below are examples of tautologies and contradictions, along with their respective truth tables to illustrate these concepts:
1. **Example of Tautology**: Consider the proposition $P \vee \neg P$. This proposition states that "P is true or not P is true,"which will always be true because a proposition is either true or false, with no other possibility.

P	**P ∨ ¬ P**
T	T
F	T

Cuadro 2.1.6: *Truth table of the tautology* $P \vee \neg P$

In this case, we observe that regardless of the value of P, the proposition $P \vee \neg P$ is always true, confirming that it is a tautology.
2. **Example of Contradiction**: Consider the proposition $P \wedge \neg P$. This proposition states that "P is true and not P is true,"which is impossible since a proposition cannot be both true and false simultaneously.

P	**P ∧ ¬ P**
T	F
F	F

Cuadro 2.1.7: *Truth table of the contradiction* $P \wedge \neg P$

Here, we see that regardless of the value of P, the proposition $P \wedge \neg P$ is always false, indicating that it is a contradiction. ■

Ⓡ Tautologies and contradictions play a key role in formal logic and Boolean algebra as they help simplify and validate logical expressions. Tautologies are useful in mathematical proofs and valid arguments, while contradictions help identify logical inconsistencies in propositions and systems of logical equations.

■ **Example 2.3** Consider additional examples of tautologies and contradictions:
- **Tautology**: The proposition $(P \to Q) \vee (Q \to P)$ is a tautology because it ensures that for any pair of propositions P and Q, at least one of them implies the other.

P	**Q**	**P → Q**	**Q → P**
T	T	T	T
T	F	F	T
F	T	T	F
F	F	T	T

Cuadro 2.1.8: *Truth table of the tautology* $(P \to Q) \vee (Q \to P)$

In all possible cases, $(P \to Q) \vee (Q \to P)$ is true, confirming it is a tautology.
- **Contradiction**: The proposition $(P \wedge Q) \wedge \neg(P \wedge Q)$ is a contradiction because it implies that $P \wedge Q$ is both true and false simultaneously, which is impossible.

P	Q	P ∧ Q	¬ (P ∧ Q)
T	T	T	F
T	F	F	T
F	T	F	T
F	F	F	T

Cuadro 2.1.9: *Truth table of the contradiction* $(P \wedge Q) \wedge \neg(P \wedge Q)$

The table shows that in any case, $(P \wedge Q) \wedge \neg(P \wedge Q)$ is always false, confirming it is a contradiction.
■

Exercise 2.2 Analyze the following propositions and determine if they are tautologies, contradictions, or neither:
1. $(P \vee Q) \rightarrow (Q \vee P)$ 2. $(P \rightarrow Q) \wedge (\neg Q \rightarrow \neg P)$ 3. $(P \vee \neg Q) \wedge (\neg P \wedge Q)$
Construct truth tables for each of them and determine whether the result is true in all cases (tautology), false in all cases (contradiction), or if the truth value depends on P and Q. ■

(R) The analysis of tautologies and contradictions is essential for verifying valid arguments and designing logical circuits. In formal logic, tautologies represent fundamental logical laws that are true in any context, while contradictions indicate impossibilities or inconsistencies in a logical system.

(R) In addition to their use in propositional logic, the concepts of tautology and contradiction are applied in other fields such as set theory, programming, and control systems, where conditions of absolute truth and falsity have direct implications in decision-making and system design.

2.2 Logical Inference: Rules of Inference (Modus Ponens, Modus Tollens).

2.2.1 Application of Modus Ponens in Logical Problems.

Definition 2.2.1 The **Modus Ponens** is a rule of logical inference that states that if a conditional proposition $P \rightarrow Q$ is given and P is known to be true, then Q can also be concluded to be true. This rule is used to make valid inferences from premises and is one of the fundamental bases of propositional logic.
Formally, the structure of Modus Ponens is:

$$\begin{array}{ll} 1. & P \rightarrow Q \\ 2. & P \\ \hline \therefore & Q \end{array}$$

That is, given a conditional and the affirmation of its antecedent, the consequent can be concluded.

■ **Example 2.4** Consider the following example:
1. If I study for the exam, then I will pass. (Proposition $P \rightarrow Q$) 2. I study for the exam. (Proposition P)
Applying Modus Ponens, we can conclude:

Therefore, I will pass. (Q)

In this case, since the antecedent P (Ï study for the exam") is true and the conditional proposition holds, we can deduce that the conclusion Q (Ï will pass") is also true. ■

■ **Example 2.5** Another example in a daily context:
1. If the car has fuel, then it can start. (Proposition $P \to Q$) 2. The car has fuel. (Proposition P)
Applying Modus Ponens, we conclude:

$$\text{Therefore, the car can start. } (Q)$$

Here, the proposition Q is valid because the condition P holds, confirming that when the antecedent of a true conditional is satisfied, the consequent can be deduced. ■

■ **Example 2.6** In mathematical logic, Modus Ponens is also used to demonstrate more abstract propositions. Suppose:
1. If $x > 5$, then $x^2 > 25$. (Proposition $P \to Q$) 2. We know that $x > 5$. (Proposition P)
By Modus Ponens, we can conclude:

$$\text{Therefore, } x^2 > 25. \ (Q)$$

This example shows how Modus Ponens allows deriving a logical conclusion in a mathematical context based on numerical conditions. ■

■ **Example 2.7** In a problem of permissions and access:
1. If a person has an access pass, then they can enter the conference room. (Proposition $P \to Q$) 2. Maria has an access pass. (Proposition P)
Applying Modus Ponens, we deduce:

$$\text{Therefore, Maria can enter the conference room. } (Q)$$

This example shows how Modus Ponens can be used in situations involving permissions and access, allowing logical inference based on authorization conditions. ■

(R) Modus Ponens is one of the most widely used rules of inference in propositional logic because it allows deriving valid and precise conclusions from conditional propositions. It is used in numerous fields, from formal mathematics to programming and logical system design, facilitating decision-making based on premises.

Exercise 2.3 Given the following propositions, use Modus Ponens to obtain valid conclusions:
1. If I practice every day, then I will improve in sports. (Proposition $P \to Q$) 2. I practice every day. (Proposition P)
Conclusion: What can be inferred?
3. If the weather is sunny, then we will go to the beach. (Proposition $P \to Q$) 4. The weather is sunny. (Proposition P)
Conclusion: What can be inferred?
For each case, indicate what the antecedent, the consequent, and how Modus Ponens is applied. ■

(R) Modus Ponens not only allows reaching valid conclusions but also verifies the coherence and consistency of arguments in formal logic. Its application in diverse fields makes this rule of inference an indispensable resource in logic and reasoning.

2.2.2 Application of Modus Tollens in Arguments.

Definition 2.2.2 The **Modus Tollens** is a rule of logical inference that states that if a conditional proposition $P \to Q$ is given and Q is known to be false ($\neg Q$), then P can also be concluded to be false ($\neg P$). This rule is used to negate the antecedent when the consequent of a conditional proposition is false, and it is a key tool in deductive reasoning.
Formally, the structure of Modus Tollens is:

$$\begin{array}{ll} 1. & P \to Q \\ 2. & \neg Q \\ \hline \therefore & \neg P \end{array}$$

That is, given a conditional and the negation of the consequent, the negation of the antecedent can be concluded.

■ **Example 2.8** Consider the following example:
1. If it rains, then the street is wet. (Proposition $P \to Q$) 2. The street is not wet. (Proposition $\neg Q$)
Applying Modus Tollens, we can conclude:

Therefore, it is not raining. ($\neg P$)

In this case, since the consequent Q ("The street is wet") is false, we can infer that the antecedent P (Ït is raining") must also be false. ■

■ **Example 2.9** Another example in a medical diagnosis context:
1. If a person has the flu, then they have a fever. (Proposition $P \to Q$) 2. The person does not have a fever. (Proposition $\neg Q$)
Applying Modus Tollens, we conclude:

Therefore, the person does not have the flu. ($\neg P$)

Here, since the consequent is false (the person does not have a fever), we infer that the initial proposition about the antecedent is also false (the person does not have the flu). ■

■ **Example 2.10** In a mathematical context, Modus Tollens is also applied for formal reasoning. Suppose:
1. If $x > 5$, then $x^2 > 25$. (Proposition $P \to Q$) 2. We know that $x^2 \leq 25$. (Proposition $\neg Q$)
By Modus Tollens, we can conclude:

Therefore, $x \leq 5$. ($\neg P$)

This example shows how Modus Tollens allows inferences about the falsity of an initial condition based on the impossibility of the consequent. ■

■ **Example 2.11** In a problem of access and restrictions:
1. If you have access to the system, then you can view confidential files. (Proposition $P \to Q$) 2. You cannot view the confidential files. (Proposition $\neg Q$)
Applying Modus Tollens, we deduce:

Therefore, you do not have access to the system. ($\neg P$)

In this example, the fact that access to the confidential files is false implies, by Modus Tollens, that the initial condition of access is also false. ■

(R) Modus Tollens is a fundamental tool in propositional logic and in solving problems of negative inference, as it allows deducing the falsity of an initial proposition based on the impossibility of the consequent. This reasoning is widely used in inconsistency proofs and logical verification in fields such as mathematics, computer science, and philosophy.

Exercise 2.4 Given the following propositions, use Modus Tollens to obtain valid conclusions:
1. If the system is online, then the database can be accessed. (Proposition $P \to Q$) 2. The database cannot be accessed. (Proposition $\neg Q$)
Conclusion: What can be inferred about the state of the system?
3. If you have a premium membership, then you can download exclusive content. (Proposition $P \to Q$) 4. You cannot download exclusive content. (Proposition $\neg Q$)
Conclusion: What can be inferred about your membership?
For each case, indicate what the antecedent, the consequent, and how Modus Tollens is applied.
■

(R) Modus Tollens is essential for analyzing consistency in arguments, as it allows deducing the falsity of an initial condition if the consequent of the conditional proposition is known to be false. Its application is useful in logical verifications and critical reasoning in logic and formal sciences.

2.3 Applications: Solving problems involving order of information and temporal relationships.

2.3.1 Problems with multiple variables.

Definition 2.3.1 In propositional logic and the resolution of complex problems, **problems with multiple variables** involve the use of several propositions or conditions that interact to derive conclusions. These problems require identifying logical relationships among the variables and using truth tables, inference rules, and diagrams to analyze all possible combinations of truth values. Solving problems with multiple variables develops skills in logical deduction and organizing complex information.

■ **Example 2.12** Consider a problem involving three variables:
1. P: "Juan arrives on time."2. Q: "The meeting takes place."3. R: "The boss is present."
Suppose we know the following logical relationships:
- If Juan arrives on time (P), then the meeting takes place (Q), i.e., $P \to Q$. - If the meeting takes place (Q), then the boss is present (R), i.e., $Q \to R$.
Given the condition that the boss is not present ($\neg R$), we can deduce, using Modus Tollens, that Juan did not arrive on time ($\neg P$). This is represented as follows:

$$P \to Q,\ Q \to R,\ \neg R \Rightarrow \neg Q \Rightarrow \neg P$$

Here, we solved a problem of logical inference with three variables by relating them through inference rules and consecutive deductions. ■

■ **Example 2.13** Let us analyze a problem of access permissions in a system with multiple conditions. Suppose the following variables:
1. A: "The user has administrator permissions."2. B: "The user is within the corporate network."3. C: "The user can access sensitive files."
The relationships are as follows:
- If the user has administrator permissions (A) and is within the corporate network (B), then they can access sensitive files (C), i.e., $(A \land B) \to C$. - If the user does not have administrator permissions ($\neg A$), then they cannot access sensitive files ($\neg C$).
Given that the user can access sensitive files (C), we can deduce that the user has administrator permissions (A) and is within the corporate network (B). Thus, we solve the problem by relating the variables and deducing their truth values.

$(A \wedge B) \rightarrow C,\ C \Rightarrow A \wedge B$

■

A	B	C	(A ∧ B) → C	Conclusion
T	T	T	T	Access granted
T	F	F	T	Access denied
F	T	F	T	Access denied
F	F	F	T	Access denied

Cuadro 2.3.1: *Truth table for an access permissions problem*

(R) Problems with multiple variables are often efficiently solved using truth tables, Venn diagrams, and inference rules such as Modus Ponens and Modus Tollens. As the number of variables increases, it is essential to organize the information systematically to visualize all possible combinations and deduce logical conclusions.

■ **Example 2.14** Consider a problem involving four variables in an alarm system:
1. S: "The motion sensor is activated."2. W: "The window is closed."3. D: "The door is closed."4. A: "The alarm is activated."
The conditions are:
- If the motion sensor is activated and the window or door is open, then the alarm is activated: $(S \wedge (\neg W \vee \neg D)) \rightarrow A$. - If the alarm is not activated ($\neg A$), we can deduce that the motion sensor is not activated or both the window and the door are closed.
Given that the alarm is not activated ($\neg A$), we apply Modus Tollens to deduce:

$$\neg A \Rightarrow \neg S \vee (W \wedge D)$$

This means that for the alarm to remain off, the motion sensor must be deactivated, or both the window and the door must be closed. Thus, we analyzed a logical problem with four variables. ■

Exercise 2.5 Given the following propositions, use truth tables and inference rules to solve the problems:
1. P: "The support team is available."2. Q: "The request is high priority."3. R: "The request is handled immediately."
The conditions are:
- If the support team is available and the request is high priority, then it is handled immediately: $(P \wedge Q) \rightarrow R$. - If the request is not high priority, then it is not handled immediately: $\neg Q \rightarrow \neg R$.
Given the information that the request is handled immediately (R), deduce the truth values of P and Q.
2. X: "The contract is signed."3. Y: "The payment is received."4. Z: "The project can begin."
Conditions:
- If the contract is signed and the payment is received, then the project can begin: $(X \wedge Y) \rightarrow Z$. - If the project cannot begin, then either the contract is not signed or the payment is not received. Based on this information, determine the possible truth values of X, Y, and Z if the project cannot begin. ■

Solving problems with multiple variables allows organizing and analyzing information logically, being especially useful in programming, mathematics, and project management. These skills help in decision-making based on multiple conditions and interdependent relationships among variables.

2.3.2 Analysis of Logical Sequences in Time.

Definition 2.3.2 Analysis of logical sequences in time refers to the study of events and conditions occurring in a specific order over time. This type of analysis is common in systems where certain events must occur before others can happen. In propositional logic, temporal analysis establishes relationships between propositions based on a chronological order, facilitating logical deduction and sequential reasoning.

■ **Example 2.15** Consider the following example of a manufacturing process with three events: 1. P: "The raw materials are ready."2. Q: "The machinery is set up."3. R: "Production can begin."
The temporal relationships between the events are:
- Production can only begin after the raw materials are ready and the machinery is set up, i.e., $(P \wedge Q) \rightarrow R$. - If production has begun (R), we can conclude that both P and Q occurred beforehand. This sequential analysis helps to understand the logical order of preparation and setup before production starts. ■

■ **Example 2.16** Another example of a logical sequence in a security system:
1. A: "The alarm is armed."2. B: "The sensor detects motion."3. C: "The alarm is triggered."
The temporal relationships are as follows:
- The alarm is only triggered if it is armed and the sensor detects motion, i.e., $(A \wedge B) \rightarrow C$. - If the alarm is triggered (C), we can infer that the alarm was armed (A) and that the sensor detected motion (B) previously.
Here, the temporal analysis ensures that the triggering of the alarm depends on a series of events occurring in sequence. ■

A	B	C	(A ∧ B) → C	Temporal Sequence
T	T	T	T	Fully triggered
T	F	F	T	Not triggered
F	T	F	T	Not triggered
F	F	F	T	Not triggered

Cuadro 2.3.2: *Truth table for a logical sequence in a security system*

The analysis of logical sequences in time is essential for organizing processes and evaluating conditions in systems where the order of events is crucial. This approach is widely used in sequential programming, project management, and control system design, where temporal dependencies impact the final outcome.

■ **Example 2.17** Consider an example in a building access process:
1. X: "The visitor registers at reception."2. Y: "The guard verifies the authorization."3. Z: .^ccess to the building is granted."
The conditions are as follows:
- Access is granted only if the visitor registers at reception and the guard verifies the authorization, i.e., $(X \wedge Y) \rightarrow Z$. - If access is granted (Z), we can conclude that registration at reception (X) and authorization verification (Y) were completed first.
This type of analysis helps to understand the sequence of events necessary for granting access. ■

Exercise 2.6 Given the following propositions and their temporal relationships, analyze the logical sequence and determine the conclusions that can be drawn:
1. P: "The payment for the order is completed."2. Q: "The order is shipped to the customer."3. R: "The customer receives the order."
Conditions:
- The order is shipped only after the payment is completed, i.e., $P \to Q$. - The customer receives the order only after it has been shipped, i.e., $Q \to R$.
Given the information that the customer has received the order (R), deduce the truth values of P and Q.
4. A: "The package arrives at the warehouse."5. B: "The package is inspected at the warehouse."6. C: "The package is dispatched to the customer."
Conditions:
- The package is inspected only if it arrives at the warehouse: $A \to B$. - The package is dispatched only after being inspected: $B \to C$.
If the package has been dispatched (C), what can be concluded about events A and B? ■

(R) The analysis of logical sequences in time allows for evaluating and organizing conditions of events in a specific order, and is particularly useful in the planning and management of complex systems. This type of analysis is applied in production lines, supply chains, and workflow programming, where events must follow a logical sequence to ensure the proper development of processes.

2.4 Solved Exercises

Exercise 2.7 Given the propositions P: Ït is raining.ªnd Q: Ï carry an umbrella,çonstruct the truth table for the following compound propositions:
1. $P \wedge Q$
2. $P \vee Q$
3. $P \to Q$
4. $P \leftrightarrow Q$
Explain in each case under what conditions they are true or false. ■

Demostración. We construct the truth table for the compound propositions:

P	Q	$P \wedge Q$	$P \vee Q$	$P \to Q$	$P \leftrightarrow Q$
T	T	T	T	T	T
T	F	F	T	F	F
F	T	F	T	T	F
F	F	F	F	T	T

1. $P \wedge Q$ is true only when both P and Q are true. 2. $P \vee Q$ is true when at least one of P or Q is true. 3. $P \to Q$ is false only when P is true and Q is false; in all other cases, it is true. 4. $P \leftrightarrow Q$ is true only when P and Q have the same truth value (both true or both false).

■

Exercise 2.8 Consider the following propositions and determine if they are tautologies, contradictions, or neither:
1. $P \vee \neg P$
2. $P \wedge \neg P$

3. $(P \to Q) \vee (Q \to P)$
For each, construct the truth table and justify your conclusion.

Demostración. We construct the truth tables for each proposition:
1. For $P \vee \neg P$:

P	$\neg P$	$P \vee \neg P$
T	F	T
F	T	T

This proposition is always true, therefore it is a tautology.
2. For $P \wedge \neg P$:

P	$\neg P$	$P \wedge \neg P$
T	F	F
F	T	F

This proposition is always false, therefore it is a contradiction.
3. For $(P \to Q) \vee (Q \to P)$:

P	Q	$P \to Q$	$Q \to P$
$(P \to Q) \vee (Q \to P)$			
T	T	T	T
T			
T	F	F	T
T			
F	T	T	F
T			
F	F	T	T
T			

This proposition is always true, therefore it is also a tautology.

■

Exercise 2.9 Use the Modus Ponens rule to deduce valid conclusions in the following cases:
1. If I study, then I will pass the exam. (Proposition $P \to Q$) I study. (Proposition P) What is the conclusion?
2. If there is traffic, I will be late. (Proposition $P \to Q$) There is traffic. (Proposition P) What is the conclusion?
Explain the use of Modus Ponens in each case.

Demostración. 1. In the first case:

1. $P \to Q$
2. P

$\therefore$ Q

Since $P \to Q$ and P are true, by applying Modus Ponens, we conclude that Q (I will pass the exam) is true.

2. In the second case:

1. $P \to Q$
2. P

$\therefore$ Q

Given that $P \to Q$ and P are true, by Modus Ponens, we conclude that Q (I will be late) is true.
In both cases, Modus Ponens allows us to infer the consequent (Q) from the truth of the antecedent (P) in a conditional proposition.

■

Exercise 2.10 Apply De Morgan's Laws to simplify the negation of the following compound propositions:
1. The negation of Ït is sunny and hot."
2. The negation of Ït is summer or I am on vacation."
3. The negation of Ï am hungry and I have no food."
Write the simplified proposition and explain the meaning of each negation.

Demostración. 1. The negation of Ït is sunny and hotïs:

$$\neg(P \land Q) \equiv \neg P \lor \neg Q$$

This means: Ït is not sunny or it is not hot."
2. The negation of Ït is summer or I am on vacationïs:

$$\neg(P \lor Q) \equiv \neg P \land \neg Q$$

This means: Ït is not summer and I am not on vacation."
3. The negation of Ï am hungry and I have no foodïs:

$$\neg(P \land \neg Q) \equiv \neg P \lor Q$$

This means: Ï am not hungry or I have food."
In each case, we apply De Morgan's Laws to distribute the negation over the compound proposition, transforming conjunctions into disjunctions and vice versa.

■

Exercise 2.11 Solve the following logical sequence problem:
1. A: "The package has been packed."
2. B: "The package has been shipped."
3. C: "The package has been delivered."
Conditions: - The package is shipped only after being packed: $A \to B$. - The package is delivered only after being shipped: $B \to C$.
Given that the package has been delivered (C), determine the truth value of A and B.

Demostración. Given that C is true (the package has been delivered), we use the implication $B \to C$.
1. If C is true, then by $B \to C$, we conclude that B must also be true (the package has been shipped).
2. Since B is true, we apply the implication $A \to B$, which leads us to conclude that A is also true (the package has been packed).
Therefore, both A and B are true in this case.

■

2.5 Proposed Exercises

2.5.1 Truth Tables: Construction and Analysis

Exercise 2.12 Construct the truth table for the compound proposition $(P \lor Q) \land \neg R$, where P, Q, and R are simple propositions. Analyze each combination of truth and falsity for P, Q, and R. ■

Exercise 2.13 Determine whether the proposition $(P \to Q) \leftrightarrow (\neg Q \to \neg P)$ is a tautology, contradiction, or neither. Justify your answer with a truth table. ■

Exercise 2.14 Construct the truth table for the proposition $\neg(P \land Q) \to (R \lor P)$. Determine under which conditions the proposition is true or false. ■

Exercise 2.15 Given the proposition $(P \lor Q) \land (P \to R)$, construct the truth table and determine whether the proposition is true or false in each case. ■

Exercise 2.16 Analyze whether the proposition $\neg P \lor (Q \to P)$ is a tautology, a contradiction, or neither. Use a truth table to justify your conclusion. ■

2.5.2 Logical Inference: Rules of Inference (Modus Ponens, Modus Tollens)

Exercise 2.17 Using Modus Ponens, deduce the conclusion in the following case: 1. If the water boils, then the temperature is at least 100 degrees Celsius. (Proposition $P \to Q$) 2. The water is boiling. (Proposition P) What can be inferred? ■

Exercise 2.18 Apply Modus Tollens in the following case: 1. If John has access permission, then he can enter the server room. (Proposition $P \to Q$) 2. John cannot enter the server room. (Proposition $\neg Q$) What can be concluded about John's access permission? ■

Exercise 2.19 Use Modus Ponens to solve: 1. If I study for the exam, then I will pass. (Proposition $P \to Q$) 2. I study for the exam. (Proposition P) What conclusion can be drawn about passing the exam? ■

Exercise 2.20 Apply Modus Tollens in this case: 1. If the equipment is on, then the screen will display images. (Proposition $P \to Q$) 2. The screen does not display images. (Proposition $\neg Q$) What can be inferred about the state of the equipment? ■

Exercise 2.21 Consider the following argument and identify whether Modus Ponens or Modus Tollens is applied: 1. If Peter has the vaccine, then he is protected against the flu. (Proposition $P \to Q$) 2. Peter is not protected against the flu. (Proposition $\neg Q$) What can be deduced about whether Peter has the vaccine or not? ■

2.5.3 Applications: Solving Problems of Information Order and Temporal Relationships

Exercise 2.22 Given the following process: 1. A: "The report has been drafted."2. B: "The report has been reviewed."3. C: "The report has been sent."

Conditions: - The report is reviewed only after being drafted: $A \rightarrow B$. - The report is sent only after being reviewed: $B \rightarrow C$.
If we know that the report has been sent (C), what can we infer about the states of A and B? ■

Exercise 2.23 Consider a security system: 1. X: "The alarm is armed."2. Y: "The sensor detects motion."3. Z: "The alarm is triggered."
Conditions: - The alarm is triggered only if it is armed and the sensor detects motion: $(X \wedge Y) \rightarrow Z$. - If the alarm has been triggered (Z), what can be inferred about the states of X and Y? ■

Exercise 2.24 In a factory, the following conditions apply: 1. P: "The raw materials have arrived."2. Q: "The machinery is set up."3. R: "Production can begin."
Relationships: - Production can begin only if the raw materials have arrived and the machinery is set up: $(P \wedge Q) \rightarrow R$. - If production has begun (R), what can be inferred about P and Q? ■

Exercise 2.25 A building access process involves: 1. A: "The visitor registers at reception."2. B: "The guard verifies the authorization."3. C: .^ccess to the building is granted."
Conditions: - Access is granted only if the visitor registers at reception and the guard verifies the authorization: $(A \wedge B) \rightarrow C$. - If access is granted (C), what can be inferred about events A and B? ■

Exercise 2.26 In a shipping protocol: 1. M: "The package has been packed."2. N: "The package has been labeled."3. O: "The package has been dispatched."
Conditions: - The package is labeled only after being packed: $M \rightarrow N$. - The package is dispatched only after being labeled: $N \rightarrow O$.
If the package has been dispatched (O), what can be concluded about M and N? ■

3. Inductive and Deductive Reasoning

3.1 Inductive Reasoning: Identifying Patterns and Generalization

3.1.1 Recognition of Numerical Series

The recognition of numerical series is fundamental in mathematics as it allows identifying patterns and predicting future terms in a sequence.

Definition 3.1.1 A **numerical series** is an ordered sequence of numbers that follows a specific law or pattern.

■ **Example 3.1** Consider the series $2,4,6,8,\ldots$. We observe that each term is the result of adding 2 to the previous term. This is an example of an *arithmetic progression* with a common difference $d = 2$. ■

Theorem 3.1.1 In an arithmetic progression, the n-th term a_n is calculated as:

$$a_n = a_1 + (n-1)d,$$

where a_1 is the first term and d is the common difference.

Demostración. Given that we are considering an arithmetic progression, each term is obtained by adding the common difference d to the previous term. This means that the first few terms can be expressed as:

$$a_1, \quad a_2 = a_1 + d, \quad a_3 = a_1 + 2d, \quad \ldots, \quad a_n = a_1 + (n-1)d.$$

We observe that to obtain the second term a_2, we add d to the first term a_1; to obtain the third term a_3, we add $2d$ to the first term, and in general, to obtain the n-th term, we add $(n-1)d$ to the first term.

Thus, the n-th term of the arithmetic progression is given by:

$$a_n = a_1 + (n-1)d.$$

This completes the proof. ■

Corollary 3.1.2 The sum of the first n terms of an arithmetic progression is:

$$S_n = \frac{n}{2}(a_1 + a_n).$$

(R) It is important to distinguish between an *arithmetic progression* and a *geometric progression.* In the former, a fixed amount is added, while in the latter, a constant ratio is multiplied.

■ **Example 3.2** Let us analyze the series $3, 9, 27, 81, \ldots$. Here, each term is the result of multiplying the previous term by 3. This is a *geometric progression* with a common ratio $r = 3$. ■

Lema 3.1.1 In a geometric progression, the n-th term a_n is calculated as:

$$a_n = a_1 \cdot r^{n-1},$$

where a_1 is the first term and r is the common ratio.

Exercise 3.1 Given the series $5, 10, 20, 40, \ldots$:

1. Identify whether it is an arithmetic or geometric progression.
2. Calculate the 7th term of the series.
3. Find the sum of the first 7 terms.

■

1. It is a **geometric progression** with a common ratio $r = 2$.
2. The 7th term is $a_7 = 5 \cdot 2^6 = 5 \cdot 64 = 320$.
3. The sum is $S_7 = a_1 \frac{r^n - 1}{r - 1} = 5 \frac{2^7 - 1}{2 - 1} = 5 \cdot (128 - 1) = 5 \cdot 127 = 635$.

(R) Recognizing the type of series is crucial to applying the appropriate formulas and solving related problems.

3.1.2 Identification of Geometric Patterns

The identification of geometric patterns is essential in mathematics and other disciplines, as it allows understanding and predicting spatial structures and relationships between shapes.

Definition 3.1.2 A **geometric pattern** is a repetitive and systematic arrangement of shapes, figures, or lines that follows a specific rule or law.

■ **Example 3.3** Consider a mosaic formed by equilateral triangles that repeat in all directions of the plane. This is an example of a geometric pattern with translational and rotational symmetry. ■

Theorem 3.1.3 In a regular geometric pattern, the internal angle θ of a regular polygon that tessellates the plane is given by:

$$\theta = \frac{(n-2) \times 180^\circ}{n},$$

where n is the number of sides of the polygon.

Demostración. To calculate the internal angle θ of a regular polygon with n sides, consider the sum of the internal angles of a polygon with n sides. This sum can be calculated using the formula:

$$\text{sum of internal angles} = (n-2) \times 180^\circ.$$

Since the polygon is regular, all its internal angles are equal. Therefore, to find the internal angle θ at each vertex, divide the total sum of the internal angles by the number of sides n:

$$\theta = \frac{(n-2) \times 180^\circ}{n}.$$

This proves that the internal angle of a regular polygon with n sides is given by the expression:

$$\theta = \frac{(n-2) \times 180^\circ}{n}.$$

■

Corollary 3.1.4 The only regular polygons that can tessellate the plane by themselves are the equilateral triangle ($n = 3$), the square ($n = 4$), and the regular hexagon ($n = 6$).

(R) Tessellation is a key example of how geometric patterns are applied in art, architecture, and nature, allowing surfaces to be covered without overlaps or gaps.

Lema 3.1.2 In a geometric pattern with radial symmetry, the number of axes of symmetry is equal to the number of repetitions of the motif around the center.

■ **Example 3.4** A snowflake exhibits radial symmetry of order 6, meaning it has 6 axes of symmetry passing through its center. ■

Exercise 3.2 Analyze the pattern formed by concentric circles with equal spacing:

1. Determine the relationship between the radius of each circle and its position in the pattern.
2. Calculate the area between the fifth and sixth circle if the distance between consecutive circles is 1 unit.

■

1. If r_n is the radius of the n-th circle, then $r_n = n \times d$, where d is the distance between consecutive circles. In this case, $d = 1$, so $r_n = n$.
2. The area between the fifth and sixth circles is the difference between the areas of both circles:

$$A = \pi r_6^2 - \pi r_5^2 = \pi(6^2 - 5^2) = \pi(36 - 25) = 11\pi \text{ square units.}$$

(R) Geometric patterns can be analyzed using concepts from Euclidean geometry, geometric transformations, and symmetries, providing a deeper understanding of their properties and applications.

3.2 Deductive Reasoning: Using Premises to Reach Conclusions

3.2.1 Deduction in Mathematical Arguments

Deduction is a logical process that allows conclusions to be inferred from given premises. In mathematics, deduction is essential to construct sound arguments and prove theorems.

Definition 3.2.1 A **deductive argument** is a sequence of propositions where, starting from premises accepted as true, a conclusion that necessarily follows from these premises is derived.

■ **Example 3.5** Consider the following premises:

1. All even numbers greater than 2 are composite.

2. 4 is an even number greater than 2.

Conclusion: 4 is a composite number.

This is a valid deductive argument because the conclusion logically follows from the premises. ■

Theorem 3.2.1 — **Hypothetical Syllogism.** If $p \implies q$ and $q \implies r$, then $p \implies r$.

Demostración. To prove the theorem, we start with the two premises:
1. $p \implies q$: If p is true, then q is true.
2. $q \implies r$: If q is true, then r is true.
We want to prove that $p \implies r$, i.e., if p is true, then r is also true.
Suppose p is true. According to the first premise ($p \implies q$), this implies that q must also be true. Now, using the second premise ($q \implies r$), since q is true, we can conclude that r must also be true. Therefore, we have proven that if p is true, then r is true. This confirms that:

$$p \implies r.$$

■

Lema 3.2.1 If a and b are even integers, then their sum $a+b$ is also an even integer.

Demostración. By definition, an integer k is even if there exists an integer m such that $k = 2m$. Let $a = 2m$ and $b = 2n$, where m and n are integers. Then:

$$a+b = 2m+2n = 2(m+n).$$

Since $m+n$ is an integer, $a+b$ is even. ■

Corollary 3.2.2 The sum of any even number of odd integers is an even number.

Demostración. An odd integer can be expressed as $2k+1$, where k is an integer. The sum of two odd numbers is:

$$(2k+1)+(2l+1) = 2(k+l+1).$$

This is an even number, as $k+l+1$ is an integer. ■

R Deduction allows us to build logical chains that lead from premises to conclusions, strengthening the validity of mathematical arguments.

Exercise 3.3 Prove that if n is a natural number and n^2 is even, then n is even. ■

Suppose n is odd. Then $n = 2k+1$ for some integer k. Calculate n^2:

$$n^2 = (2k+1)^2 = 4k^2+4k+1 = 2(2k^2+2k)+1.$$

This shows that n^2 is odd, which contradicts the premise that n^2 is even. Therefore, n must be even.

Theorem 3.2.3 — Contrapositive. For any propositions p and q, $p \implies q$ is logically equivalent to $\neg q \implies \neg p$.

Demostración. We want to prove that $p \implies q$ is logically equivalent to $\neg q \implies \neg p$.
Recall that an implication $p \implies q$ is false only when p is true and q is false; in all other cases, it is true. Now, consider the contrapositive $\neg q \implies \neg p$:
1. If q is false, then for $\neg q \implies \neg p$ to be true, p must also be false (otherwise, the implication would be false).
2. If q is true, then $\neg q$ is false, and an implication with a false antecedent (like $\neg q$) is true regardless of the value of $\neg p$.
We observe that the truth values of $p \implies q$ and $\neg q \implies \neg p$ coincide in all possible cases for p and q, meaning both expressions are logically equivalent.
Therefore:

$$p \implies q \iff \neg q \implies \neg p.$$

This concludes the proof. ∎

▪ **Example 3.6** Prove that if an integer n is not divisible by 3, then n^2 is not divisible by 9.
Proof: Suppose n is not divisible by 3. Then n can be of the form $3k+1$ or $3k+2$. Calculate n^2 in both cases:

- If $n = 3k+1$:

$$n^2 = (3k+1)^2 = 9k^2 + 6k + 1.$$

- If $n = 3k+2$:

$$n^2 = (3k+2)^2 = 9k^2 + 12k + 4.$$

In both cases, n^2 is not divisible by 9, as the constant term is not a multiple of 9. ▪

(R) Using the contrapositive is an effective technique in proving implications, especially when the direct implication is difficult to prove.

Exercise 3.4 Let n be an integer. Prove that if n^2 is divisible by 4, then n is divisible by 2. ▪

Suppose n is not divisible by 2, i.e., n is odd and can be written as $n = 2k+1$. Then:

$$n^2 = (2k+1)^2 = 4k^2 + 4k + 1.$$

The result n^2 is odd and therefore not divisible by 4, which contradicts the premise. Thus, n must be even and divisible by 2.

Theorem 3.2.4 — Proof by Contradiction. If assuming a proposition is false leads to a contradiction, then the proposition must be true.

Demostración. To prove this theorem, suppose we want to show that a proposition P is true. We proceed by **proof by contradiction**.
1. Assume, with the intent of reaching a contradiction, that P is false. That is, assume $\neg P$ is true.
2. From this assumption, if we derive a contradiction, it means that our initial assumption $\neg P$ cannot be correct because a contradiction implies a logical impossibility.
3. Since $\neg P$ leads to a contradiction, $\neg P$ must be false.

4. If $\neg P$ is false, then P must be true.
Thus, we have shown that if assuming P is false leads to a contradiction, then P must be true. This confirms the principle of proof by contradiction.

■

■ **Example 3.7** Prove that $\sqrt{2}$ is an irrational number.
Proof by Contradiction: Suppose that $\sqrt{2}$ is rational. Then it can be expressed as $\sqrt{2} = \frac{a}{b}$, where a and b are positive integers that are coprime. Then:

$$2 = \left(\frac{a}{b}\right)^2 \implies 2b^2 = a^2.$$

This implies that a^2 is even, and by the previous lemma, a is even. Thus $a = 2k$ for some integer k. Substituting:

$$2b^2 = (2k)^2 \implies 2b^2 = 4k^2 \implies b^2 = 2k^2.$$

This implies that b^2 is even and, therefore, b is even. But if both a and b are even, then they share a common factor, which contradicts the assumption that they are coprime. Therefore, $\sqrt{2}$ is irrational.
■

(R) Proof by contradiction is a powerful tool in mathematics for establishing the truth of a proposition by showing that its negation leads to a contradiction.

Exercise 3.5 Using proof by contradiction, show that there do not exist positive integers x and y such that $x^2 = 3y^2$. ■

Suppose that there exist positive integers x and y such that $x^2 = 3y^2$. Without loss of generality, suppose x and y are coprime. Then:

$$x^2 = 3y^2 \implies \left(\frac{x}{y}\right)^2 = 3.$$

This implies that $\frac{x}{y} = \sqrt{3}$, which is irrational, contradicting the fact that x and y are integers. Therefore, such positive integers x and y do not exist.

(R) Deduction in mathematical arguments is fundamental to the advancement of mathematical knowledge, enabling rigorous and reliable proofs.

3.2.2 Premise Analysis in Proofs

The analysis of premises is an essential part of constructing and understanding mathematical proofs. It involves carefully examining the initial assumptions to ensure that the derived conclusions are logically valid and well-founded.

Definition 3.2.2 A **premise** is a proposition or statement assumed to be true, from which conclusions are drawn in a logical argument.

■ **Example 3.8** In the proof that the sum of two even numbers is even, the premises are:

1. An even number can be expressed as $2k$, where k is an integer.
2. The sum of two numbers of the form $2k$ and $2m$ is $2k + 2m$.

These premises allow us to deduce that $2k + 2m = 2(k + m)$, which is a multiple of 2 and, therefore, even. ■

Theorem 3.2.5 If all the premises in an argument are true and the reasoning is valid, then the conclusion is necessarily true.

Demostración. To prove this theorem, consider an argument with a set of premises $P_1, P_2, \ldots, P_n$ and a conclusion C.
1. Assume that all the premises $P_1, P_2, \ldots, P_n$ are true.
2. Additionally, assume that the reasoning is valid, meaning there exists a logical relationship such that if all the premises are true, then the conclusion C must also be true.
3. Since the reasoning is valid and all the premises are true, by the definition of logical validity, the conclusion C cannot be false unless at least one of the premises is false.
4. Therefore, the conclusion C is necessarily true.
In conclusion, if all the premises in an argument are true and the reasoning is valid, then the conclusion is necessarily true. ■

Lema 3.2.2 In a logical argument, if a premise is false, the argument may still be valid, but the conclusion is not guaranteed to be true.

Demostración. An argument is valid if its logical structure is correct, regardless of the truth of the premises. However, if one or more premises are false, the conclusion may be true or false, but its truth cannot be guaranteed based solely on the argument. ■

Corollary 3.2.6 To ensure the truth of a conclusion, it is necessary not only for the reasoning to be valid but also for all the premises to be true.

R It is crucial to distinguish between the validity of an argument and the truth of its premises. An argument can be logically valid but lead to a false conclusion if any premise is false.

■ **Example 3.9** Consider the following argument:
1. Premise 1: All mammals can fly.
2. Premise 2: Dolphins are mammals.
3. Conclusion: Dolphins can fly.

The argument is logically valid, but Premise 1 is false, leading to a false conclusion. ■

Exercise 3.6 Analyze the premises and determine the validity of the following argument:
1. Premise 1: If a figure is a square, then it has four equal sides.
2. Premise 2: Figure F has four equal sides.
3. Conclusion: Figure F is a square.

■

The argument commits the *affirming the consequent* fallacy. While all squares have four equal sides, not all figures with four equal sides are squares (e.g., a rhombus). Therefore, the premises are insufficient to conclude that F is a square. The argument is invalid.

R This example illustrates the importance of carefully analyzing premises and understanding logical implications to avoid errors in proofs.

Theorem 3.2.7 — **Disjunctive Syllogism.** Given the premises:
1. $p \vee q$ (p or q is true)
2. $\neg p$ (p is false)

We can conclude that q is true.

Demostración. To prove the theorem, we start with the two given premises:
1. $p \vee q$: At least one of p or q is true. 2. $\neg p$: p is false.
We want to prove that q must be true.
Since the first premise tells us that at least one of p or q is true, there are two possible cases:
- If p is true, then $p \vee q$ would be true regardless of the value of q. - If p is false, then for $p \vee q$ to be true, q must be true (otherwise, $p \vee q$ would be false).
The second premise tells us that p is false, which implies that q must be true for $p \vee q$ to hold.
Therefore, we conclude that q is true. ■

■ **Example 3.10** Applying disjunctive syllogism:

1. Premise 1: Either the number n is even, or it is odd.
2. Premise 2: n is not even.
3. Conclusion: n is odd.

This argument is valid, and the conclusion is correct if the premises are true. ■

Exercise 3.7 Determine if the following argument is valid and analyze its premises:

1. Premise 1: If n is divisible by 4, then n is divisible by 2.
2. Premise 2: n is divisible by 2.
3. Conclusion: n is divisible by 4.

■

The argument commits the *affirming the consequent* fallacy. While n being divisible by 4 implies that it is divisible by 2, the fact that n is divisible by 2 does not necessarily mean it is divisible by 4 (e.g., $n = 6$ is divisible by 2 but not by 4). Therefore, the argument is invalid.

(R) It is essential not only to verify the logical validity of the argument but also to carefully examine the premises and their relationship to the conclusion to avoid common errors in proofs.

Theorem 3.2.8 — **Modus Tollens.** Given the premises:

1. $p \implies q$ (if p then q)
2. $\neg q$ (q is false)

We can conclude that $\neg p$ (p is false).

Demostración. To prove the theorem, we start with the two premises:
1. $p \implies q$: If p is true, then q is true. 2. $\neg q$: q is false.
We want to prove $\neg p$, i.e., that p is false.
Consider the implication $p \implies q$. The implication is false only when p is true and q is false. However, this contradicts the first premise, as $p \implies q$ is true.
Since q is false ($\neg q$), the only way to maintain the truth of $p \implies q$ is if p is also false. In other words, if q is false, then p cannot be true, implying $\neg p$ must be true.
Therefore, we have proven that $\neg q$ and $p \implies q$ imply $\neg p$. ■

■ **Example 3.11** Applying Modus Tollens:

1. Premise 1: If a number is divisible by 6, then it is divisible by 3.
2. Premise 2: The number n is not divisible by 3.
3. Conclusion: n is not divisible by 6.

The argument is valid, and the conclusion logically follows from the premises. ■

Exercise 3.8 Analyze the following argument and determine if it is valid:

1. Premise 1: If x is a prime number greater than 2, then x is odd.
2. Premise 2: x is odd.
3. Conclusion: x is a prime number greater than 2.

The argument commits the *affirming the consequent* fallacy. While all prime numbers greater than 2 are odd, not all odd numbers are prime (e.g., 9 is odd but not prime). Therefore, the argument is invalid.

R Premise analysis helps identify logical fallacies and strengthen the soundness of mathematical proofs through a deep understanding of the implications and relationships between propositions.

3.3 Applications: Numerical Series and Graphical Analogies

3.3.1 Analogies Based on Arithmetic Progressions

Arithmetic progressions are numerical sequences where each term is obtained by adding a constant amount to the previous term. Analogies based on these progressions allow solving problems and establishing relationships between different mathematical concepts.

Definition 3.3.1 An **arithmetic progression** is a sequence of numbers (a_n) such that the difference between consecutive terms is constant. That is, there exists a constant d (common difference) such that:

$$a_n = a_{n-1} + d \quad \text{for all } n \geq 2.$$

Example 3.12 The sequence $5, 8, 11, 14, 17, \ldots$ is an arithmetic progression with the first term $a_1 = 5$ and common difference $d = 3$. ■

Theorem 3.3.1 The n-th term of an arithmetic progression can be expressed as:

$$a_n = a_1 + (n-1)d,$$

where a_1 is the first term and d is the common difference.

Demostración. To prove the formula for the n-th term of an arithmetic progression, consider the definition of an arithmetic progression: each term is obtained by adding a constant difference d to the previous term.
Given the first term a_1, the first terms of the arithmetic progression are:

$$a_1, \quad a_2 = a_1 + d, \quad a_3 = a_1 + 2d, \quad \ldots, \quad a_n = a_1 + (n-1)d.$$

We observe the pattern in the terms: - For a_2, d has been added once to a_1. - For a_3, d has been added twice to a_1. - In general, for the n-th term a_n, d has been added $(n-1)$ times to a_1.
Thus, the n-th term a_n is given by the formula:

$$a_n = a_1 + (n-1)d.$$

This concludes the proof. ■

Lema 3.3.1 The sum of the first n terms of an arithmetic progression is:

$$S_n = \frac{n}{2}(a_1 + a_n).$$

Demostración. The sum of the first n terms is:

$$S_n = a_1 + a_2 + a_3 + \cdots + a_n.$$

We can write the sum in reverse order:

$$S_n = a_n + a_{n-1} + a_{n-2} + \cdots + a_1.$$

Adding both expressions term by term:

$$2S_n = (a_1 + a_n) + (a_2 + a_{n-1}) + \cdots + (a_n + a_1).$$

There are n pairs, and each pair sums to $(a_1 + a_n)$, so:

$$2S_n = n(a_1 + a_n) \implies S_n = \frac{n}{2}(a_1 + a_n).$$

■

■ **Example 3.13** Calculate the sum of the first 20 terms of the arithmetic progression $7, 10, 13, \ldots$.
Solution: Here, $a_1 = 7$, $d = 3$, and $n = 20$.
First, find a_{20}:

$$a_{20} = a_1 + (20-1)d = 7 + 19 \times 3 = 7 + 57 = 64.$$

Now, apply the sum formula:

$$S_{20} = \frac{20}{2}(7 + 64) = 10 \times 71 = 710.$$

■

(R) Arithmetic progressions are useful in various fields, including finance, physics, and logical reasoning problem-solving.

Exercise 3.9 Based on numerical analogies derived from arithmetic progressions, complete the following exercises:

1. If $2 \to 5$, $5 \to 11$, $11 \to 23$, what is the pattern, and what is the next pair?
2. Find the missing term in the sequence: 15, _, 27, 33, knowing it is an arithmetic progression.
3. Determine the sum of all even integers between 1 and 100.

■

1. Observe that from 2 to 5, the difference is 3; from 5 to 11, the difference is 6; from 11 to 23, the difference is 12. The differences form the sequence $3, 6, 12$, which doubles each time. Following the pattern, the next difference should be 24. Thus, the next number after 23 is $23 + 24 = 47$. Therefore, the next pair is $23 \to 47$.
2. The common difference d can be found using the known terms. Between 27 and 33, the difference is 6, so $d = 6$. The missing term before 27 is $27 - 6 = 21$. Therefore, the sequence is 15, 21, 27, 33.
3. The even integers between 1 and 100 are $2, 4, 6, \ldots, 100$. This is an arithmetic progression with $a_1 = 2$, $d = 2$, and $a_n = 100$. The number of terms n is:

$$n = \frac{(a_n - a_1)}{d} + 1 = \frac{(100-2)}{2} + 1 = 49 + 1 = 50.$$

The sum is:

$$S_n = \frac{n}{2}(a_1 + a_n) = 25 \times (2 + 100) = 25 \times 102 = 2550.$$

Theorem 3.3.2 If a sequence of numbers satisfies the recurrence relation $a_n = a_{n-1} + d$ for all $n \geq 2$, then the sequence is an arithmetic progression with a common difference d.

Demostración. Assume a sequence of numbers $\{a_n\}$ satisfies the recurrence relation

$$a_n = a_{n-1} + d$$

for all $n \geq 2$, where d is a constant. We aim to prove that this sequence is an arithmetic progression with a common difference d.
1. From the recurrence relation, we observe that each term a_n is obtained by adding a constant d to the previous term a_{n-1}. This implies that the difference between consecutive terms is always d, i.e.,

$$a_n - a_{n-1} = d \quad \text{for all } n \geq 2.$$

2. By the definition of an arithmetic progression, a sequence is arithmetic if the difference between consecutive terms is constant. Since the difference between consecutive terms in this sequence is d, the sequence is, by definition, an arithmetic progression with a common difference d.
Therefore, we have demonstrated that any sequence satisfying the recurrence relation $a_n = a_{n-1} + d$ is an arithmetic progression with a common difference d.

■

Lema 3.3.2 In an arithmetic progression, the average of any pair of terms equidistant from the extremes is equal to the average of the first and last terms.

Demostración. Consider an arithmetic progression with terms $a_1, a_2, \ldots, a_n$. Let a_k and a_{n-k+1} be terms equidistant from the extremes, where $1 \leq k \leq n$.
The average is calculated as:

$$\frac{a_k + a_{n-k+1}}{2} = \frac{[a_1 + (k-1)d] + [a_1 + (n-k)d]}{2} = \frac{2a_1 + (n-1)d}{2} = \frac{a_1 + a_n}{2}.$$

■

■ **Example 3.14** In the arithmetic progression $3, 7, 11, 15, 19$, verify that the average of equidistant terms is constant.

- a_1 and a_5: $\frac{3+19}{2} = 11$
- a_2 and a_4: $\frac{7+15}{2} = 11$
- a_3 and a_3: $\frac{11+11}{2} = 11$

The average is always 11, which is the average of a_1 and a_5. ■

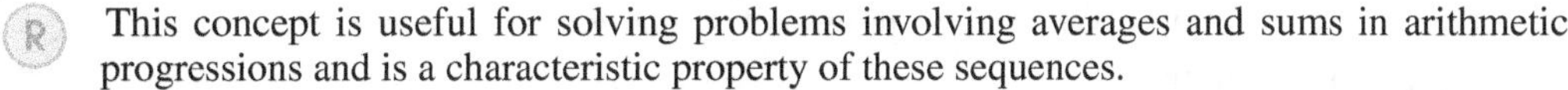

This concept is useful for solving problems involving averages and sums in arithmetic progressions and is a characteristic property of these sequences.

3.3.2 Geometric Series Applied to Visual Problems

Geometric series are sequences where each term is obtained by multiplying the previous term by a constant called the common ratio. These series have broad applications in visual problems, particularly in figures that exhibit repetitive patterns or decreasing scales.

Definition 3.3.2 A **geometric series** is a sequence of numbers (a_n) where each term is obtained by multiplying the previous term by a common ratio r:

$$a_n = a_{n-1} \cdot r, \quad \text{for all } n \geq 2.$$

Example 3.15 Consider a figure where each level contains a number of squares that is half the number of squares in the previous level. If the first level contains 16 squares, the subsequent levels will have 8, 4, 2, and 1 square, respectively. This is a geometric series with the first term $a_1 = 16$ and common ratio $r = \frac{1}{2}$. ■

Theorem 3.3.3 The sum of the first n terms of a geometric series is:

$$S_n = a_1 \frac{1-r^n}{1-r}, \quad \text{if } r \neq 1.$$

Demostración. Consider a geometric series with the first term a_1, ratio r, and n terms. We want to find the sum S_n of the first n terms:

$$S_n = a_1 + a_1 r + a_1 r^2 + \cdots + a_1 r^{n-1}.$$

Multiply both sides of the equation by r:

$$rS_n = a_1 r + a_1 r^2 + a_1 r^3 + \cdots + a_1 r^n.$$

Subtract this equation from the first:

$$S_n - rS_n = (a_1 + a_1 r + a_1 r^2 + \cdots + a_1 r^{n-1}) - (a_1 r + a_1 r^2 + a_1 r^3 + \cdots + a_1 r^n),$$

which simplifies to:

$$S_n(1-r) = a_1(1-r^n).$$

Assuming $r \neq 1$, divide both sides by $1-r$ to obtain:

$$S_n = a_1 \frac{1-r^n}{1-r}.$$

This completes the proof. ■

Corollary 3.3.4 If $|r| < 1$, the infinite sum of the geometric series is:

$$S = \lim_{n\to\infty} S_n = \frac{a_1}{1-r}.$$

Demostración. For a geometric series with the first term a_1 and ratio r such that $|r| < 1$, the sum of the first n terms is given by:

$$S_n = a_1 \frac{1-r^n}{1-r}.$$

We want to find the infinite sum of the series, which is the limit of S_n as $n \to \infty$:

$$S = \lim_{n\to\infty} S_n = \lim_{n\to\infty} a_1 \frac{1-r^n}{1-r}.$$

Since $|r| < 1$, the term r^n approaches 0 as $n \to \infty$. Thus:

$$\lim_{n\to\infty} r^n = 0.$$

Substituting into the expression for S:

$$S = a_1 \frac{1-0}{1-r} = \frac{a_1}{1-r}.$$

This shows that the infinite sum of the geometric series is:

$$S = \frac{a_1}{1-r}.$$

■

R This property is fundamental in visual problems involving fractals or infinite patterns, where the total sum converges to a finite value.

■ **Example 3.16** Consider a square with side length 1. A circle is inscribed within the square, followed by another square inscribed within the circle, and so on infinitely. The total area of all the inscribed squares can be calculated using a geometric series.
The area of the first square is $A_1 = 1$. The ratio between the areas of two consecutive squares is $r = \left(\frac{1}{\sqrt{2}}\right)^2 = \frac{1}{2}$, since each inscribed square's side is scaled by $\frac{1}{\sqrt{2}}$.
Thus, the total sum of the areas is:

$$S = \frac{A_1}{1-r} = \frac{1}{1-\frac{1}{2}} = 2.$$

■

Lema 3.3.3 In a visual problem where each iteration reduces a linear dimension by a factor k, the area is reduced by a factor k^2, and the volume by a factor k^3.

Demostración. Let L be the initial linear dimension. After a reduction by a factor k, the new dimension is $L' = kL$. The area is calculated as $A = (L')^2 = (kL)^2 = k^2L^2$. Similarly, the volume is $V = (L')^3 = k^3L^3$. ■

Exercise 3.10 In an equilateral triangle with side length 1, an inverted triangle is drawn inside it by connecting the midpoints of its sides, forming four smaller equilateral triangles. This process is repeated infinitely with the central inverted triangles. Calculate the total area of all the removed triangles. ■

The area of the original triangle is:

$$A_1 = \frac{\sqrt{3}}{4} \times 1^2 = \frac{\sqrt{3}}{4}.$$

In each iteration, a triangle is removed whose area is $\frac{1}{4}$ of the area of the triangle in the previous iteration.
The total sum of the removed areas is:

$$S = A_1\left(\frac{1}{4} + \left(\frac{1}{4}\right)^2 + \left(\frac{1}{4}\right)^3 + \dots\right) = A_1\left(\frac{1/4}{1-1/4}\right) = A_1\left(\frac{1/4}{3/4}\right) = \frac{A_1}{3}.$$

Thus, the total removed area is $\frac{\sqrt{3}}{12}$.

R This is an example of how geometric series allow the calculation of total areas and volumes in figures exhibiting self-similarity and infinite patterns.

Theorem 3.3.5 In a fractal figure generated by iterations of reduced scaling by a factor r, the total sum of a geometric property (such as length, area, or volume) that decreases geometrically is finite if $|r| < 1$.

Demostración. The geometric property in the n-th iteration is $P_n = P_1 r^{n-1}$. The total sum is:

$$S = \sum_{n=1}^{\infty} P_n = P_1 \sum_{n=1}^{\infty} r^{n-1} = P_1 \left(\frac{1}{1-r}\right), \quad \text{if } |r| < 1.$$

■

■ **Example 3.17** Calculate the total length of a broken line formed by dividing a segment of length 1 into two equal parts and adding perpendicular segments of length $\frac{1}{2}$ at each division, repeating this process infinitely.
Solution:
In each iteration, segments are added with a total length of:

$$L_n = 2^{n-1}\left(\frac{1}{2^n}\right) = \frac{1}{2^{n-1}}.$$

The total added length is:

$$S = \sum_{n=1}^{\infty} L_n = \sum_{n=1}^{\infty} \frac{1}{2^{n-1}} = 2.$$

■

Exercise 3.11 An infinite mirror is formed by placing two parallel mirrors facing each other. If a ray of light loses a fixed percentage of its intensity with each reflection, what is the total perceived intensity after summing all reflections? ■

If the initial intensity is I_0, and the reflection coefficient is r (with $0 < r < 1$), then the intensity after n reflections is:

$$I_n = I_0 r^n.$$

The total perceived intensity is:

$$S = I_0 \sum_{n=0}^{\infty} r^n = I_0 \left(\frac{1}{1-r}\right).$$

(R) This model applies in optics and physics, demonstrating how geometric series describe attenuation and multiple reflection phenomena.

Lema 3.3.4 In a decreasing geometric series with $0 < r < 1$, the n-th term approaches zero as $n \to \infty$.

Demostración. Since $0 < r < 1$, raising r to increasing powers leads to $r^n \to 0$ as $n \to \infty$. ■

Exercise 3.12 In a painting, an artist draws a circle of radius 1. Within this circle, another circle with radius $\frac{1}{2}$ is drawn, and circles continue to be nested with radii halving each time. Calculate the total sum of the areas of all the circles drawn. ■

■

The area of the first circle is $A_1 = \pi(1)^2 = \pi$. Each subsequent circle's area is $\left(\frac{1}{2}\right)^2 = \frac{1}{4}$ of the previous one.
The total sum of the areas is:

$$S = \pi\left(1 + \frac{1}{4} + \left(\frac{1}{4}\right)^2 + \ldots\right) = \pi\left(\frac{1}{1-\frac{1}{4}}\right) = \pi\left(\frac{1}{\frac{3}{4}}\right) = \frac{4\pi}{3}.$$

(R) Visual problems involving geometric series illustrate how infinite quantities can sum to a finite value, a key concept in calculus and mathematical analysis.

3.4 Solved Exercises

Exercise 3.13 Consider an arithmetic progression with the first term $a_1 = 3$ and common difference $d = 5$. Calculate the tenth term of the progression and the sum of the first ten terms. ■

Demostración. To find the tenth term, we use the formula for the n-th term of an arithmetic progression:

$$a_n = a_1 + (n-1)d.$$

Substituting $a_1 = 3$, $d = 5$, and $n = 10$:

$$a_{10} = 3 + (10-1)\cdot 5 = 3 + 9\cdot 5 = 3 + 45 = 48.$$

Now, we calculate the sum of the first ten terms using the formula for the sum of the first n terms:

$$S_n = \frac{n}{2}(a_1 + a_n).$$

Substituting $n = 10$, $a_1 = 3$, and $a_{10} = 48$:

$$S_{10} = \frac{10}{2}(3 + 48) = 5\cdot 51 = 255.$$

Thus, the tenth term is 48, and the sum of the first ten terms is 255. ■

—

Exercise 3.14 Prove that the sum of the first n odd numbers equals n^2. ■

Demostración. Let the series of the first n odd numbers be $1, 3, 5, \ldots, (2n-1)$. We aim to prove:

$$1 + 3 + 5 + \cdots + (2n-1) = n^2.$$

We proceed by mathematical induction.
Base case: For $n = 1$:

$$1 = 1^2,$$

which is true.
Inductive step: Assume the formula is true for $n = k$, i.e.,

$$1 + 3 + 5 + \cdots + (2k-1) = k^2.$$

We need to prove that the formula is true for $n = k+1$:

$$1+3+5+\cdots+(2k-1)+(2(k+1)-1) = (k+1)^2.$$

By the induction hypothesis, we know $1+3+5+\cdots+(2k-1) = k^2$. Thus:

$$1+3+5+\cdots+(2k-1)+(2(k+1)-1) = k^2+(2k+1).$$

Simplifying the right-hand side:

$$k^2+2k+1 = (k+1)^2.$$

Therefore, the formula holds for $n = k+1$.
By the principle of mathematical induction, the formula is true for all $n \in \mathbb{N}$. Hence, the sum of the first n odd numbers equals n^2. ■

—

Exercise 3.15 Prove that the square root of 3 is an irrational number.

Demostración. We proceed by contradiction. Suppose $\sqrt{3}$ is rational. Then, we can write $\sqrt{3} = \frac{a}{b}$, where a and b are positive integers that are coprime (i.e., their greatest common divisor is 1). Squaring both sides, we get:

$$3 = \frac{a^2}{b^2} \Rightarrow a^2 = 3b^2.$$

This implies that a^2 is divisible by 3, which in turn implies that a is divisible by 3 (because if the square of a number is divisible by 3, the number itself must also be divisible by 3).
Thus, we can write $a = 3k$ for some integer k. Substituting into the equation above, we get:

$$(3k)^2 = 3b^2 \Rightarrow 9k^2 = 3b^2 \Rightarrow b^2 = 3k^2.$$

This implies that b^2 is divisible by 3, and hence b is also divisible by 3.
We have now reached a contradiction, as both a and b are divisible by 3, which contradicts the assumption that they are coprime. Therefore, $\sqrt{3}$ cannot be rational, and it must be irrational. ■

—

Exercise 3.16 Calculate the infinite sum of the geometric series $\sum_{n=0}^{\infty} \frac{1}{2^n}$.

Demostración. The given geometric series is:

$$\sum_{n=0}^{\infty} \frac{1}{2^n}.$$

This is an infinite geometric series with the first term $a = 1$ and common ratio $r = \frac{1}{2}$.
The formula for the sum of an infinite geometric series with $|r| < 1$ is:

$$S = \frac{a}{1-r}.$$

Substituting the values of a and r:

$$S = \frac{1}{1-\frac{1}{2}} = \frac{1}{\frac{1}{2}} = 2.$$

Therefore, the infinite sum of the series is 2. ■

—

Exercise 3.17 Prove that if n^2 is even, then n is even. ■

Demostración. We proceed by contradiction. Suppose n^2 is even, but n is odd. If n is odd, then n can be written as $n = 2k+1$ for some integer k.
Calculating n^2:

$$n^2 = (2k+1)^2 = 4k^2 + 4k + 1 = 2(2k^2 + 2k) + 1.$$

We observe that n^2 is of the form $2m+1$ for some integer m, which implies that n^2 is odd.
This contradicts the assumption that n^2 is even. Therefore, our assumption that n is odd must be incorrect. This implies that n is even. ■

3.5 Proposed Exercises

3.5.1 Inductive Reasoning: Pattern Identification and Generalization

Exercise 3.18 Identify the pattern in the numerical sequence $2, 5, 10, 17, 26, \ldots$ and find the next term. ■

Exercise 3.19 Determine a general formula for the n-th term of the series $1, 4, 9, 16, 25, \ldots$, which represents the squares of natural numbers. ■

Exercise 3.20 Observe the sequence $3, 6, 12, 24, \ldots$ and determine whether it is arithmetic, geometric, or of another type. Find the seventh term of the sequence. ■

Exercise 3.21 Given the sequence $1, 1, 2, 3, 5, 8, \ldots$, which follows the Fibonacci sequence, find the tenth term of the sequence. ■

Exercise 3.22 In the sequence of even numbers $2, 4, 6, 8, \ldots$, find the sum of the first twenty terms. ■

3.5.2 Deductive Reasoning: Using Premises to Reach Conclusions

Exercise 3.23 Given the premises: 1. If it rains, then the street is wet. 2. The street is not wet. Use deductive reasoning to conclude whether it is raining or not. ■

Exercise 3.24 Consider the following premises: 1. If a number is even, then it is divisible by 2. 2. The number 10 is even.
Use the premises to deduce whether the number 10 is divisible by 2. ■

Exercise 3.25 Given the premises: 1. All cats are mammals. 2. All mammals have hearts. What can be concluded about cats using deductive reasoning? ■

Exercise 3.26 Given the premises: 1. If you study, then you will pass the exam. 2. You did not pass the exam.
Use deductive reasoning to determine whether you studied or not. ■

Exercise 3.27 Given the following premises: 1. If a number is a multiple of 4, then it is a multiple of 2. 2. The number 18 is a multiple of 2.
Can you conclude that the number 18 is a multiple of 4? Justify your answer. ■

3.5.3 Applications: Numerical Series and Graphic Analogies

Exercise 3.28 Given the geometric series $5, 10, 20, 40, \ldots$, determine the eighth term of the series. ■

Exercise 3.29 In a geometric figure, each level of a triangle has twice as many triangles as the previous level. If the first level has one triangle, how many triangles will the fifth level have? ■

Exercise 3.30 Calculate the sum of the first ten terms of the series $3, 6, 9, 12, \ldots$. ■

Exercise 3.31 Observe the series $100, 50, 25, 12{,}5, \ldots$. Determine whether it is an arithmetic or geometric progression, and calculate the infinite sum of the series if possible. ■

Exercise 3.32 Draw a pattern of squares where the area of the first square is 1, the area of the second is $\frac{1}{4}$, the area of the third is $\frac{1}{16}$, and so on. Calculate the infinite sum of the areas of all the squares. ■

4. Propositional Functions and Quantifiers

4.1 Predicates: Definition and Examples.

4.1.1 Predicates of One Variable

A **predicate** is a propositional function that depends on one or more variables. When specific values are assigned to these variables, the predicate becomes a proposition that can be either true or false.

Definition 4.1.1 A **predicate of one variable** is a function $P(x)$ that assigns each element x of a set D (domain) a proposition. That is, for each $x \in D$, $P(x)$ is a proposition that can be either true or false.

■ **Example 4.1** Let $P(x)$ be the predicate "x is an even number,"where the domain D is the set of integers $\mathbb{Z}$. Then, for $x = 2$, $P(2)$ is "2 is an even number,"which is a true proposition. For $x = 3$, $P(3)$ is "3 is an even number,"which is a false proposition. ■

Theorem 4.1.1 If $P(x)$ is a predicate of one variable over a finite domain $D = \{x_1, x_2, \ldots, x_n\}$, then the proposition "For all $x \in D$, $P(x)$ is trueïs equivalent to the conjunction $P(x_1) \wedge P(x_2) \wedge \cdots \wedge P(x_n)$.

Demostración. We aim to demonstrate that the proposition "For all $x \in D$, $P(x)$ is trueïs equivalent to the conjunction $P(x_1) \wedge P(x_2) \wedge \cdots \wedge P(x_n)$, where the domain D is finite and is given by $D = \{x_1, x_2, \ldots, x_n\}$.

1. The proposition "For all $x \in D$, $P(x)$ is true"means that $P(x)$ is true for every element x in the set D. Since D is finite, it contains exactly the elements $x_1, x_2, \ldots, x_n$.
2. By the definition of "for allïn a finite domain, the proposition "For all $x \in D$, $P(x)$ is trueïs equivalent to the statement that $P(x_1)$, $P(x_2)$, ..., $P(x_n)$ are all true.
3. In propositional logic, stating that all propositions $P(x_1), P(x_2), \ldots, P(x_n)$ are true is equivalent to stating that their conjunction is true, that is:

$$P(x_1) \wedge P(x_2) \wedge \cdots \wedge P(x_n).$$

Therefore, we have demonstrated that the proposition "For all $x \in D$, $P(x)$ is trueïs equivalent to the conjunction $P(x_1) \wedge P(x_2) \wedge \cdots \wedge P(x_n)$.

■

Lema 4.1.1 The negation of a universal proposition $\forall x \in D, P(x)$ is equivalent to the existence of a counterexample, that is:

$$\neg(\forall x \in D, P(x)) \equiv \exists x \in D \text{ such that } \neg P(x).$$

Demostración. By the definition of universal and existential quantification, the negation of "for all x, $P(x)$ïs "there exists some x such that $P(x)$ is false." ■

■ **Example 4.2** Consider the predicate $P(x)$: "x is greater than 0,"with domain $D = \mathbb{Z}$.
- The proposition $\forall x \in \mathbb{Z}, P(x)$ is .ᴬll integers are greater than 0,"which is false. - The negation is $\exists x \in \mathbb{Z}$ such that $\neg P(x)$, that is, "There exists an integer that is not greater than 0,"which is true (for example, $x = -1$).

■

(R) Predicates allow the generalization of propositions and working with them in terms of variables, which is fundamental in mathematical logic and in the formalization of mathematical theories.

Exercise 4.1 Let $P(x)$ be the predicate "x is divisible by 3,"with domain $D = \mathbb{Z}$. Determine whether the proposition $\exists x \in D$ such that $P(x)$ is true or false. Justify your answer.

■

The proposition $\exists x \in \mathbb{Z}$ such that $P(x)$ is true means "There exists an integer that is divisible by 3."This is true because $x = 3$ is an example (3 is divisible by 3). Therefore, the proposition is true.

Theorem 4.1.2 If $P(x)$ is a predicate of one variable and $Q(x)$ is another predicate of one variable over the same domain D, then:
1. $\forall x \in D, [P(x) \wedge Q(x)] \equiv [\forall x \in D, P(x)] \wedge [\forall x \in D, Q(x)]$.
2. $\exists x \in D, [P(x) \vee Q(x)] \equiv [\exists x \in D, P(x)] \vee [\exists x \in D, Q(x)]$.

Demostración. We will prove each part of the theorem separately.
Part 1 We want to prove that:

$$\forall x \in D, [P(x) \wedge Q(x)] \equiv [\forall x \in D, P(x)] \wedge [\forall x \in D, Q(x)].$$

1. ($\Rightarrow$) Suppose $\forall x \in D, [P(x) \wedge Q(x)]$ is true. This means that for every $x \in D$, both $P(x)$ and $Q(x)$ are true. Therefore: - $\forall x \in D, P(x)$ is true because $P(x)$ is true for each $x \in D$. - $\forall x \in D, Q(x)$ is also true because $Q(x)$ is true for each $x \in D$.
Thus, $[\forall x \in D, P(x)] \wedge [\forall x \in D, Q(x)]$ is true.
2. ($\Leftarrow$) Suppose $[\forall x \in D, P(x)] \wedge [\forall x \in D, Q(x)]$ is true. This implies that: - $\forall x \in D, P(x)$ is true. - $\forall x \in D, Q(x)$ is true.
Since $P(x)$ and $Q(x)$ are true for every $x \in D$, $P(x) \wedge Q(x)$ is also true for every $x \in D$. This means that $\forall x \in D, [P(x) \wedge Q(x)]$ is true.
Therefore, we have demonstrated that:

$$\forall x \in D, [P(x) \wedge Q(x)] \equiv [\forall x \in D, P(x)] \wedge [\forall x \in D, Q(x)].$$

Part 2 We want to prove that:

$$\exists x \in D, [P(x) \vee Q(x)] \equiv [\exists x \in D, P(x)] \vee [\exists x \in D, Q(x)].$$

1. ($\Rightarrow$) Suppose $\exists x \in D, [P(x) \vee Q(x)]$ is true. This means that there exists at least one $x \in D$ such that $P(x) \vee Q(x)$ is true. By the definition of disjunction, this implies that: - $P(x)$ is true for some $x \in D$, or - $Q(x)$ is true for some $x \in D$.
Therefore, $\exists x \in D, P(x)$ is true, or $\exists x \in D, Q(x)$ is true, which implies $[\exists x \in D, P(x)] \vee [\exists x \in D, Q(x)]$ is true.
2. ($\Leftarrow$) Suppose $[\exists x \in D, P(x)] \vee [\exists x \in D, Q(x)]$ is true. This means that: - $\exists x \in D, P(x)$ is true, or - $\exists x \in D, Q(x)$ is true.
In either case, there exists at least one $x \in D$ such that $P(x) \vee Q(x)$ is true. Thus, $\exists x \in D, [P(x) \vee Q(x)]$ is true.
With this, we have demonstrated that:

$$\exists x \in D, [P(x) \vee Q(x)] \equiv [\exists x \in D, P(x)] \vee [\exists x \in D, Q(x)].$$

This completes the proof. ■

■ **Example 4.3** Let $P(x)$: "x is even.ªnd $Q(x)$: "x is a multiple of 3,"with domain $D = \{1,2,3,4,5,6\}$.
- Evaluate $\forall x \in D, [P(x) \wedge Q(x)]$:
- Is it true that for all $x \in D$, x is even and x is a multiple of 3? No, because, for example, $x = 2$ is even but not a multiple of 3, and $x = 3$ is a multiple of 3 but not even. Therefore, the proposition is false.
- Evaluate $[\forall x \in D, P(x)] \wedge [\forall x \in D, Q(x)]$:
- $\forall x \in D, P(x)$: .ᴬll numbers in D are evenïs false. - $\forall x \in D, Q(x)$: .ᴬll numbers in D are multiples of 3ïs false. - The conjunction of two false propositions is false, so $[\forall x \in D, P(x)] \wedge [\forall x \in D, Q(x)]$ is false.
- Both results coincide, confirming the equivalence.

■

Exercise 4.2 Use the predicates $P(x)$: "x is prime.ªnd $Q(x)$: "x is odd,"with domain $D = \{2,3,4,5,6,7\}$. Verify the validity of the equivalence:

$$\exists x \in D, [P(x) \wedge Q(x)] \equiv [\exists x \in D, P(x)] \wedge [\exists x \in D, Q(x)].$$

■

- Evaluate $\exists x \in D, [P(x) \wedge Q(x)]$:
- Find $x \in D$ such that x is prime and odd. - The numbers that satisfy this are $x = 3,5,7$. - Therefore, the proposition is true.
- Evaluate $[\exists x \in D, P(x)] \wedge [\exists x \in D, Q(x)]$:
- $\exists x \in D, P(x)$: "There exists x in D that is primeïs true (e.g., $x = 2$). - $\exists x \in D, Q(x)$: "There exists x in D that is oddïs true (e.g., $x = 3$). - The conjunction of two true propositions is true.
- Both propositions are true, confirming the equivalence.

(R) The handling of predicates and quantifiers is fundamental for formalizing mathematical and logical arguments, enabling the precise expression of complex propositions.

Theorem 4.1.3 — **De Morgan's Laws for Quantifiers.** For any predicate $P(x)$ and domain D:
1. $\neg(\forall x \in D, P(x)) \equiv \exists x \in D$ such that $\neg P(x)$.
2. $\neg(\exists x \in D, P(x)) \equiv \forall x \in D$ such that $\neg P(x)$.

Demostración. We will prove each part of De Morgan's Laws for quantifiers separately.
Part 1 We want to prove that:

$$\neg(\forall x \in D, P(x)) \equiv \exists x \in D \text{ such that } \neg P(x).$$

1. ($\Rightarrow$) Suppose $\neg(\forall x \in D, P(x))$ is true. This means that the proposition $\forall x \in D, P(x)$ is false. If $\forall x \in D, P(x)$ is false, then there must exist at least one element $x \in D$ for which $P(x)$ is false. Therefore, there exists $x \in D$ such that $\neg P(x)$ is true. This implies that $\exists x \in D$ such that $\neg P(x)$ is true.
2. ($\Leftarrow$) Suppose $\exists x \in D$ such that $\neg P(x)$ is true. This means that there is at least one element $x \in D$ for which $P(x)$ is false. Therefore, $\forall x \in D, P(x)$ is not true, which implies that $\neg(\forall x \in D, P(x))$ is true.
Thus, we have proven that:

$$\neg(\forall x \in D, P(x)) \equiv \exists x \in D \text{ such that } \neg P(x).$$

Part 2 We want to prove that:

$$\neg(\exists x \in D, P(x)) \equiv \forall x \in D \text{ such that } \neg P(x).$$

1. ($\Rightarrow$) Suppose $\neg(\exists x \in D, P(x))$ is true. This means that the proposition $\exists x \in D, P(x)$ is false. If $\exists x \in D, P(x)$ is false, then there does not exist any $x \in D$ for which $P(x)$ is true. This implies that for all $x \in D$, $P(x)$ is false, or equivalently, that $\neg P(x)$ is true for all $x \in D$. Therefore, $\forall x \in D$ such that $\neg P(x)$ is true.
2. ($\Leftarrow$) Suppose $\forall x \in D$ such that $\neg P(x)$ is true. This means that $P(x)$ is false for every $x \in D$, which implies that there does not exist any $x \in D$ for which $P(x)$ is true. Therefore, $\exists x \in D, P(x)$ is false, which implies that $\neg(\exists x \in D, P(x))$ is true.
Thus, we have proven that:

$$\neg(\exists x \in D, P(x)) \equiv \forall x \in D \text{ such that } \neg P(x).$$

This completes the proof of De Morgan's Laws for quantifiers. ■

■ **Example 4.4** Let $P(x)$: "x is greater than 10,"with domain $D = \{8, 9, 10, 11, 12\}$.
- Evaluate $\neg(\forall x \in D, P(x))$:
- Ït is not true that for all $x \in D$, x is greater than 10. This is equivalent to "There exists $x \in D$ such that x is not greater than 10. Indeed, $x = 8$ satisfies that x is not greater than 10.

■

Exercise 4.3 Formulate the negation of the proposition .Ąll students in the class passed the examüsing quantifiers and predicates.

■

First, define the predicate $P(x)$: "x passed the exam,"where x belongs to the domain D of all students in the class.
The original proposition is $\forall x \in D, P(x)$.
The negation is $\neg(\forall x \in D, P(x))$, which by De Morgan's Laws is equivalent to $\exists x \in D$ such that $\neg P(x)$.
Interpretation: "There exists at least one student in the class who did not pass the exam."

R Understanding how to negate propositions with quantifiers is essential for logical analysis and to avoid common errors in mathematical reasoning.

4.1.2 Predicates of Two Variables

Predicates of two variables are propositional functions that depend on two variables and become propositions when specific values are assigned to both variables. These predicates allow us to express relationships between elements of one or more sets.

Definition 4.1.2 A **predicate of two variables** is a function $P(x,y)$ that assigns each ordered pair (x,y) of elements from a domain $D_x \times D_y$ a proposition. That is, for each $x \in D_x$ and $y \in D_y$, $P(x,y)$ is a proposition that can be either true or false.

Example 4.5 Let $P(x,y)$ be the predicate "x is greater than y,"where $D_x = D_y = \mathbb{N}$ (the set of natural numbers). Then:

- $P(5,3)$ is "5 is greater than 3,"which is true.
- $P(2,4)$ is "2 is greater than 4,"which is false.

■

Theorem 4.1.4 If $P(x,y)$ is a predicate of two variables over finite domains D_x and D_y, then the proposition $\forall x \in D_x, \forall y \in D_y, P(x,y)$ is equivalent to the conjunction of all possible propositions $P(x,y)$ for each $x \in D_x$ and $y \in D_y$.

Demostración. We want to demonstrate that the proposition

$$\forall x \in D_x, \forall y \in D_y, P(x,y)$$

is equivalent to the conjunction of all propositions $P(x,y)$ for each pair of values $x \in D_x$ and $y \in D_y$.
The proposition $\forall x \in D_x, \forall y \in D_y, P(x,y)$ means that $P(x,y)$ is true for every combination of x and y within the domains D_x and D_y.
Since D_x and D_y are finite, we can list all possible values of x in D_x and y in D_y. Then, this proposition reduces to verifying that $P(x,y)$ is true for each specific combination of x and y in these domains.
Saying that $P(x,y)$ is true for all $x \in D_x$ and $y \in D_y$ is the same as saying that each $P(x,y)$ is individually true for every pair. This can be expressed as the conjunction of all these propositions:

$$\forall x \in D_x, \forall y \in D_y, P(x,y) \iff P(x_1,y_1) \wedge P(x_1,y_2) \wedge \cdots \wedge P(x_m,y_n)$$

Thus, we have demonstrated that the proposition is equivalent to the conjunction of all possible propositions $P(x,y)$ for each $x \in D_x$ and $y \in D_y$. ■

Lema 4.1.2 The negation of a double universal proposition is equivalent to the existence of at least one pair for which the predicate is false:

$$\neg(\forall x \in D_x, \forall y \in D_y, P(x,y)) \equiv \exists x \in D_x, \exists y \in D_y \text{ such that } \neg P(x,y).$$

Demostración. Applying De Morgan's laws for quantifiers:

$$\neg(\forall x \in D_x, \forall y \in D_y, P(x,y)) \equiv \exists x \in D_x, \neg(\forall y \in D_y, P(x,y)) \equiv \exists x \in D_x, \exists y \in D_y \text{ such that } \neg P(x,y).$$

■

■ **Example 4.6** Let $P(x,y)$ be the predicate "x is divisible by y,"with $D_x = \{2,4,6\}$ and $D_y = \{1,2,3\}$.
- Evaluate $\forall x \in D_x, \exists y \in D_y$ such that $P(x,y)$:
- For $x = 2$, $P(2,1)$ is true (2 is divisible by 1). - For $x = 4$, $P(4,1)$ is true. - For $x = 6$, $P(6,1)$ is true.
Therefore, the proposition is true.
- Evaluate $\exists y \in D_y, \forall x \in D_x$ such that $P(x,y)$:
- Does there exist a y such that for all x, x is divisible by y? - Test $y = 1$: all x are divisible by 1, proposition is true. - Therefore, the proposition is true.

■

R The order of quantifiers is crucial in predicates with multiple variables, as changing the order can alter the meaning of the proposition.

Theorem 4.1.5 In general, for predicates of two variables, it holds that:

$$\forall x \in D_x, \exists y \in D_y, P(x,y) \not\equiv \exists y \in D_y, \forall x \in D_x, P(x,y).$$

Demostración. To demonstrate that

$$\forall x \in D_x, \exists y \in D_y, P(x,y) \not\equiv \exists y \in D_y, \forall x \in D_x, P(x,y),$$

that is, that these two expressions are not equivalent, it suffices to find a counterexample where one of the expressions is true and the other false.
Consider the domains $D_x = \{1,2\}$ and $D_y = \{1,2\}$, and the predicate $P(x,y)$ defined by:

$$P(x,y) = \begin{cases} \text{true} & \text{if } x = y, \\ \text{false} & \text{if } x \neq y. \end{cases}$$

Evaluate each expression separately:
1. **For $\forall x \in D_x, \exists y \in D_y, P(x,y)$:** For each x in D_x, we can find a y in D_y such that $P(x,y)$ is true: - If $x = 1$, choose $y = 1$, where $P(1,1)$ is true. - If $x = 2$, choose $y = 2$, where $P(2,2)$ is true. Therefore, $\forall x \in D_x, \exists y \in D_y, P(x,y)$ is true.
2. **For $\exists y \in D_y, \forall x \in D_x, P(x,y)$:** Find a single y in D_y such that $P(x,y)$ is true for all x in D_x. - If we choose $y = 1$, then $P(2,1)$ is false. - If we choose $y = 2$, then $P(1,2)$ is false.
There is no y in D_y such that $P(x,y)$ is true for all $x \in D_x$. Therefore, $\exists y \in D_y, \forall x \in D_x, P(x,y)$ is false.
This counterexample shows that $\forall x \in D_x, \exists y \in D_y, P(x,y)$ is true, while $\exists y \in D_y, \forall x \in D_x, P(x,y)$ is false. Thus, the two expressions are not equivalent.

■

Exercise 4.4 Let $P(x,y)$ be the predicate "x plus y equals 5,"with $D_x = D_y = \{1,2,3,4\}$. Determine whether the following propositions are true or false:

1. $\exists x \in D_x, \exists y \in D_y$ such that $P(x,y)$.
2. $\forall x \in D_x, \exists y \in D_y$ such that $P(x,y)$.
3. $\exists y \in D_y, \forall x \in D_x$ such that $P(x,y)$.
4. $\forall x \in D_x, \forall y \in D_y, P(x,y)$.

■

Demostración.

1. **True**. There are multiple pairs (x,y) that satisfy $x+y=5$, for example, $(1,4)$, $(2,3)$.
2. **True**. For each $x \in D_x$, we can find a $y \in D_y$ such that $x+y=5$:
   - For $x=1$, $y=4$.
   - For $x=2$, $y=3$.
   - For $x=3$, $y=2$.
   - For $x=4$, $y=1$.
3. **False**. Is there a y that works for all x?
   - If $y=1, x+1=5 \implies x=4$.
   - $y=1$ does not work for all x.
   - No y satisfies $x+y=5$ for all x.
4. **False**. Not all pairs (x,y) satisfy $x+y=5$. For example, $x=1$, $y=1$ gives $1+1=2\neq 5$.

■

(R) This exercise illustrates how changes in the order and type of quantifiers affect the truth value of propositions in predicates with two variables.

Theorem 4.1.6 — **Quantifier Exchange.** For certain domains and specific predicates, the following equivalence holds:

$$\forall x \in D_x, \forall y \in D_y, P(x,y) \equiv \forall y \in D_y, \forall x \in D_x, P(x,y).$$

Demostración. To prove that

$$\forall x \in D_x, \forall y \in D_y, P(x,y) \equiv \forall y \in D_y, \forall x \in D_x, P(x,y),$$

that is, that the quantifiers can be exchanged in this case, note that each expression states that $P(x,y)$ is true for all pairs (x,y) in the domains D_x and D_y.
1. For $\forall x \in D_x, \forall y \in D_y, P(x,y)$: This means that for each $x \in D_x$, the predicate $P(x,y)$ is true for all $y \in D_y$.
2. For $\forall y \in D_y, \forall x \in D_x, P(x,y)$: This means that for each $y \in D_y$, the predicate $P(x,y)$ is true for all $x \in D_x$.
Since both expressions require that $P(x,y)$ be true for all pairs $(x,y) \in D_x \times D_y$, the two expressions are equivalent, as both assert that $P(x,y)$ is true over the entire combination of the domains.
Therefore,

$$\forall x \in D_x, \forall y \in D_y, P(x,y) \equiv \forall y \in D_y, \forall x \in D_x, P(x,y).$$

■

Lema 4.1.3 If $P(x,y)$ is a predicate and $D_x = D_y$, then:

$$\forall x \in D, P(x,x) \implies \forall x \in D, \exists y \in D, P(x,y).$$

Demostración. If for all $x \in D$, $P(x,x)$ is true, then for each x, there exists at least one y (in this case $y = x$) such that $P(x,y)$ is true. ■

■ **Example 4.7** Let $P(x,y)$ be the predicate "x is equal to y,"with $D = \{1,2,3\}$.
- $\forall x \in D, P(x,x)$ is true, since each element is equal to itself. - By the lemma, $\forall x \in D, \exists y \in D$ such that $P(x,y)$ is true, which is correct.

■

Exercise 4.5 Define the predicate $P(x,y)$: "x is a friend of yïn a set of people $D = \{A,B,C\}$. If it is known that the friendship relation is symmetric (if x is a friend of y, then y is a friend of x), and that A is a friend of B, and B is a friend of C, is it necessarily true that A is a friend of C? Analyze using predicates and quantifiers.

■

Demostración. Symmetry gives us:

$$\forall x,y \in D, P(x,y) \implies P(y,x).$$

We know that:

$$P(A,B) \text{ is true}, \quad P(B,C) \text{ is true}.$$

However, we cannot conclude that $P(A,C)$ is true without additional information. Friendship in this case is not necessarily transitive. Therefore, it is not true that A is a friend of C based solely on the given information.

■

(R) This exercise demonstrates the importance of understanding the properties of relations (symmetry, reflexivity, transitivity) when working with predicates of two variables.

Theorem 4.1.7 If $P(x,y)$ is a reflexive and symmetric relation on a finite set D, then:

$$\forall x,y \in D, P(x,y) \implies P(y,x) \wedge P(x,x).$$

Demostración. To prove that

$$\forall x,y \in D, P(x,y) \implies P(y,x) \wedge P(x,x),$$

we start with the given properties of the relation $P(x,y)$ on the finite set D: $P(x,y)$ is reflexive and symmetric.
1. **Reflexivity**: Since $P(x,y)$ is reflexive, we have $P(x,x)$ is true for all $x \in D$.
2. **Symmetry**: Since $P(x,y)$ is symmetric, if $P(x,y)$ is true, then $P(y,x)$ is also true for all $x,y \in D$.
Therefore, if $P(x,y)$ is true, then $P(y,x)$ is true (by symmetry) and $P(x,x)$ is also true (by reflexivity). Thus, we have proven that

$$\forall x,y \in D, P(x,y) \implies P(y,x) \wedge P(x,x).$$

■

■ **Example 4.8** Consider the relation of .equality modulo n"defined by the predicate $P(x,y)$: "$x \equiv y \mod n$"with $D = \mathbb{Z}$.
- The relation is reflexive: $\forall x \in \mathbb{Z}, x \equiv x \mod n$. - The relation is symmetric: If $x \equiv y \mod n$, then $y \equiv x \mod n$. - The relation is also transitive: If $x \equiv y \mod n$ and $y \equiv z \mod n$, then $x \equiv z \mod n$.

■

Exercise 4.6 Let $P(x,y)$ be the predicate "x is an ancestor of y,"with D being the set of all people. If the ancestor relation is transitive, what can we say about the proposition $\forall x,y,z \in D, [P(x,y) \wedge P(y,z)] \implies P(x,z)$?

■

Demostración. The proposition is true by the definition of transitivity in the ancestor relation. If x is an ancestor of y and y is an ancestor of z, then x is an ancestor of z.

■

(R) Relations between elements of a set can be formalized using predicates with two variables, and analyzing their properties (reflexivity, symmetry, transitivity) is fundamental in relation theory and mathematical structures.

Theorem 4.1.8 If $P(x,y)$ is a predicate on $D_x \times D_y$, then:

$$\neg(\exists x \in D_x, \forall y \in D_y, P(x,y)) \equiv \forall x \in D_x, \exists y \in D_y \text{ such that } \neg P(x,y).$$

Demostración. We want to prove that

$$\neg(\exists x \in D_x, \forall y \in D_y, P(x,y)) \equiv \forall x \in D_x, \exists y \in D_y \text{ such that } \neg P(x,y).$$

1. **Left to Right ($\Rightarrow$)**: Suppose $\neg(\exists x \in D_x, \forall y \in D_y, P(x,y))$ is true. This means that there does not exist any $x \in D_x$ such that $P(x,y)$ is true for all $y \in D_y$. In other words, for each $x \in D_x$, there exists at least one $y \in D_y$ such that $P(x,y)$ is false. This implies that:

$$\forall x \in D_x, \exists y \in D_y \text{ such that } \neg P(x,y).$$

2. **Right to Left ($\Leftarrow$)**: Suppose $\forall x \in D_x, \exists y \in D_y$ such that $\neg P(x,y)$ is true. This means that for each $x \in D_x$, there exists at least one $y \in D_y$ such that $P(x,y)$ is false. Therefore, it is not possible for there to exist an $x \in D_x$ such that $P(x,y)$ is true for all $y \in D_y$. This implies:

$$\neg(\exists x \in D_x, \forall y \in D_y, P(x,y)).$$

Thus, we have shown that:

$$\neg(\exists x \in D_x, \forall y \in D_y, P(x,y)) \equiv \forall x \in D_x, \exists y \in D_y \text{ such that } \neg P(x,y).$$

■

Exercise 4.7 Formulate the negation of the proposition "There exists a student who knows all the teachersüsing predicates and quantifiers.

Demostración. Define:
- D_s: the set of students. - D_p: the set of teachers. - $P(x,y)$: "Student x knows teacher y."
The original proposition is:

$$\exists x \in D_s, \forall y \in D_p, P(x,y).$$

The negation is:

$$\neg(\exists x \in D_s, \forall y \in D_p, P(x,y)) \equiv \forall x \in D_s, \exists y \in D_p \text{ such that } \neg P(x,y).$$

Interpretation: "For every student x, there exists at least one teacher y whom x does not know." ■

R Understanding how to negate propositions with multiple quantifiers is essential for logical analysis and solving problems in mathematical logic.

4.2 Quantifiers: Universal Quantifier () and Existential Quantifier ()

4.2.1 Use of Quantifiers in Universal Propositions

In mathematical logic, **quantifiers** allow us to express properties and propositions involving variables that can take multiple values within a domain. The **universal quantifier** $\forall$ is used to indicate that a proposition is true for all elements of a given domain.

Definition 4.2.1 The **universal quantifier** $\forall$ is an operator that, applied to a propositional function $P(x)$, produces the proposition $\forall x \in D, P(x)$, which reads as "for all x in the domain D, $P(x)$ is true."

■ **Example 4.9** Let $P(x)$ be the predicate "x is divisible by 2,"with domain $D = \{2,4,6,8\}$. The proposition $\forall x \in D, P(x)$ means .all elements in D are divisible by 2..Evaluating:
- $P(2)$: 2 is divisible by 2 (true).
- $P(4)$: 4 is divisible by 2 (true).
- $P(6)$: 6 is divisible by 2 (true).
- $P(8)$: 8 is divisible by 2 (true).

Since $P(x)$ is true for all $x \in D$, the proposition $\forall x \in D, P(x)$ is true. ■

Theorem 4.2.1 If $P(x)$ is a predicate over a finite domain D, then the universal proposition $\forall x \in D, P(x)$ is true if and only if $P(x)$ is true for every element of D.

Demostración. To prove that

$$\forall x \in D, P(x) \text{ is true if and only if } P(x) \text{ is true for every element of } D,$$

we proceed in both directions.
($\Rightarrow$) Suppose $\forall x \in D, P(x)$ is true. This means $P(x)$ is true for all x in the domain D. Since D is finite, this implies $P(x)$ is true for each specific element of D.
($\Leftarrow$) Suppose $P(x)$ is true for each element of D. Since D is finite and $P(x)$ is true for each element of D, this implies $P(x)$ is true for all $x \in D$, i.e., $\forall x \in D, P(x)$ is true.

Thus, we have shown that

$$\forall x \in D, P(x) \text{ is true if and only if } P(x) \text{ is true for every element of } D.$$

■

Lema 4.2.1 The negation of a universal proposition is equivalent to an existential proposition:

$$\neg(\forall x \in D, P(x)) \equiv \exists x \in D \text{ such that } \neg P(x).$$

Demostración. According to De Morgan's laws for quantifiers, the negation of the universal quantifier becomes the existential quantifier, and the inner proposition is negated:

$$\neg(\forall x \in D, P(x)) = \exists x \in D \text{ such that } \neg P(x).$$

This means that if it is not true that $P(x)$ holds for every x, then there exists at least one x in D for which $P(x)$ is false. ■

■ **Example 4.10** Consider the proposition $\forall x \in \mathbb{N}, x > 0$, where $\mathbb{N}$ is the set of natural numbers. This proposition is true since every natural number is greater than 0.
The negation would be:

$$\neg(\forall x \in \mathbb{N}, x > 0) \equiv \exists x \in \mathbb{N} \text{ such that } x \leq 0.$$

This is false because there is no natural number that is less than or equal to 0. ■

(R) It is important to understand that using the universal quantifier allows us to generalize statements about all elements of a set, and its negation implies the existence of a counterexample.

Theorem 4.2.2 Let $P(x)$ be a predicate and D a non-empty domain. If $\forall x \in D, P(x)$ is true, then for any $x \in D$, $P(x)$ is true.

Demostración. Suppose $\forall x \in D, P(x)$ is true. This means $P(x)$ is true for all elements x in the domain D.
Since D is a non-empty domain, any element $x \in D$ belongs to this set, and by the assertion that $P(x)$ is true for all $x \in D$, we conclude that $P(x)$ is true for any $x \in D$.
Thus, if $\forall x \in D, P(x)$ is true, then for any $x \in D$, $P(x)$ is true.

■

Exercise 4.8 Let $D = \{1,2,3,4\}$ and $P(x)$ be the predicate "$x^2 \geq x$". Determine whether the proposition $\forall x \in D, P(x)$ is true or false.

■

Demostración. We evaluate $P(x)$ for each $x \in D$:
- $x = 1$: $1^2 = 1 \geq 1$ (true).
- $x = 2$: $2^2 = 4 \geq 2$ (true).
- $x = 3$: $3^2 = 9 \geq 3$ (true).
- $x = 4$: $4^2 = 16 \geq 4$ (true).

Since $P(x)$ is true for all $x \in D$, the proposition $\forall x \in D, P(x)$ is true. ■

In finite domains, we can verify the validity of a universal proposition by evaluating the predicate for each element in the domain. However, in infinite domains, general arguments or proofs are needed to establish the truth of the proposition.

Here's the translation of your LaTeX content to English, maintaining the original structure:

—

Theorem 4.2.3 In the context of real numbers, the proposition $\forall x \in \mathbb{R}, x^2 \geq 0$ is true.

Demostración. Consider an arbitrary real number $x \in \mathbb{R}$. We want to prove that $x^2 \geq 0$.
Recall that the square of any real number is always non-negative since:

$$x^2 = x \cdot x.$$

If $x = 0$, then $x^2 = 0$.
If $x > 0$ or $x < 0$, then $x^2 > 0$, since the product of two positive numbers or two negative numbers is positive.
Therefore, for any $x \in \mathbb{R}$, it always holds that $x^2 \geq 0$.
This proves that:

$$\forall x \in \mathbb{R}, x^2 \geq 0.$$

■

Exercise 4.9 Determine whether the proposition $\forall x \in \mathbb{R}, x^3 \geq 0$ is true or false. ■

We evaluate the cube of real numbers:

- If $x > 0$, then $x^3 > 0$.
- If $x = 0$, then $x^3 = 0$.
- If $x < 0$, then $x^3 < 0$ (because the product of three negative numbers is negative).

For $x < 0$, x^3 is negative, so $x^3 \geq 0$ is false. Therefore, the proposition $\forall x \in \mathbb{R}, x^3 \geq 0$ is false.
The negation is:

$$\neg(\forall x \in \mathbb{R}, x^3 \geq 0) \equiv \exists x \in \mathbb{R} \text{ such that } x^3 < 0.$$

Indeed, for $x = -1$, $(-1)^3 = -1 < 0$.

R This example shows that to refute a universal proposition, it is sufficient to find a counterexample.

Theorem 4.2.4 Universal propositions can be combined using logical connectives, and quantifiers can distribute over some of these connectives:

$$\forall x \in D, [P(x) \land Q(x)] \equiv [\forall x \in D, P(x)] \land [\forall x \in D, Q(x)].$$

Demostración. We want to prove that:

$$\forall x \in D, [P(x) \land Q(x)] \equiv [\forall x \in D, P(x)] \land [\forall x \in D, Q(x)].$$

($\Rightarrow$) Suppose that $\forall x \in D, [P(x) \land Q(x)]$ is true. This means that for each $x \in D$, both $P(x)$ and $Q(x)$ are true. Therefore: - $\forall x \in D, P(x)$ is true since $P(x)$ is true for every $x \in D$. - $\forall x \in D, Q(x)$ is also true since $Q(x)$ is true for every $x \in D$.

Thus, $[\forall x \in D, P(x)] \wedge [\forall x \in D, Q(x)]$ is true.
($\Leftarrow$) Suppose that $[\forall x \in D, P(x)] \wedge [\forall x \in D, Q(x)]$ is true. This implies that: - $\forall x \in D, P(x)$ is true. - $\forall x \in D, Q(x)$ is true.
Since $P(x)$ and $Q(x)$ are true for all $x \in D$, $P(x) \wedge Q(x)$ is also true for all $x \in D$. This means that $\forall x \in D, [P(x) \wedge Q(x)]$ is true.
Therefore, we have proven that:

$$\forall x \in D, [P(x) \wedge Q(x)] \equiv [\forall x \in D, P(x)] \wedge [\forall x \in D, Q(x)].$$

■

■ **Example 4.11** Let $D = \mathbb{N}$, $P(x)$: "x is even", and $Q(x)$: "x is divisible by 2".
The proposition $\forall x \in D, [P(x) \leftrightarrow Q(x)]$ is true since "being even.ªnd "being divisible by 2.ªre equivalent for natural numbers. ■

Exercise 4.10 Verify the equivalence:

$$\forall x \in D, P(x) \vee Q(x) \not\equiv [\forall x \in D, P(x)] \vee [\forall x \in D, Q(x)].$$

Provide an example demonstrating that the equivalence does not hold. ■

Consider $D = \{1,2\}$, $P(x)$: "x is even", $Q(x)$: "x is odd".
We evaluate:

- $\forall x \in D, P(x) \vee Q(x)$: For each $x \in D$, $P(x) \vee Q(x)$ is true because every number is either even or odd. Therefore, $\forall x \in D, P(x) \vee Q(x)$ is true.
- $[\forall x \in D, P(x)] \vee [\forall x \in D, Q(x)]$:
  - $\forall x \in D, P(x)$: Are all x in D even? $P(1)$ is false because 1 is not even. Therefore, $\forall x \in D, P(x)$ is false.
  - $\forall x \in D, Q(x)$: Are all x in D odd? $Q(2)$ is false because 2 is not odd. Therefore, $\forall x \in D, Q(x)$ is false.
- Therefore, $[\forall x \in D, P(x)] \vee [\forall x \in D, Q(x)]$ is false.

We conclude that:

$$\forall x \in D, P(x) \vee Q(x) \text{ is true, but } [\forall x \in D, P(x)] \vee [\forall x \in D, Q(x)] \text{ is false.}$$

Therefore, they are not equivalent.

(R) This difference shows that the distribution of quantifiers over disjunction is not generally equivalent, and we must be careful when manipulating propositions with quantifiers and logical connectives.

Theorem 4.2.5 — **De Morgan's Laws for Universal Quantifiers.** The negation of a universal proposition with a conjunction or disjunction transforms in the following way:

$$\neg(\forall x \in D, [P(x) \wedge Q(x)]) \equiv \exists x \in D \text{ such that } \neg P(x) \vee \neg Q(x),$$
$$\neg(\forall x \in D, [P(x) \vee Q(x)]) \equiv \exists x \in D \text{ such that } \neg P(x) \wedge \neg Q(x).$$

Demostración. First Law We want to show that:

$$\neg(\forall x \in D, [P(x) \wedge Q(x)]) \equiv \exists x \in D \text{ such that } \neg P(x) \vee \neg Q(x).$$

($\Rightarrow$) Suppose that $\neg(\forall x \in D, [P(x) \wedge Q(x)])$ is true. This means that it is not the case that $P(x) \wedge Q(x)$ is true for all $x \in D$. Therefore, there must exist at least one $x \in D$ such that $P(x) \wedge Q(x)$ is false. The negation of $P(x) \wedge Q(x)$ is $\neg P(x) \vee \neg Q(x)$, which implies that:

$$\exists x \in D \text{ such that } \neg P(x) \vee \neg Q(x).$$

($\Leftarrow$) Suppose that $\exists x \in D$ such that $\neg P(x) \vee \neg Q(x)$ is true. This means that there is at least one $x \in D$ for which $P(x) \wedge Q(x)$ is false. Therefore, it is not the case that $P(x) \wedge Q(x)$ is true for all $x \in D$, which implies that:

$$\neg(\forall x \in D, [P(x) \wedge Q(x)]).$$

Therefore, we have shown that:

$$\neg(\forall x \in D, [P(x) \wedge Q(x)]) \equiv \exists x \in D \text{ such that } \neg P(x) \vee \neg Q(x).$$

Second Law We want to show that:

$$\neg(\forall x \in D, [P(x) \vee Q(x)]) \equiv \exists x \in D \text{ such that } \neg P(x) \wedge \neg Q(x).$$

($\Rightarrow$) Suppose that $\neg(\forall x \in D, [P(x) \vee Q(x)])$ is true. This means that it is not the case that $P(x) \vee Q(x)$ is true for all $x \in D$. Therefore, there must exist at least one $x \in D$ such that $P(x) \vee Q(x)$ is false. The negation of $P(x) \vee Q(x)$ is $\neg P(x) \wedge \neg Q(x)$, which implies that:

$$\exists x \in D \text{ such that } \neg P(x) \wedge \neg Q(x).$$

($\Leftarrow$) Suppose that $\exists x \in D$ such that $\neg P(x) \wedge \neg Q(x)$ is true. This means that there is at least one $x \in D$ for which $P(x) \vee Q(x)$ is false. Therefore, it is not the case that $P(x) \vee Q(x)$ is true for all $x \in D$, which implies that:

$$\neg(\forall x \in D, [P(x) \vee Q(x)]).$$

Therefore, we have shown that:

$$\neg(\forall x \in D, [P(x) \vee Q(x)]) \equiv \exists x \in D \text{ such that } \neg P(x) \wedge \neg Q(x).$$

This completes the proof of De Morgan's Laws for Universal Quantifiers.

■

■ **Example 4.12** Let $D = \{1, 2, 3\}$, $P(x)$: "x is even", $Q(x)$: "x is prime".
Consider the proposition $\forall x \in D, [P(x) \vee Q(x)]$.
The negation is:

$$\neg(\forall x \in D, [P(x) \vee Q(x)]) \equiv \exists x \in D \text{ such that } \neg P(x) \wedge \neg Q(x).$$

We evaluate $\neg P(x) \wedge \neg Q(x)$ for each $x \in D$:

- $x = 1$: $P(1)$ is false, $Q(1)$ is false $\Longrightarrow \neg P(1) \wedge \neg Q(1)$ is true.

Therefore, there exists an $x \in D$ such that $\neg P(x) \wedge \neg Q(x)$ is true, confirming the negation.

■

Exercise 4.11 Formulate and prove the negation of the following universal proposition:

$$\forall x \in \mathbb{R}, [x \geq 0 \implies \sqrt{x} \geq 0].$$

■

First, we note that the proposition is true, since the square root of a non-negative number is always non-negative.
The negation is:

$$\neg \left(\forall x \in \mathbb{R}, [x \geq 0 \implies \sqrt{x} \geq 0]\right) \equiv \exists x \in \mathbb{R} \text{ such that } \neg[x \geq 0 \implies \sqrt{x} \geq 0].$$

We simplify the negation of the conditional:

$$\neg[x \geq 0 \implies \sqrt{x} \geq 0] \equiv x \geq 0 \wedge \sqrt{x} < 0.$$

However, the square root of a non-negative number is always greater than or equal to zero, so $\sqrt{x} < 0$ is impossible for $x \geq 0$.
Therefore, the negation is false, and the original proposition is true.

(R) This exercise demonstrates how to apply negation to universal propositions with implications and how to analyze the validity of the resulting propositions.

Exercise 4.12 Let $P(x)$: "x is a multiple of 3", with $D = \mathbb{Z}$. Write the negation of the proposition $\forall x \in D, P(x)$ and explain its meaning. ■

The negation is:

$$\neg(\forall x \in D, P(x)) \equiv \exists x \in D \text{ such that } \neg P(x).$$

Meaning: "There exists at least one integer that is not a multiple of 3."
This is true; for example, $x = 1$ is not a multiple of 3.

(R) Understanding how to negate universal propositions is fundamental in logic and mathematics, especially when working with proofs by contradiction or counterexamples.

Theorem 4.2.6 If a universal proposition $\forall x \in D, P(x)$ is false, then its negation $\exists x \in D$ such that $\neg P(x)$ is true, and vice versa.

Demostración. We want to show that if $\forall x \in D, P(x)$ is false, then $\exists x \in D$ such that $\neg P(x)$ is true, and vice versa.
1. ($\Rightarrow$) Suppose that $\forall x \in D, P(x)$ is false. This means it is not true that $P(x)$ holds for all $x \in D$. Therefore, there must exist at least one element $x \in D$ for which $P(x)$ is false, i.e., $\neg P(x)$ is true. This implies that $\exists x \in D$ such that $\neg P(x)$ is true.
2. ($\Leftarrow$) Suppose that $\exists x \in D$ such that $\neg P(x)$ is true. This means there exists at least one $x \in D$ for which $P(x)$ is false. Therefore, it is not true that $P(x)$ holds for all $x \in D$, which implies that $\forall x \in D, P(x)$ is false.
Thus, we have proven that $\forall x \in D, P(x)$ is false if and only if $\exists x \in D$ such that $\neg P(x)$ is true.

■

Exercise 4.13 Prove that the proposition $\forall x \in \mathbb{R}, x^2 + 1 > 0$ is true. ■

For any real number x, $x^2 \geq 0$ (since the square of a real number is non-negative). Therefore, $x^2 + 1 \geq 0 + 1 = 1 > 0$.
Hence, $\forall x \in \mathbb{R}, x^2 + 1 > 0$ is true.

(R) Universal propositions are fundamental for establishing general properties in mathematics, and the proper use of quantifiers is essential for precise and rigorous communication.

4.2.2 Examples of Existential Quantifiers in Logical Problems

The **existential quantifier** $\exists$ is used in logic to express that there exists at least one element in a domain that satisfies a given property. It is fundamental in the formulation and resolution of logical and mathematical problems.

Definition 4.2.2 The **existential quantifier** $\exists$ is an operator that, when applied to a propositional function $P(x)$, produces the proposition $\exists x \in D, P(x)$, which is read as "there exists at least one x in the domain D such that $P(x)$ is true."

■ **Example 4.13** Let $P(x)$ be the predicate "x is an even prime number,"with the domain $D = \mathbb{N}$ (natural numbers). The proposition $\exists x \in D, P(x)$ means "there exists at least one natural number that is an even prime number."We evaluate:

- The number 2 is an even prime number.

Therefore, the proposition $\exists x \in D, P(x)$ is true. ■

Theorem 4.2.7 The existential proposition $\exists x \in D, P(x)$ is true if and only if the set $\{x \in D \mid P(x) \text{ is true}\}$ is non-empty.

Demostración. We want to prove that:

$$\exists x \in D, P(x) \text{ is true if and only if the set } \{x \in D \mid P(x) \text{ is true}\} \text{ is non-empty.}$$

($\Rightarrow$) Suppose $\exists x \in D, P(x)$ is true. This means there is at least one element $x \in D$ for which $P(x)$ is true. Therefore, the set $\{x \in D \mid P(x) \text{ is true}\}$ contains at least one element, i.e., it is non-empty.
($\Leftarrow$) Suppose the set $\{x \in D \mid P(x) \text{ is true}\}$ is non-empty. This means there exists at least one $x \in D$ such that $P(x)$ is true. Therefore, $\exists x \in D, P(x)$ is true.
Thus, we have proven that:

$$\exists x \in D, P(x) \text{ is true if and only if the set } \{x \in D \mid P(x) \text{ is true}\} \text{ is non-empty.}$$

■

Lema 4.2.2 The negation of an existential proposition is equivalent to a universal negative proposition:

$$\neg(\exists x \in D, P(x)) \equiv \forall x \in D, \neg P(x).$$

Demostración. Applying De Morgan's laws for quantifiers:

$$\neg(\exists x \in D, P(x)) \equiv \forall x \in D, \neg P(x).$$

This means that if there does not exist any x in D such that $P(x)$ is true, then $P(x)$ is false for all $x \in D$. ■

■ **Example 4.14** Consider the predicate $P(x)$: "x is divisible by 5,"with the domain $D = \{1,2,3,4\}$. The proposition $\exists x \in D, P(x)$ means "there exists at least one number in D that is divisible by 5."We evaluate:

- No number in D is divisible by 5.

Therefore, $\exists x \in D, P(x)$ is false.
The negation is:

$$\neg(\exists x \in D, P(x)) \equiv \forall x \in D, \neg P(x).$$

This means "for all $x \in D$, x is not divisible by 5,"which is true in this case. ■

R The existential quantifier is useful for expressing the existence of solutions or particular cases in logical and mathematical problems.

Theorem 4.2.8 If $P(x)$ and $Q(x)$ are predicates over the same domain D, then:
1. $\exists x \in D, P(x) \vee Q(x) \equiv (\exists x \in D, P(x)) \vee (\exists x \in D, Q(x))$.
2. $\exists x \in D, P(x) \wedge Q(x) \implies (\exists x \in D, P(x)) \wedge (\exists x \in D, Q(x))$.

Demostración. We will prove each part of the theorem separately.
Part 1 We want to prove that:

$$\exists x \in D, P(x) \vee Q(x) \equiv (\exists x \in D, P(x)) \vee (\exists x \in D, Q(x)).$$

($\Rightarrow$) Suppose $\exists x \in D, P(x) \vee Q(x)$ is true. This means there exists at least one $x \in D$ such that $P(x) \vee Q(x)$ is true. By the definition of disjunction, this implies that: - $P(x)$ is true for some $x \in D$, or - $Q(x)$ is true for some $x \in D$.
Thus, $(\exists x \in D, P(x)) \vee (\exists x \in D, Q(x))$ is true.
($\Leftarrow$) Suppose $(\exists x \in D, P(x)) \vee (\exists x \in D, Q(x))$ is true. This means that: - $\exists x \in D, P(x)$ is true, or - $\exists x \in D, Q(x)$ is true.
In either case, there exists at least one $x \in D$ such that $P(x) \vee Q(x)$ is true. Therefore, $\exists x \in D, P(x) \vee Q(x)$ is true.
Thus, we have proven that:

$$\exists x \in D, P(x) \vee Q(x) \equiv (\exists x \in D, P(x)) \vee (\exists x \in D, Q(x)).$$

Part 2 We want to prove that:

$$\exists x \in D, P(x) \wedge Q(x) \implies (\exists x \in D, P(x)) \wedge (\exists x \in D, Q(x)).$$

Suppose $\exists x \in D, P(x) \wedge Q(x)$ is true. This means there exists at least one $x \in D$ such that $P(x)$ is true and $Q(x)$ is also true simultaneously.
- Since $P(x)$ is true for that x, we have $\exists x \in D, P(x)$ is true. - Since $Q(x)$ is also true for that same x, we have $\exists x \in D, Q(x)$ is true.
Therefore, $(\exists x \in D, P(x)) \wedge (\exists x \in D, Q(x))$ is true.
This proves that:

$$\exists x \in D, P(x) \wedge Q(x) \implies (\exists x \in D, P(x)) \wedge (\exists x \in D, Q(x)).$$

■

■ **Example 4.15** Let $D = \{1,2,3,4,5\}$, $P(x)$: "x is even", $Q(x)$: "x is greater than 4".
We evaluate $\exists x \in D, P(x) \wedge Q(x)$:

- We look for an x that is even and greater than 4.
- $x = 2$: even, but not greater than 4.
- $x = 4$: even, but not greater than 4.
- $x = 5$: not even.

There is no such x in D, so $\exists x \in D, P(x) \wedge Q(x)$ is false.
However, $\exists x \in D, P(x)$ is true (for $x = 2$), and $\exists x \in D, Q(x)$ is true (for $x = 5$). This shows that although
both exist, the conjunction $\exists x \in D, P(x) \wedge Q(x)$ can still be false. ■

Exercise 4.14 Let $D = \mathbb{Z}$ (the set of integers), and $P(x)$: "$x^2 = 4$". Determine whether the proposition $\exists x \in D, P(x)$ is true or false. Find all values of x that satisfy $P(x)$. ■

Demostración. We are looking for $x \in \mathbb{Z}$ such that $x^2 = 4$.
We compute:

- $x = -2$: $(-2)^2 = 4$.
- $x = 2$: $(2)^2 = 4$.

Therefore, $\exists x \in \mathbb{Z}, P(x)$ is true. The values of x that satisfy $P(x)$ are $x = -2$ and $x = 2$. ■

(R) Existential quantifiers allow us to assert the existence of specific solutions in mathematical problems, which is crucial in many proofs and applications.

■ **Example 4.16** In graph theory, consider the predicate $P(x)$: "Vertex x has an odd degree,"with D being the set of vertices of a graph.
The proposition $\exists x \in D, P(x)$ means "There exists a vertex with an odd degree."This is relevant in the context of Eulerian graphs. ■

Theorem 4.2.9 A connected graph has an Eulerian path if and only if it has exactly zero or two vertices with an odd degree.

Demostración. (Sketch) If all vertices have even degree, there exists an Eulerian cycle (a path that starts and ends at the same vertex and traverses each edge exactly once). If exactly two vertices have an odd degree, there exists an Eulerian path that starts at one of the vertices with an odd degree and ends at the other.
The existence of vertices with an odd degree can be expressed using the existential quantifier: $\exists x, y \in D, x \neq y$, such that $P(x)$ and $P(y)$ are true, and for all other vertices z, $\neg P(z)$ is true. ■

Exercise 4.15 Let $D = \mathbb{N}$ (natural numbers), and $P(x)$: "x is prime and x is even."Determine if $\exists x \in D, P(x)$ is true or false. ■

Demostración. The only even prime number is 2. Therefore, $\exists x \in D, P(x)$ is true, and the value of x is 2. ■

■ **Example 4.17** In predicate logic, we can use existential quantifiers to express statements like: "The function f has a root in the interval $[a,b]$çan be expressed as:

$$\exists x \in [a,b] \text{ such that } f(x) = 0.$$

This proposition is essential in theorems like the Intermediate Value Theorem (Bolzano's Theorem). ■

Theorem 4.2.10 — **Bolzano's Theorem.** Let f be a continuous function on the closed interval $[a,b]$, and suppose that $f(a)$ and $f(b)$ have opposite signs, i.e., $f(a) \cdot f(b) < 0$. Then, there exists at least one $c \in (a,b)$ such that $f(c) = 0$.

Demostración. The theorem guarantees the existence of a root of the function in the interval, which is expressed by the existential quantifier: $\exists c \in (a,b)$ such that $f(c) = 0$.
The proof uses the continuity of f and the intermediate value theorem. ■

Exercise 4.16 In a group of people, let $P(x)$ be the predicate "x is older than 30 years.Ïf we know that there is at least one person older than 30 in the group, how can this information be expressed using quantifiers? What is the negation of this proposition? ■

Demostración. The information is expressed as:

$$\exists x \in \text{Group}, P(x).$$

The negation of this proposition is:

$$\neg(\exists x \in \text{Group}, P(x)) \equiv \forall x \in \text{Group}, \neg P(x).$$

This means "For all x in the group, x is not older than 30,ï.e., .Everyone in the group is 30 years old or younger." ■

(R) Understanding how to formulate and negate propositions with existential quantifiers is essential for analyzing and solving logical problems in various contexts.

■ **Example 4.18** In algebra, consider the predicate $P(a,b)$: "There exists a real number x such that $ax+b=0$,"with $a,b \in \mathbb{R}$.
The proposition $\exists x \in \mathbb{R}, ax+b=0$ is true if $a \neq 0$, since we can solve for $x = -\frac{b}{a}$.
If $a = 0$, the equation reduces to $b = 0$, and the existence of a solution depends on the value of b. ■

Exercise 4.17 Determine for which values of a and b the proposition $\exists x \in \mathbb{R}, ax+b=0$ is true. ■

Demostración.
- If $a \neq 0$, then there always exists $x = -\frac{b}{a}$ that satisfies the equation.
- If $a = 0$:
  - If $b = 0$, the equation is $0 = 0$, which is true for any x, so $\exists x \in \mathbb{R}, 0x+0=0$ is true.
  - If $b \neq 0$, the equation is $0 = -b$, which is false, and there is no x that satisfies it. Therefore, $\exists x \in \mathbb{R}, 0x+b=0$ is false.

In summary, the proposition is true for all $a \in \mathbb{R}$ and $b \in \mathbb{R}$, except when $a = 0$ and $b \neq 0$. ■

■ **Example 4.19** In set theory, the proposition "There exist sets A and B such that $A \subset B$ and $A = B$çan be analyzed.
The proposition is:

$$\exists A, B \text{ such that } A \subset B \wedge A = B.$$

This implies that A is equal to B, and hence trivially $A \subset B$ is true. ■

Exercise 4.18 Determine whether the following proposition is true or false:

$$\exists A, B \text{ such that } A \subset B \text{ and } A \neq B.$$

■

Demostración. The proposition is true. For example, take $A = \{1\}$ and $B = \{1, 2\}$. Then, $A \subset B$ and $A \neq B$. ■

(R) This exercise shows how existential quantifiers can express the existence of elements or sets with specific properties, and how examples satisfying the given conditions can be found.

■ **Example 4.20** In geometry, the proposition "There exists a right triangle whose legs have integer lengthsçan be expressed as:

$$\exists a, b, c \in \mathbb{N}, a^2 + b^2 = c^2.$$

This is the definition of Pythagorean triples. ■

Exercise 4.19 Find a Pythagorean triple that satisfies the previous proposition. ■

Demostración. A known Pythagorean triple is $(a, b, c) = (3, 4, 5)$.
We verify:

$$3^2 + 4^2 = 9 + 16 = 25 = 5^2.$$

Therefore, $\exists a = 3, b = 4, c = 5$ in $\mathbb{N}$, such that $a^2 + b^2 = c^2$. ■

(R) Pythagorean triples are classic examples of how the existential quantifier is used to affirm the existence of integer solutions to Diophantine equations.

■ **Example 4.21** In logic, the Witness Principle states that if there exists an element with a certain property, then we can introduce a symbol or name to refer to that particular element.
For example, if we know that:

$$\exists x \in D, P(x),$$

then we can say "Let $c \in D$ such that $P(c)$."

■

Theorem 4.2.11 — **Witness Principle.** If $\exists x \in D, P(x)$ is true, then there exists a specific element $c \in D$ for which $P(c)$ is true.

Demostración. By the definition of the existential quantifier, if $\exists x \in D, P(x)$ is true, then there is at least one element $c \in D$ such that $P(c)$ is true. We can refer to this element in our proofs.

■

Exercise 4.20 Using the Witness Principle, prove that if a function $f : \mathbb{R} \to \mathbb{R}$ is continuous and positive at some point, then there exists a $\delta > 0$ such that $f(x) > 0$ in an interval around that point.

■

Demostración. Let $x_0 \in \mathbb{R}$ such that $f(x_0) > 0$ (by the existential quantifier). Since f is continuous at x_0, for $\varepsilon = \frac{f(x_0)}{2} > 0$, there exists $\delta > 0$ such that if $|x - x_0| < \delta$, then $|f(x) - f(x_0)| < \varepsilon$.
Therefore, for x in this interval:

$$|f(x) - f(x_0)| < \varepsilon \implies f(x) > f(x_0) - \varepsilon = f(x_0) - \frac{f(x_0)}{2} = \frac{f(x_0)}{2} > 0.$$

Hence, $f(x) > 0$ in this interval around x_0.

■

(R) The use of the existential quantifier and the Witness Principle is fundamental in analysis and in formulating proofs that involve the existence of elements with specific properties.

■ **Example 4.22** In programming, the existential quantifier relates to searching for elements that meet certain conditions. For example, in an array of numbers, "There exists a negative element."

$$\exists i \in \{1, 2, \ldots, n\}, \text{ such that } a_i < 0.$$

■

Exercise 4.21 Write an algorithm to determine if there is at least one negative element in an array of integers.

■

Demostración. A simple algorithm in pseudocode:

```
For i from 1 to n do:
    If a_i < 0 then:
        Write "There is a negative element"
        Stop
End For
Write "All elements are non-negative"
```

■

R This example shows the practical application of the existential quantifier in algorithms and programming, where we search for the existence of elements that satisfy certain conditions.

■ **Example 4.23** In number theory, the proposition "There exists a prime number greater than any given natural numberïs fundamental.

■

Theorem 4.2.12 — **Euclid's Theorem.** There are infinitely many prime numbers.

Demostración. (Sketch) Suppose there are only finitely many primes $p_1, p_2, \ldots, p_n$. Consider the number $Q = p_1 p_2 \ldots p_n + 1$. This number is not divisible by any of the primes p_i, since dividing by p_i leaves a remainder of 1. Therefore, Q is prime or has prime factors not included in the list, contradicting our assumption. Therefore, there are more primes.
This argument implies that for any natural number, there exists a prime number greater than it, i.e.,

$$\forall N \in \mathbb{N}, \exists p \in \mathbb{P}, \text{ such that } p > N.$$

■

R The use of the existential quantifier in this context is essential to express the infinitude of prime numbers and to formulate fundamental propositions in number theory.

4.3 Transformation of Propositions: Analysis of the Negation of Quantifiers

4.3.1 Negation of Propositions with Universal Quantifiers

The negation of propositions involving universal quantifiers is a fundamental aspect of mathematical logic and rigorous argument construction. Understanding how to correctly negate these propositions is essential to avoid errors in reasoning and proofs.

Definition 4.3.1 A **universal proposition** is a statement that uses the universal quantifier $\forall$, indicating that a certain property holds for all elements of a domain D. It is expressed as:

$$\forall x \in D, P(x),$$

where $P(x)$ is a predicate over the domain D.

■ **Example 4.24** Consider the universal proposition:

$$\forall n \in \mathbb{N}, n + 0 = n.$$

This proposition states that "for every natural number n, n plus zero equals n."This is a fundamental truth in arithmetic. ■

Theorem 4.3.1 — **Negation of a Universal Proposition.** The negation of a universal proposition $\forall x \in D, P(x)$ is logically equivalent to an existential proposition where the predicate is negated:

$$\neg(\forall x \in D, P(x)) \equiv \exists x \in D \text{ such that } \neg P(x).$$

Demostración. We want to prove that

$$\neg(\forall x \in D, P(x)) \equiv \exists x \in D \text{ such that } \neg P(x).$$

($\Rightarrow$) Suppose that $\neg(\forall x \in D, P(x))$ is true. This means that the proposition $\forall x \in D, P(x)$ is false. If $\forall x \in D, P(x)$ is false, then there must exist at least one element $x \in D$ such that $P(x)$ is false. This implies that there exists an $x \in D$ such that $\neg P(x)$ is true, i.e., $\exists x \in D$ such that $\neg P(x)$ is true.
($\Leftarrow$) Suppose that $\exists x \in D$ such that $\neg P(x)$ is true. This means there is at least one $x \in D$ for which $P(x)$ is false. Therefore, it is not true that $P(x)$ holds for all $x \in D$, which implies that $\forall x \in D, P(x)$ is false, i.e., $\neg(\forall x \in D, P(x))$ is true.
Thus, we have shown that

$$\neg(\forall x \in D, P(x)) \equiv \exists x \in D \text{ such that } \neg P(x).$$

■

Lema 4.3.1 The negation of the universal quantifier affects only the quantifier and the predicate, not the domain. The domain remains unchanged when negating.

Demostración. The domain D is the set of all elements being quantified. When negating the universal proposition, we change the statement about the elements of the domain, but the domain itself remains unchanged. ■

■ **Example 4.25** Let $P(x)$ be the predicate "x is greater than 0,"with domain $D = \mathbb{R}$ (real numbers). The universal proposition is:

$$\forall x \in \mathbb{R}, x > 0.$$

This proposition is false because there are real numbers that are not greater than zero (for example, $x = -1$).
The negation of this proposition is:

$$\neg(\forall x \in \mathbb{R}, x > 0) \equiv \exists x \in \mathbb{R} \text{ such that } x \leq 0.$$

This means "there exists a real number such that x is less than or equal to zero,"which is true. ■

(R) The negation of a universal proposition does not imply that the opposite proposition is true for all elements; it means that there exists at least one case where the original proposition does not hold.

Theorem 4.3.2 — **General Process for Negating Universal Propositions.** To negate a compound universal proposition, one must:

1. Change the universal quantifier $\forall$ to an existential quantifier $\exists$.
2. Negate the internal predicate or proposition.

Mathematically, this is expressed as:

$$\neg(\forall x \in D, P(x)) \equiv \exists x \in D \text{ such that } \neg P(x).$$

Demostración. This is a direct application of De Morgan's laws for quantifiers and logical connectives. By negating a universal proposition, we are stating that it is not true that $P(x)$ holds for all x in D, which means there is at least one x in D for which $P(x)$ is false. ■

■ **Example 4.26** Consider the proposition:

$$\forall x \in \mathbb{N}, x^2 \geq x.$$

This proposition is true because for all natural numbers x, x^2 is greater than or equal to x.
The negation of this proposition is:

$$\neg(\forall x \in \mathbb{N}, x^2 \geq x) \equiv \exists x \in \mathbb{N} \text{ such that } x^2 < x.$$

This means "there exists a natural number x such that x^2 is less than x."
Let's evaluate this proposition:

- For $x = 0$, $0^2 = 0$, and $0^2 \geq 0$ (true).
- For $x = 1$, $1^2 = 1$, and $1^2 \geq 1$ (true).
- For $x = 2$, $2^2 = 4$, and $4 \geq 2$ (true).

There is no natural number x for which $x^2 < x$. Therefore, the negation is false, and the original proposition is true. ■

Exercise 4.22 Let $P(x)$ be the predicate "x is prime,"with $D = \{2,3,4,5\}$. Write the negation of the universal proposition:

$$\forall x \in D, P(x),$$

and determine whether the negation is true or false. ■

Demostración. The negation of the proposition is:

$$\neg(\forall x \in D, P(x)) \equiv \exists x \in D \text{ such that } \neg P(x).$$

This means "there exists an x in D such that x is not prime."
We evaluate $P(x)$ for each $x \in D$:

- $x = 2$: prime (true).
- $x = 3$: prime (true).
- $x = 4$: not prime (false).
- $x = 5$: prime (true).

We see that $x = 4$ is an element in D such that $P(4)$ is false.
Therefore, the negation is true because there exists an x in D (specifically $x = 4$) for which $P(x)$ is false. ■

(R) This exercise shows how the negation of a universal proposition can be true even if the original proposition is false, and vice versa. Careful evaluation of each case is essential.

Theorem 4.3.3 — De Morgan's Laws for Quantifiers. De Morgan's laws apply to quantifiers as follows:

$$\neg(\forall x \in D, P(x)) \equiv \exists x \in D \text{ such that } \neg P(x),$$
$$\neg(\exists x \in D, P(x)) \equiv \forall x \in D, \neg P(x).$$

Demostración. We will prove each of De Morgan's laws for quantifiers separately.
First Law We want to prove that:

$$\neg(\forall x \in D, P(x)) \equiv \exists x \in D \text{ such that } \neg P(x).$$

($\Rightarrow$) Suppose $\neg(\forall x \in D, P(x))$ is true. This means that the proposition $\forall x \in D, P(x)$ is false. If $\forall x \in D, P(x)$ is false, then there must exist at least one element $x \in D$ such that $P(x)$ is false, i.e., $\neg P(x)$ is true for some $x \in D$. This implies that $\exists x \in D$ such that $\neg P(x)$ is true.
($\Leftarrow$) Suppose $\exists x \in D$ such that $\neg P(x)$ is true. This means that there exists at least one $x \in D$ for which $P(x)$ is false. Therefore, it is not true that $P(x)$ holds for all $x \in D$, which implies that $\forall x \in D, P(x)$ is false, i.e., $\neg(\forall x \in D, P(x))$ is true.
Hence, we have shown that:

$$\neg(\forall x \in D, P(x)) \equiv \exists x \in D \text{ such that } \neg P(x).$$

Second Law We want to prove that:

$$\neg(\exists x \in D, P(x)) \equiv \forall x \in D, \neg P(x).$$

($\Rightarrow$) Suppose $\neg(\exists x \in D, P(x))$ is true. This means that the proposition $\exists x \in D, P(x)$ is false. If $\exists x \in D, P(x)$ is false, then there does not exist any $x \in D$ such that $P(x)$ is true. This implies that $P(x)$ is false for all $x \in D$, i.e., $\forall x \in D, \neg P(x)$ is true.
($\Leftarrow$) Suppose $\forall x \in D, \neg P(x)$ is true. This means that $P(x)$ is false for each $x \in D$. Therefore, there does not exist any $x \in D$ such that $P(x)$ is true, which implies that $\exists x \in D, P(x)$ is false, i.e., $\neg(\exists x \in D, P(x))$ is true.
Hence, we have shown that:

$$\neg(\exists x \in D, P(x)) \equiv \forall x \in D, \neg P(x).$$

This completes the proof of De Morgan's laws for quantifiers. ■

■ **Example 4.27** Consider the proposition:

$$\forall x \in \mathbb{R}, x^2 \geq 0.$$

This is a true proposition because the square of any real number is non-negative.
The negation is:

$$\neg(\forall x \in \mathbb{R}, x^2 \geq 0) \equiv \exists x \in \mathbb{R} \text{ such that } x^2 < 0.$$

This means "there exists a real number x such that x^2 is less than zero."
However, no such real number exists because the square of any real number is always greater than or equal to zero. Therefore, the negation is false, and the original proposition is true. ■

Exercise 4.23 Write the negation of the following universal proposition and determine its truth value:

$$\forall x \in \mathbb{Z}, x \text{ is even.}$$

Demostración. The negation is:

$$\neg(\forall x \in \mathbb{Z}, x \text{ is even}) \equiv \exists x \in \mathbb{Z} \text{ such that } x \text{ is not even.}$$

This means "there exists an integer that is not even,ï.e., "there exists an odd integer."
Since odd integers exist (for example, $x = 1$), the negation is true.
The original proposition .^All integers are evenïs false because not all integers are even. ■

(R) Negating a universal proposition is useful for finding counterexamples that show a general statement is not valid.

Theorem 4.3.4 The negation of a universal proposition that includes a conditional is transformed as follows:

$$\neg(\forall x \in D, P(x) \implies Q(x)) \equiv \exists x \in D \text{ such that } P(x) \wedge \neg Q(x).$$

Demostración. We want to prove that

$$\neg(\forall x \in D, P(x) \implies Q(x)) \equiv \exists x \in D \text{ such that } P(x) \wedge \neg Q(x).$$

($\Rightarrow$) Suppose $\neg(\forall x \in D, P(x) \implies Q(x))$ is true. This means that the proposition $\forall x \in D, P(x) \implies Q(x)$ is false. If $\forall x \in D, P(x) \implies Q(x)$ is false, then there must exist at least one $x \in D$ such that $P(x) \implies Q(x)$ is false.
Recall that $P(x) \implies Q(x)$ is false if and only if $P(x)$ is true and $Q(x)$ is false, i.e., $P(x) \wedge \neg Q(x)$ is true. Therefore, there exists at least one $x \in D$ such that $P(x) \wedge \neg Q(x)$ is true. This implies that:

$$\exists x \in D \text{ such that } P(x) \wedge \neg Q(x).$$

($\Leftarrow$) Suppose $\exists x \in D$ such that $P(x) \wedge \neg Q(x)$ is true. This means there is at least one $x \in D$ for which $P(x)$ is true and $Q(x)$ is false. For this x, $P(x) \implies Q(x)$ is false. Therefore, $\forall x \in D, P(x) \implies Q(x)$ is not true, which implies that $\neg(\forall x \in D, P(x) \implies Q(x))$ is true.
Hence, we have shown that:

$$\neg(\forall x \in D, P(x) \implies Q(x)) \equiv \exists x \in D \text{ such that } P(x) \wedge \neg Q(x).$$

■

Example 4.28 Consider the proposition:

$$\forall x \in \mathbb{R}, x > 0 \implies \frac{1}{x} > 0.$$

This proposition is true because the reciprocal of a positive number is positive.
The negation is:

$$\neg(\forall x \in \mathbb{R}, x > 0 \implies \frac{1}{x} > 0) \equiv \exists x \in \mathbb{R} \text{ such that } x > 0 \wedge \frac{1}{x} \leq 0.$$

We are looking for a real number x such that $x > 0$ and $\frac{1}{x} \leq 0$. No such number exists because if $x > 0$, then $\frac{1}{x} > 0$.
Therefore, the negation is false, and the original proposition is true. ■

Exercise 4.24 Write the negation of the following proposition and determine whether the negation is true or false:

$$\forall x \in \mathbb{R}, x \neq 0 \implies x^2 > 0.$$

■

Demostración. The negation is:

$$\neg(\forall x \in \mathbb{R}, x \neq 0 \implies x^2 > 0) \equiv \exists x \in \mathbb{R} \text{ such that } x \neq 0 \wedge x^2 \leq 0.$$

We are looking for a real number x such that $x \neq 0$ and $x^2 \leq 0$.
We know that $x^2 \geq 0$ for all real numbers, and $x^2 = 0$ if and only if $x = 0$.
Therefore, no $x \neq 0$ exists such that $x^2 \leq 0$.
The negation is false, and the original proposition is true. ■

R When working with propositions that include conditionals, it is important to remember the logical equivalences when negating implications.

Theorem 4.3.5 The negation of a universal proposition with a conjunction transforms to:

$$\neg(\forall x \in D, P(x) \wedge Q(x)) \equiv \exists x \in D \text{ such that } \neg(P(x) \wedge Q(x)) \equiv \exists x \in D \text{ such that } \neg P(x) \vee \neg Q(x).$$

Demostración. We want to prove that

$$\neg(\forall x \in D, P(x) \wedge Q(x)) \equiv \exists x \in D \text{ such that } \neg(P(x) \wedge Q(x)) \equiv \exists x \in D \text{ such that } \neg P(x) \vee \neg Q(x).$$

First, observe that:

$$\neg(\forall x \in D, P(x) \wedge Q(x)) \equiv \exists x \in D \text{ such that } \neg(P(x) \wedge Q(x)).$$

This follows from De Morgan's law for the negation of a universal proposition: the negation of "for all $x \in D$, $P(x) \wedge Q(x)$ïs "there exists an $x \in D$ such that $P(x) \wedge Q(x)$ is false."

Now, apply De Morgan's law to the negation of the conjunction within the predicate:

$$\neg(P(x) \wedge Q(x)) \equiv \neg P(x) \vee \neg Q(x).$$

Therefore,

$$\exists x \in D \text{ such that } \neg(P(x) \wedge Q(x)) \equiv \exists x \in D \text{ such that } \neg P(x) \vee \neg Q(x).$$

Thus, we have proven that:

$$\neg(\forall x \in D, P(x) \wedge Q(x)) \equiv \exists x \in D \text{ such that } \neg P(x) \vee \neg Q(x).$$

■

■ **Example 4.29** Consider the proposition:

$$\forall x \in \mathbb{Z}, x \text{ is even} \wedge x \text{ is a multiple of } 3.$$

This proposition is false because not all integers are simultaneously even and multiples of 3.
The negation is:

$\neg(\forall x \in \mathbb{Z}, x \text{ is even} \wedge x \text{ is a multiple of } 3) \equiv \exists x \in \mathbb{Z} \text{ such that } \neg(x \text{ is even} \wedge x \text{ is a multiple of } 3) \equiv \exists x \in \mathbb{Z}$ such that x is not even $\vee x$ is not a multiple of 3.
This means there exists an integer that is not even or not a multiple of 3.
Since many integers do not satisfy both conditions simultaneously (for example, $x = 1$ is neither even nor a multiple of 3), the negation is true. ■

Exercise 4.25 Negate the following proposition and determine its truth value:

$$\forall x \in \mathbb{N}, x > 5 \wedge x \text{ is prime}.$$

■

Demostración. The negation is:

$\neg(\forall x \in \mathbb{N}, x > 5 \wedge x \text{ is prime}) \equiv \exists x \in \mathbb{N} \text{ such that } \neg(x > 5 \wedge x \text{ is prime}) \equiv \exists x \in \mathbb{N}$ such that $x \leq 5$ $\vee x$ is not prime.

This means there exists a natural number that is less than or equal to 5, or that is not prime.
Since numbers like $x = 4$ satisfy this condition ($x \leq 5$ and x is not prime), the negation is true.
The original proposition is false, since not all natural numbers are greater than 5 and prime. ■

(R) Careful application of De Morgan's laws is crucial when negating propositions that contain conjunctions or disjunctions.

Theorem 4.3.6 The negation of a universal proposition with a disjunction transforms to:

$$\neg(\forall x \in D, P(x) \vee Q(x)) \equiv \exists x \in D \text{ such that } \neg(P(x) \vee Q(x)) \equiv \exists x \in D \text{ such that } \neg P(x) \wedge \neg Q(x).$$

Demostración. We want to prove that

$$\neg(\forall x \in D, P(x) \vee Q(x)) \equiv \exists x \in D \text{ such that } \neg(P(x) \vee Q(x)) \equiv \exists x \in D \text{ such that } \neg P(x) \wedge \neg Q(x).$$

First, apply De Morgan's law for the negation of a universal proposition. The negation of "for all $x \in D$, $P(x) \vee Q(x)$ïs "there exists an $x \in D$ such that $P(x) \vee Q(x)$ is false."Therefore,

$$\neg(\forall x \in D, P(x) \vee Q(x)) \equiv \exists x \in D \text{ such that } \neg(P(x) \vee Q(x)).$$

Now, apply De Morgan's law to the negation of the disjunction within the predicate:

$$\neg(P(x) \vee Q(x)) \equiv \neg P(x) \wedge \neg Q(x).$$

This allows us to write:

$$\exists x \in D \text{ such that } \neg(P(x) \vee Q(x)) \equiv \exists x \in D \text{ such that } \neg P(x) \wedge \neg Q(x).$$

Thus, we have proven that:

$$\neg(\forall x \in D, P(x) \vee Q(x)) \equiv \exists x \in D \text{ such that } \neg P(x) \wedge \neg Q(x).$$

■

■ **Example 4.30** Consider the proposition:

$\forall x \in \mathbb{N}, x$ is even $\vee\, x$ is odd.

This proposition is true because every natural number is either even or odd.
The negation is:
$\neg(\forall x \in \mathbb{N}, x \text{ is even} \vee x \text{ is odd}) \equiv \exists x \in \mathbb{N} \text{ such that } \neg(x \text{ is even} \vee x \text{ is odd}) \equiv \exists x \in \mathbb{N} \text{ such that } \neg x$ is even $\wedge \neg x$ is odd.
This means there exists a natural number that is neither even nor odd.
No such number exists in the natural numbers; therefore, the negation is false, and the original proposition is true. ■

Exercise 4.26 Negate the following proposition and determine its truth value:

$\forall x \in \mathbb{Z}, x$ is a multiple of $4 \vee x$ is a multiple of 5.

■

Demostración. The negation is:

$\neg(\forall x \in \mathbb{Z}, x \text{ is a multiple of } 4 \vee x \text{ is a multiple of } 5) \equiv \exists x \in \mathbb{Z} \text{ such that } \neg(x \text{ is a multiple of } 4 \vee$

$x \text{ is a multiple of } 5) \equiv \exists x \in \mathbb{Z}$ such that x is not a multiple of $4 \wedge x$ is not a multiple of 5.

This means there exists an integer that is neither a multiple of 4 nor a multiple of 5.

For example, $x = 1$ satisfies this condition. Therefore, the negation is true.
The original proposition is false since not all integers are multiples of 4 or 5. ■

(R) This exercise emphasizes the importance of understanding how disjunctions transform when negating universal propositions.

Theorem 4.3.7 The negation of a nested universal proposition (with multiple universal quantifiers) transforms by changing each universal quantifier to an existential quantifier and negating the predicate:

$$\neg(\forall x \in D, \forall y \in D, P(x,y)) \equiv \exists x \in D, \exists y \in D \text{ such that } \neg P(x,y).$$

Demostración. We want to show that

$$\neg(\forall x \in D, \forall y \in D, P(x,y)) \equiv \exists x \in D, \exists y \in D \text{ such that } \neg P(x,y).$$

According to De Morgan's law for negating a universal proposition, the negation of "for all $x \in D$, for all $y \in D$, $P(x,y)$ïs "there exists at least one $x \in D$ and one $y \in D$ such that $P(x,y)$ is false." Thus,

$$\neg(\forall x \in D, \forall y \in D, P(x,y)) \equiv \exists x \in D, \exists y \in D \text{ such that } \neg P(x,y).$$

This equivalence tells us that if $P(x,y)$ is not true for all pairs $(x,y) \in D \times D$, then there must exist at least one pair (x,y) such that $P(x,y)$ is false.
Therefore, we have proven that:

$$\neg(\forall x \in D, \forall y \in D, P(x,y)) \equiv \exists x \in D, \exists y \in D \text{ such that } \neg P(x,y).$$

■

■ **Example 4.31** Consider the proposition:

$$\forall x \in \mathbb{N}, \forall y \in \mathbb{N}, x+y = y+x.$$

This proposition is true because the addition of natural numbers is commutative.
The negation is:

$$\neg(\forall x \in \mathbb{N}, \forall y \in \mathbb{N}, x+y = y+x) \equiv \exists x \in \mathbb{N}, \exists y \in \mathbb{N} \text{ such that } x+y \neq y+x.$$

No such pair (x,y) exists in the natural numbers; therefore, the negation is false, and the original proposition is true. ■

Exercise 4.27 Negate the following proposition and determine its truth value:

$$\forall x \in \mathbb{R}, \forall y \in \mathbb{R}, xy = yx.$$

■

Demostración. The negation is:

$$\neg(\forall x \in \mathbb{R}, \forall y \in \mathbb{R}, xy = yx) \equiv \exists x \in \mathbb{R}, \exists y \in \mathbb{R} \text{ such that } xy \neq yx.$$

However, in the real numbers, multiplication is commutative, and $xy = yx$ for any $x, y \in \mathbb{R}$. No such pair (x,y) exists where $xy \neq yx$.
Therefore, the negation is false, and the original proposition is true. ■

(R) When dealing with nested quantifiers, it is essential to apply the negation to each quantifier and understand how it affects the composite predicate.

Exercise 4.28 Consider the proposition:

$$\forall x \in \mathbb{N}, \forall y \in \mathbb{N}, x \leq y \implies x^2 \leq y^2.$$

Write the negation of this proposition and determine if the negation is true or false.

Demostración. The negation is:

$$\neg(\forall x \in \mathbb{N}, \forall y \in \mathbb{N}, x \leq y \implies x^2 \leq y^2) \equiv \exists x \in \mathbb{N}, \exists y \in \mathbb{N} \text{ such that } x \leq y \wedge x^2 > y^2.$$

We are looking for natural numbers x and y such that $x \leq y$ and $x^2 > y^2$.
Consider $x = 2$ and $y = 3$:

$$2 \leq 3 \text{ (true)}, \quad 2^2 = 4, \quad 3^2 = 9, \quad 4 \leq 9 \text{ (true)}.$$

Try $x = 3$ and $y = 2$:

$$3 \leq 2 \text{ (false)}, \quad \text{Does not satisfy } x \leq y.$$

Now, try $x = 1$ and $y = 1$:

$$1 \leq 1 \text{ (true)}, \quad 1^2 = 1, \quad 1^2 = 1, \quad 1 \leq 1 \text{ (true)}.$$

We do not find any pair (x, y) that satisfies the negation. If we consider $x = 0$ and $y = 1$ (if 0 is included in $\mathbb{N}$):

$$0 \leq 1 \text{ (true)}, \quad 0^2 = 0, \quad 1^2 = 1, \quad 0 \leq 1 \text{ (true)}.$$

It appears that the negation is false. Therefore, the original proposition is true in $\mathbb{N}$. ■

(R) This exercise demonstrates that, even when relationships seem obvious, it is important to carefully analyze the negation and look for counterexamples that might invalidate the proposition.

Understanding the negation of propositions with universal quantifiers is fundamental in mathematical logic and in constructing rigorous proofs. By correctly applying De Morgan's laws and logical equivalences, we can analyze and refute propositions, identify counterexamples, and strengthen our logical reasoning.

4.3.2 Negation of Propositions with Existential Quantifiers

The **existential quantifier** $\exists$ is used in logic to indicate that there exists at least one element in a domain that satisfies a certain property. The negation of propositions involving the existential quantifier is fundamental for logical analysis and the construction of mathematical arguments.

Definition 4.3.2 An **existential proposition** is a statement of the form:

$$\exists x \in D, P(x),$$

where $P(x)$ is a predicate over the domain D. This proposition reads as "there exists at least one x in D such that $P(x)$ is true."

Theorem 4.3.8 — **Negation of an Existential Proposition.** The negation of an existential proposition $\exists x \in D, P(x)$ is logically equivalent to a universal proposition where the predicate is negated:

$$\neg(\exists x \in D, P(x)) \equiv \forall x \in D, \neg P(x).$$

Demostración. We want to prove that

$$\neg(\exists x \in D, P(x)) \equiv \forall x \in D, \neg P(x).$$

($\Rightarrow$) Suppose $\neg(\exists x \in D, P(x))$ is true. This means that the proposition $\exists x \in D, P(x)$ is false. If $\exists x \in D, P(x)$ is false, then there does not exist any $x \in D$ such that $P(x)$ is true. This implies that $P(x)$ is false for all $x \in D$, that is, $\forall x \in D, \neg P(x)$ is true.
($\Leftarrow$) Suppose $\forall x \in D, \neg P(x)$ is true. This means that $P(x)$ is false for each $x \in D$. Therefore, there does not exist any $x \in D$ such that $P(x)$ is true, which implies that $\exists x \in D, P(x)$ is false, that is, $\neg(\exists x \in D, P(x))$ is true.
Thus, we have shown that

$$\neg(\exists x \in D, P(x)) \equiv \forall x \in D, \neg P(x).$$

■

Lema 4.3.2 The negation of the existential quantifier affects only the quantifier and the predicate, while the domain remains unchanged.

Demostración. The domain D is the set of elements over which we quantify. When negating an existential proposition, we change the quantifier and negate the predicate, but the domain remains the same. The negation transforms the statement "there exists at least one x in D such that $P(x)$ is trueïnto "for all x in D, $P(x)$ is false." ■

■ **Example 4.32** Let $P(x)$ be the predicate "x is an odd number,"with domain $D = \{2,4,6\}$. The existential proposition is:

$$\exists x \in D, P(x).$$

Let's evaluate $P(x)$ for each x in D:

- $P(2)$: 2 is odd (false).
- $P(4)$: 4 is odd (false).
- $P(6)$: 6 is odd (false).

Therefore, $\exists x \in D, P(x)$ is false.
The negation of the existential proposition is:

$$\neg(\exists x \in D, P(x)) \equiv \forall x \in D, \neg P(x).$$

This means "for all x in D, x is not odd,"which is true since all elements in D are even numbers. ■

(R) The negation of an existential proposition does not imply that the opposite proposition is true for some element; rather, it asserts that the predicate is false for all elements in the domain.

Theorem 4.3.9 — **General Process for Negating Existential Propositions.** To negate a compound existential proposition, one must:

1. Change the existential quantifier $\exists$ to a universal quantifier $\forall$.
2. Negate the predicate or internal proposition.

Mathematically, this is expressed as:

$$\neg(\exists x \in D, P(x)) \equiv \forall x \in D, \neg P(x).$$

Demostración. This is a direct application of De Morgan's laws for quantifiers and logical connectives. By negating an existential proposition, we assert that no x in D makes $P(x)$ true, meaning $P(x)$ is false for all x in D. ■

■ **Example 4.33** Consider the proposition:

$$\exists x \in \mathbb{N}, x < 0.$$

This proposition is false because there are no natural numbers less than zero.
The negation of the existential proposition is:

$$\neg(\exists x \in \mathbb{N}, x < 0) \equiv \forall x \in \mathbb{N}, x \geq 0.$$

This means "for all natural numbers x, x is greater than or equal to zero,"which is true. ■

Exercise 4.29 Let $P(x)$ be the predicate "x is an even prime number greater than 2,"with domain $D = \mathbb{N}$. Write the negation of the existential proposition:

$$\exists x \in D, P(x),$$

and determine if the negation is true or false. ■

Demostración. The negation of the proposition is:

$$\neg(\exists x \in D, P(x)) \equiv \forall x \in D, \neg P(x).$$

This means "for all x in $\mathbb{N}$, x is not an even prime number greater than 2."
Let's evaluate $P(x)$ for natural numbers:
We know that the only even prime number is 2, and 2 is not greater than 2. Therefore, there does not exist any natural number x such that $P(x)$ is true.
Hence, the existential proposition $\exists x \in D, P(x)$ is false, and its negation $\forall x \in D, \neg P(x)$ is true. ■

(R) This exercise illustrates how the negation of an existential proposition leads to a universal statement about all elements in the domain.

Theorem 4.3.10 — **De Morgan's Laws for Quantifiers.** De Morgan's laws apply to quantifiers as follows:

$$\neg(\exists x \in D, P(x)) \equiv \forall x \in D, \neg P(x),$$
$$\neg(\forall x \in D, P(x)) \equiv \exists x \in D, \neg P(x).$$

Demostración. We will prove each of De Morgan's laws for quantifiers separately.
First Law We want to prove:

$$\neg(\exists x \in D, P(x)) \equiv \forall x \in D, \neg P(x).$$

($\Rightarrow$) Suppose $\neg(\exists x \in D, P(x))$ is true. This means the proposition $\exists x \in D, P(x)$ is false. If $\exists x \in D, P(x)$ is false, then there does not exist any $x \in D$ such that $P(x)$ is true. This implies $P(x)$ is false for all $x \in D$, i.e., $\forall x \in D, \neg P(x)$ is true.
($\Leftarrow$) Suppose $\forall x \in D, \neg P(x)$ is true. This means $P(x)$ is false for each $x \in D$. Therefore, there does not exist any $x \in D$ such that $P(x)$ is true, implying $\neg(\exists x \in D, P(x))$ is true.
Second Law We want to prove:

$$\neg(\forall x \in D, P(x)) \equiv \exists x \in D, \neg P(x).$$

($\Rightarrow$) Suppose $\neg(\forall x \in D, P(x))$ is true. This means $\forall x \in D, P(x)$ is false. Therefore, there exists at least one $x \in D$ such that $P(x)$ is false, i.e.,
exists x $\in D$ such that $\neg P(x)$ is true.
($\Leftarrow$) Suppose $\exists x \in D$ such that $\neg P(x)$ is true. This means there exists at least one $x \in D$ for which $P(x)$ is false. Therefore, $\forall x \in D, P(x)$ is false, i.e., $\neg(\forall x \in D, P(x))$ is true.
Thus, we have proven:

$$\neg(\exists x \in D, P(x)) \equiv \forall x \in D, \neg P(x), \quad \text{and} \quad \neg(\forall x \in D, P(x)) \equiv \exists x \in D, \neg P(x).$$

■

■ **Example 4.34** Consider the proposition:

$$\exists x \in \mathbb{R}, x^2 = -1.$$

In the set of real numbers, this proposition is false because there is no real number whose square is -1.
The negation is:

$$\neg(\exists x \in \mathbb{R}, x^2 = -1) \equiv \forall x \in \mathbb{R}, x^2 \neq -1.$$

This means "for all real numbers x, x^2 is not equal to -1,"which is true.
However, if we consider complex numbers, the original proposition is true since i is a complex number such that $i^2 = -1$. ■

Exercise 4.30 Write the negation of the following existential proposition and determine its truth value in the integers:

$$\exists x \in \mathbb{Z}, x^2 = 2.$$

■

Demostración. The negation is:

$$\neg(\exists x \in \mathbb{Z}, x^2 = 2) \equiv \forall x \in \mathbb{Z}, x^2 \neq 2.$$

This means "for all integers x, x^2 is not equal to 2."
In the set of integers, there is no x such that $x^2 = 2$ since $\sqrt{2}$ is not an integer. Therefore, the existential proposition is false, and its negation is true. ■

(R) The negation of an existential proposition is useful for stating that no solution or specific case satisfies the given property within the considered domain.

Theorem 4.3.11 The negation of an existential proposition involving a conditional transforms as follows:

$$\neg(\exists x \in D, P(x) \implies Q(x)) \equiv \forall x \in D, P(x) \land \neg Q(x).$$

Demostración. We want to prove:

$$\neg(\exists x \in D, P(x) \implies Q(x)) \equiv \forall x \in D, P(x) \land \neg Q(x).$$

1. **Left-Hand Side (⇒)**: Suppose $\neg(\exists x \in D, P(x) \implies Q(x))$ is true. This means that the proposition $\exists x \in D, P(x) \implies Q(x)$ is false. If $\exists x \in D, P(x) \implies Q(x)$ is false, then for all $x \in D$, the implication $P(x) \implies Q(x)$ must be false.
2. **Evaluating the Falsehood of the Implication**: Recall that an implication $P(x) \implies Q(x)$ is false if and only if $P(x)$ is true and $Q(x)$ is false, i.e., $P(x) \land \neg Q(x)$ is true. Since we want the implication to be false for all $x \in D$, it follows that:

$$\forall x \in D, P(x) \land \neg Q(x).$$

3. **Right-Hand Side (⇐)**: Suppose $\forall x \in D, P(x) \land \neg Q(x)$ is true. This means that for each $x \in D$, $P(x)$ is true and $Q(x)$ is false. If this is true for all $x \in D$, then there does not exist any $x \in D$ such that $P(x) \implies Q(x)$ is true, which implies that $\neg(\exists x \in D, P(x) \implies Q(x))$ is true.
Thus, we have proven:

$$\neg(\exists x \in D, P(x) \implies Q(x)) \equiv \forall x \in D, P(x) \land \neg Q(x).$$

■

▪ **Example 4.35** Consider the proposition:

$$\exists x \in \mathbb{Z}, x > 5 \implies x^2 > 25.$$

The negation is:

$$\neg(\exists x \in \mathbb{Z}, x > 5 \implies x^2 > 25) \equiv \forall x \in \mathbb{Z}, x > 5 \land x^2 \leq 25.$$

This means that for all x in $\mathbb{Z}$, if $x > 5$, then $x^2 \leq 25$. However, this is false because for $x = 6$, we have $x^2 = 36 > 25$.
This indicates that the original proposition $\exists x \in \mathbb{Z}, x > 5 \implies x^2 > 25$ is true (because there exists at least one x that satisfies the conditional), and its negation is false. ▪

Exercise 4.31 Write the negation of the following existential proposition and determine its truth value:

$$\exists x \in \mathbb{R}, x \neq 0 \implies \frac{1}{x} = 0.$$

▪

Demostración. The negation is:

$$\neg(\exists x \in \mathbb{R}, x \neq 0 \implies \frac{1}{x} = 0) \equiv \forall x \in \mathbb{R}, x \neq 0 \land \frac{1}{x} \neq 0.$$

This means "for all x in $\mathbb{R}$, $x \neq 0$ and $\frac{1}{x} \neq 0$."
We know that for all $x \neq 0$, $\frac{1}{x} \neq 0$ because the reciprocal of any non-zero number is not zero.
Therefore, the negation is true, and the original existential proposition is false, as there does not exist an $x \neq 0$ such that its reciprocal is zero. ■

(R) When dealing with propositions involving conditionals, it is important to remember the logical equivalences when negating implications and how they affect quantifiers.

Theorem 4.3.12 The negation of an existential proposition with a conjunction transforms as:

$$\neg(\exists x \in D, P(x) \wedge Q(x)) \equiv \forall x \in D, \neg(P(x) \wedge Q(x)) \equiv \forall x \in D, \neg P(x) \vee \neg Q(x).$$

Demostración. We want to prove:

$$\neg(\exists x \in D, P(x) \wedge Q(x)) \equiv \forall x \in D, \neg(P(x) \wedge Q(x)) \equiv \forall x \in D, \neg P(x) \vee \neg Q(x).$$

We begin by applying De Morgan's law for the negation of an existential proposition. The negation of "there exists an $x \in D$ such that $P(x) \wedge Q(x)$ïs "for all $x \in D$, $P(x) \wedge Q(x)$ is false."This gives:

$$\neg(\exists x \in D, P(x) \wedge Q(x)) \equiv \forall x \in D, \neg(P(x) \wedge Q(x)).$$

Then, applying De Morgan's law to the negation of the conjunction within the predicate, the negation of $P(x) \wedge Q(x)$ is $\neg P(x) \vee \neg Q(x)$. Thus:

$$\forall x \in D, \neg(P(x) \wedge Q(x)) \equiv \forall x \in D, \neg P(x) \vee \neg Q(x).$$

Therefore, we have proven:

$$\neg(\exists x \in D, P(x) \wedge Q(x)) \equiv \forall x \in D, \neg P(x) \vee \neg Q(x).$$

■

■ **Example 4.36** Consider the proposition:

$$\exists x \in \mathbb{N}, x \text{ is even} \wedge x \text{ is prime}.$$

We know that the only natural number that is both even and prime is $x = 2$.
The negation is:

$$\neg(\exists x \in \mathbb{N}, x \text{ is even} \wedge x \text{ is prime}) \equiv \forall x \in \mathbb{N}, \neg(x \text{ is even} \wedge x \text{ is prime}) \equiv \forall x \in \mathbb{N}, (x \text{ is not even} \vee x \text{ is not prime}).$$

This means that for all natural numbers x, x is not even or x is not prime.
However, $x = 2$ is a natural number that is both even and prime, so the negation is false, and the original existential proposition is true. ■

Exercise 4.32 Negate the following existential proposition and determine its truth value:

$$\exists x \in \mathbb{Z}, x \text{ is odd} \wedge x \text{ is divisible by 2.}$$

■

Demostración. The negation is:

$\neg(\exists x \in \mathbb{Z}, x \text{ is odd} \wedge x \text{ is divisible by } 2) \equiv \forall x \in \mathbb{Z}, \neg(x \text{ is odd} \wedge x \text{ is divisible by } 2) \equiv \forall x \in \mathbb{Z},$
$(x \text{ is not odd} \vee x \text{ is not divisible by } 2)$.

We know that an integer cannot be both odd and divisible by 2, since numbers divisible by 2 are even.
Therefore, the original existential proposition is false (no such x exists), and its negation is true. ■

R This exercise shows how the negation of an existential proposition with a conjunction leads to a universal statement with a disjunction.

Theorem 4.3.13 The negation of an existential proposition with a disjunction transforms as:

$$\neg(\exists x \in D, P(x) \vee Q(x)) \equiv \forall x \in D, \neg(P(x) \vee Q(x)) \equiv \forall x \in D, \neg P(x) \wedge \neg Q(x).$$

Demostración. We want to prove:

$$\neg(\exists x \in D, P(x) \vee Q(x)) \equiv \forall x \in D, \neg(P(x) \vee Q(x)) \equiv \forall x \in D, \neg P(x) \wedge \neg Q(x).$$

We start by applying De Morgan's law to the negation of an existential proposition. The negation of "there exists an $x \in D$ such that $P(x) \vee Q(x)$ïs "for all $x \in D$, $P(x) \vee Q(x)$ is false."This gives:

$$\neg(\exists x \in D, P(x) \vee Q(x)) \equiv \forall x \in D, \neg(P(x) \vee Q(x)).$$

Next, apply De Morgan's law to the negation of the disjunction within the predicate. The negation of $P(x) \vee Q(x)$ is $\neg P(x) \wedge \neg Q(x)$, so:

$$\forall x \in D, \neg(P(x) \vee Q(x)) \equiv \forall x \in D, \neg P(x) \wedge \neg Q(x).$$

Thus, we have proven:

$$\neg(\exists x \in D, P(x) \vee Q(x)) \equiv \forall x \in D, \neg P(x) \wedge \neg Q(x).$$

■

■ **Example 4.37** Consider the proposition:

$$\exists x \in \mathbb{Z}, x \text{ is a multiple of } 2 \vee x \text{ is a multiple of } 3.$$

This proposition is true since there exist integers that are multiples of 2 or 3 (for example, $x = 2$, $x = 3$).

The negation is:

$$\neg(\exists x \in \mathbb{Z}, x \text{ is a multiple of } 2 \vee x \text{ is a multiple of } 3) \equiv \forall x \in \mathbb{Z}, \neg P(x) \wedge \neg Q(x),$$

where $P(x)$ is "x is a multiple of 2.ªnd $Q(x)$ is "x is a multiple of 3."
This means that for all x in $\mathbb{Z}$, x is not a multiple of 2 and x is not a multiple of 3. This is false since there exist numbers that are multiples of 2 or 3.
Therefore, the negation is false, and the original existential proposition is true. ■

Exercise 4.33 Negate the following existential proposition and determine its truth value:

$$\exists x \in \mathbb{N}, x \text{ is divisible by } 5 \vee x \text{ is divisible by } 7.$$

■

Demostración. The negation is:

$\neg(\exists x \in \mathbb{N}, x \text{ is divisible by } 5 \vee x \text{ is divisible by } 7) \equiv \forall x \in \mathbb{N}, \neg(x \text{ is divisible by } 5 \vee x \text{ is divisible by } 7)$
$\equiv \forall x \in \mathbb{N}, (x \text{ is not divisible by } 5 \wedge x \text{ is not divisible by } 7).$

This means that all natural numbers are neither divisible by 5 nor by 7. This is false since there are natural numbers that are divisible by 5 (e.g., $x = 5$) or by 7 (e.g., $x = 7$).
Therefore, the negation is false, and the original existential proposition is true. ■

(R) This exercise highlights how the negation of an existential proposition with a disjunction leads to a universal statement with a conjunction.

Theorem 4.3.14 The negation of a nested existential proposition (with multiple existential quantifiers) is transformed by replacing each existential quantifier with a universal quantifier and negating the predicate:

$$\neg(\exists x \in D, \exists y \in D, P(x,y)) \equiv \forall x \in D, \forall y \in D, \neg P(x,y).$$

Demostración. We want to prove:

$$\neg(\exists x \in D, \exists y \in D, P(x,y)) \equiv \forall x \in D, \forall y \in D, \neg P(x,y).$$

Applying De Morgan's law for the negation of a nested existential proposition, the negation of "there exists an $x \in D$ and a $y \in D$ such that $P(x,y)$ïs "for all $x \in D$ and for all $y \in D$, $P(x,y)$ is false."This translates to:

$$\neg(\exists x \in D, \exists y \in D, P(x,y)) \equiv \forall x \in D, \forall y \in D, \neg P(x,y).$$

Thus, we have proven:

$$\neg(\exists x \in D, \exists y \in D, P(x,y)) \equiv \forall x \in D, \forall y \in D, \neg P(x,y).$$

■

■ **Example 4.38** Consider the proposition:

$$\exists x \in \mathbb{N}, \exists y \in \mathbb{N}, x+y=5.$$

This proposition is true since there exist natural numbers x and y that sum to 5 (for example, $x=2$, $y=3$).
The negation is:

$$\neg(\exists x \in \mathbb{N}, \exists y \in \mathbb{N}, x+y=5) \equiv \forall x \in \mathbb{N}, \forall y \in \mathbb{N}, x+y \neq 5.$$

This means that for all natural numbers x and y, the sum $x+y$ is not equal to 5, which is false since we have found pairs that sum to 5.
Therefore, the negation is false, and the original existential proposition is true. ■

Exercise 4.34 Negate the following existential proposition and determine its truth value:

$$\exists x \in \mathbb{R}, \exists y \in \mathbb{R}, x^2+y^2=-1.$$

■

Demostración. The negation is:

$$\neg(\exists x \in \mathbb{R}, \exists y \in \mathbb{R}, x^2+y^2=-1) \equiv \forall x \in \mathbb{R}, \forall y \in \mathbb{R}, x^2+y^2 \neq -1.$$

We know that for any real number x, $x^2 \geq 0$. The sum of two non-negative squares is always non-negative. Therefore, $x^2+y^2 \geq 0$.
There is no pair of real numbers x, y such that $x^2+y^2=-1$.
Hence, the original existential proposition is false, and its negation is true. ■

(R) When dealing with nested quantifiers, it is essential to apply the negation to each quantifier and understand how it affects the compound predicate.

Exercise 4.35 Consider the proposition:

$$\exists x \in \mathbb{N}, \exists y \in \mathbb{N}, x > y \wedge y > x.$$

Write the negation of this proposition and determine its truth value. ■

Demostración. The negation is:

$$\neg(\exists x \in \mathbb{N}, \exists y \in \mathbb{N}, x > y \wedge y > x) \equiv \forall x \in \mathbb{N}, \forall y \in \mathbb{N}, \neg(x > y \wedge y > x) \equiv \forall x \in \mathbb{N}, \forall y \in \mathbb{N}, (x \leq y \vee y \leq x).$$

Simplifying, this means that for all natural numbers x and y, x is not simultaneously greater than y and y greater than x.
Since it is impossible for $x > y$ and $y > x$ to hold simultaneously, the original existential proposition is false, and its negation is true. ■

(R) This exercise demonstrates how the negation of an existential proposition that is logically impossible results in a universal statement that is true.

Understanding the negation of propositions with existential quantifiers is essential in mathematical logic and the construction of rigorous arguments. By correctly applying De Morgan's laws and logical equivalences, we can analyze and refute propositions, identify the absence of solutions, and strengthen our logical reasoning.

4.4 Solved Exercises

Exercise 4.36 Let $P(x)$ be the proposition "x is an even number." Express the statement "There exist odd numbers" using propositional functions and quantifiers. ■

Demostración. To express "There exist odd numbers" using $P(x)$, we first need to define the propositional function for "x is odd." A number x is odd if it is not even, so we can define the function $Q(x)$ as:

$$Q(x) = \neg P(x)$$

Thus, the statement "There exist odd numbers" is expressed as:

$$\exists x\, Q(x) = \exists x \neg P(x)$$

This means there is at least one value of x such that x is not even, which fulfills the original statement. ■

Exercise 4.37 Let $P(x) : x > 5$ and $Q(x) : x < 10$ be propositions, both defined over the domain of natural numbers. Express and determine the truth value of the proposition: "For all x, x is between 5 and 10." ■

Demostración. The statement "x is between 5 and 10" can be expressed as:

$$R(x) : P(x) \wedge Q(x)$$

Therefore, the statement "For all x, x is between 5 and 10" is represented as:

$$\forall x\, (P(x) \wedge Q(x))$$

We verify whether this statement is true for all natural numbers. However, this statement is false because there are values of x in the natural numbers that are not between 5 and 10 (for example, $x = 4$ or $x = 11$). ■

Exercise 4.38 Let $P(x)$ be the proposition "x is a multiple of 3" and $Q(x)$ the proposition "x is a multiple of 5," both defined over the domain of natural numbers. Express the statement "There exist numbers that are multiples of 3 and 5" and determine its truth value. ■

Demostración. The statement "There exist numbers that are multiples of 3 and 5" is represented as:

$$\exists x\, (P(x) \wedge Q(x))$$

This means there is at least one value of x such that x is a multiple of 3 and also a multiple of 5. In the natural numbers, any multiple of 15 satisfies this condition (for example, $x = 15$). Therefore, this statement is true. ■

Exercise 4.39 Define the proposition $P(x,y)$: "$x+y = 10$," where x and y are natural numbers. Express the statement "There exists a number x such that for all y, $x+y \neq 10$" and determine its truth value. ■

Demostración. The statement is represented as:

$$\exists x \forall y \neg P(x,y) = \exists x \forall y (x+y \neq 10)$$

This means there exists at least one number x such that, regardless of the value of y, $x+y$ is not equal to 10. If we choose $x = 11$, for example, no matter what value y takes, the sum $11+y$ will never be equal to 10. Therefore, this statement is true. ■

Exercise 4.40 Let $P(x)$: "x^2 is an even number," where x is an integer. Express and determine the truth value of the statement "For all x, if x is even, then x^2 is even." ■

Demostración. The statement is represented as:

$$\forall x (x \text{ is even} \rightarrow P(x))$$

This is an implication that states if x is even, then x^2 is also even. Indeed, if x is an even number, it can be expressed as $x = 2k$ for some integer k. Then:

$$x^2 = (2k)^2 = 4k^2 = 2(2k^2)$$

This shows that x^2 is a multiple of 2, that is, it is even. Therefore, this statement is true. ■

4.5 Proposed Exercises

4.5.1 Predicates: Definition and Examples

Exercise 4.41 Let $P(x)$ be the proposition "x is a prime number." Determine whether $P(7)$ and $P(10)$ are true or false. ■

Exercise 4.42 Define a predicate $Q(x)$ that represents the proposition "x is an odd number" and evaluate $Q(3)$ and $Q(8)$. ■

Exercise 4.43 Let $R(x,y)$ be the proposition "$x+y = 10$," where x and y are natural numbers. Determine whether $R(4,6)$ and $R(3,7)$ are true or false. ■

Exercise 4.44 Formulate a predicate $S(x)$ to express the proposition "x is a multiple of 5" and determine whether $S(15)$ and $S(13)$ are true or false. ■

Exercise 4.45 Define a predicate $T(x,y)$ for the proposition "x is greater than y" and determine whether $T(7,3)$ and $T(2,5)$ are true or false. ■

4.5.2 Quantifiers: Universal Quantifier (∀) and Existential Quantifier (∃)

Exercise 4.46 Express the statement "All natural numbers are greater than or equal to 0" using the universal quantifier. ■

Exercise 4.47 Write the statement "There exists a natural number that is a multiple of 7 and less than 10" using the existential quantifier. ■

Exercise 4.48 Use quantifiers to express the statement "For every real number x, if x is positive, then x^2 is also positive." ■

Exercise 4.49 Express the statement "There exists an integer that is odd and greater than 15" using propositional functions and quantifiers. ■

Exercise 4.50 Formalize the statement "All multiples of 4 are even" using the universal quantifier and a propositional function. ■

4.5.3 Proposition Transformation: Analysis of the Negation of Quantifiers

Exercise 4.51 Find the negation of the statement "For every integer x, x is positive or x is negative." ■

Exercise 4.52 Write the negation of the proposition "There exists a real number y such that $y^2 = -1$." ■

Exercise 4.53 Transform the negation of the statement "For every natural number x, $x+1$ is even" into an equivalent statement using quantifiers. ■

Exercise 4.54 Find the negation of the proposition "For every integer x, if x is odd, then x^2 is odd." ■

Exercise 4.55 Determine the negation of the statement "There exists a real number z such that z is less than 0 and greater than 5." ■

5. Sets and Set Algebra

5.1 Set Operations: Union, Intersection, Complement

5.1.1 Finite and Infinite Sets

In mathematics, especially in set theory, the distinction between finite and infinite sets is fundamental. A finite set is one that has a limited number of elements, while an infinite set has an unlimited number of elements. Understanding these concepts is essential for the study of cardinality, infinity, and number properties.

Definition 5.1.1 A set A is **finite** if there exists a bijection (a bijective function) between A and a set of the form $\{1,2,3,\ldots,n\}$ for some natural number n. If no such n exists, i.e., if it is not possible to establish a bijection with any finite set of natural numbers, then the set A is **infinite**.

■ **Example 5.1**
- The set $A = \{a,b,c\}$ is finite because it has exactly three elements.
- The set of natural numbers $\mathbb{N} = \{1,2,3,4,\ldots\}$ is infinite because there is no natural number n such that $\mathbb{N}$ can be put into a one-to-one correspondence with $\{1,2,\ldots,n\}$.

■

Theorem 5.1.1 A subset of a finite set is finite.

Demostración. Let A be a finite set and $B \subseteq A$ a subset of A. By definition, a set is finite if it has a finite number of elements.
Since A is finite, the number of elements in A is finite. Because $B \subseteq A$, the number of elements in B cannot exceed the number of elements in A. Therefore, B also has a finite number of elements.
Thus, B is finite, and we have shown that any subset of a finite set is also finite. ■

Theorem 5.1.2 The union of a finite number of finite sets is finite.

Demostración. Let $\{A_1, A_2, \ldots, A_n\}$ be a finite collection of finite sets. We want to show that the union $U = A_1 \cup A_2 \cup \cdots \cup A_n$ is finite.
Since each A_i is finite, it has a finite number of elements, say $|A_i| = k_i$, where k_i is a natural number.
The total number of elements in U does not exceed the sum of the numbers of elements in the A_i:

$$|U| \leq |A_1| + |A_2| + \cdots + |A_n| = k_1 + k_2 + \cdots + k_n.$$

Since the sum of a finite number of finite numbers is finite, it follows that $|U|$ is finite. Therefore, the union of a finite number of finite sets is finite. ■

Definition 5.1.2 An infinite set is called **countable** or **denumerable** if its elements can be put into a one-to-one correspondence with the natural numbers $\mathbb{N}$. If it is not possible to establish such a correspondence, the set is called **uncountable**.

Example 5.2

- The set of integers $\mathbb{Z}$ is countable. We can establish a bijection with $\mathbb{N}$ by ordering the integers as follows: $0, 1, -1, 2, -2, 3, -3, \ldots$
- The set of rational numbers $\mathbb{Q}$ is countable. Although there may seem to be "more" rational numbers than natural numbers, a bijection with $\mathbb{N}$ can be established using enumeration techniques.
- The set of real numbers $\mathbb{R}$ is uncountable, meaning its cardinality is greater than that of $\mathbb{N}$. This is demonstrated using Cantor's diagonal argument.

■

Theorem 5.1.3 — **The Set of Rational Numbers is Countable.** The set $\mathbb{Q}$ of rational numbers is an infinite countable set.

Demostración. To show that the set of rational numbers $\mathbb{Q}$ is countable, we need to demonstrate that its elements can be put into a one-to-one correspondence with the natural numbers $\mathbb{N}$. Consider the set of positive rational numbers, which can be written as:

$$\mathbb{Q}^+ = \left\{ \frac{m}{n} \mid m, n \in \mathbb{N}, n \neq 0 \right\}.$$

We can organize the elements of $\mathbb{Q}^+$ in an infinite matrix where the entry in row m and column n represents the rational number $\frac{m}{n}$. By traversing this matrix along ascending diagonals (using a zigzag path), we can list all positive rational numbers.
In this traversal, every positive rational number will eventually appear in the list, though some numbers may be repeated (e.g., $\frac{2}{4} = \frac{1}{2}$). By eliminating duplicates, we obtain a list of all positive rational numbers, establishing a one-to-one correspondence with $\mathbb{N}$.
To include negative rational numbers, we simply alternate each positive number $\frac{m}{n}$ with its negative counterpart $-\frac{m}{n}$, and also include zero. This gives us an enumeration of all rational numbers $\mathbb{Q}$. Therefore, $\mathbb{Q}$ is countable because we have established a bijection with $\mathbb{N}$. ■

Theorem 5.1.4 — **The Set of Real Numbers is Uncountable.** The set $\mathbb{R}$ of real numbers is uncountable.

Demostración. To show that the set of real numbers $\mathbb{R}$ is uncountable, we use Cantor's diagonal argument.
Suppose, for the sake of contradiction, that $\mathbb{R}$ is countable. This would imply that the real numbers in the interval $[0,1]$ are also countable, as any countable infinite set contains countable subsets. If $[0,1]$ were countable, we could enumerate all its elements as an infinite sequence $x_1, x_2, x_3, \ldots$, where each x_i represents a real number in $[0,1]$. Write each x_i in its infinite decimal expansion:

$$x_1 = 0.a_{11}a_{12}a_{13}\ldots,$$

$$x_2 = 0.a_{21}a_{22}a_{23}\ldots,$$

$$x_3 = 0.a_{31}a_{32}a_{33}\ldots,$$

and so forth, where a_{ij} represents the j-th decimal digit of x_i.
Now construct a real number $y = 0.b_1b_2b_3\ldots$ in $[0,1]$ such that y differs from each x_i in at least the i-th decimal place. Define the digits b_i of y as follows: for each i, let $b_i = 5$ if $a_{ii} \neq 5$, and $b_i = 6$ if $a_{ii} = 5$. This ensures that y differs from x_i in the i-th digit.
Thus, y cannot be equal to any x_i in the list, as it differs from each x_i in at least one digit. This contradicts our assumption that we had listed all real numbers in $[0,1]$.
Therefore, the set of real numbers $\mathbb{R}$ is uncountable. ■

Definition 5.1.3 The **cardinality** of a set is a measure of the "number of elements" in the set. For finite sets, the cardinality is simply the number of elements. For infinite sets, the cardinality is defined in terms of bijections with known sets.

- The cardinality of the natural numbers $\mathbb{N}$ is denoted as $\aleph_0$ (aleph-null).
- A set is **countable** if its cardinality is $\aleph_0$.
- The cardinality of the real numbers $\mathbb{R}$ is $\mathfrak{c}$ (the cardinality of the continuum).

Theorem 5.1.5 There are no sets with cardinality strictly between $\aleph_0$ and $\mathfrak{c}$.

Demostración. This theorem is a formulation of the **Continuum Hypothesis**, which states that there are no sets with cardinality strictly between $\aleph_0$ (the cardinality of the natural numbers) and $\mathfrak{c}$ (the cardinality of the real numbers).
The Continuum Hypothesis, proposed by Georg Cantor, asserts that the set of real numbers $\mathbb{R}$ is the smallest infinite set larger than $\mathbb{N}$, in the sense that there is no set with intermediate cardinality between $\mathbb{N}$ and $\mathbb{R}$. This means that any infinite set that can be injected into $\mathbb{R}$ or $\mathbb{N}$ must have a cardinality of either $\aleph_0$ or $\mathfrak{c}$.
However, the Continuum Hypothesis is independent of the axioms of Zermelo-Fraenkel set theory with the Axiom of Choice (ZFC). This implies that within ZFC, it cannot be proved or disproved whether sets with cardinality strictly between $\aleph_0$ and $\mathfrak{c}$ exist; it is consistent to assume either their existence or non-existence.
Therefore, the truth of the theorem depends on accepting the Continuum Hypothesis as an additional axiom since it cannot be proven or disproven within the standard axioms of ZFC. ■

Theorem 5.1.6 — **Cantor's Theorem.** For any set A, the cardinality of the power set $\mathscr{P}(A)$ is strictly greater than the cardinality of A.

Demostración. Let A be any set. We want to prove that the cardinality of the power set $\mathscr{P}(A)$ is strictly greater than the cardinality of A.
Suppose, for contradiction, that there exists a bijection $f: A \to \mathscr{P}(A)$, meaning f assigns each element of A a unique subset of A, and each subset of A is the image of some element of A under f.
Consider the subset

$$B = \{x \in A \mid x \notin f(x)\}.$$

The set B is the set of all elements $x \in A$ such that x does not belong to its own image $f(x)$.
Since $B \subseteq A$, there should exist some $a \in A$ such that $f(a) = B$, because f is a bijection. Now, we analyze whether $a \in B$:

- If $a \in B$, then by the definition of B, $a \notin f(a) = B$, which is a contradiction. - If $a \notin B$, then by the definition of B, $a \in f(a) = B$, which is also a contradiction.
In both cases, we reach a contradiction, which implies that no bijection between A and $\mathscr{P}(A)$ can exist. Therefore, the cardinality of $\mathscr{P}(A)$ is strictly greater than the cardinality of A. ■

R This theorem shows that there is no universal set containing all sets, as one can always construct a set with greater cardinality. This has profound implications in set theory and the understanding of infinity.

Exercise 5.1 Show that the open interval $(0,1)$ has the same cardinality as $\mathbb{R}$.

Demostración. To show that two sets have the same cardinality, it suffices to find a bijection between them.
Define the function $f : (0,1) \to \mathbb{R}$ by:

$$f(x) = \tan\left(\pi x - \frac{\pi}{2}\right).$$

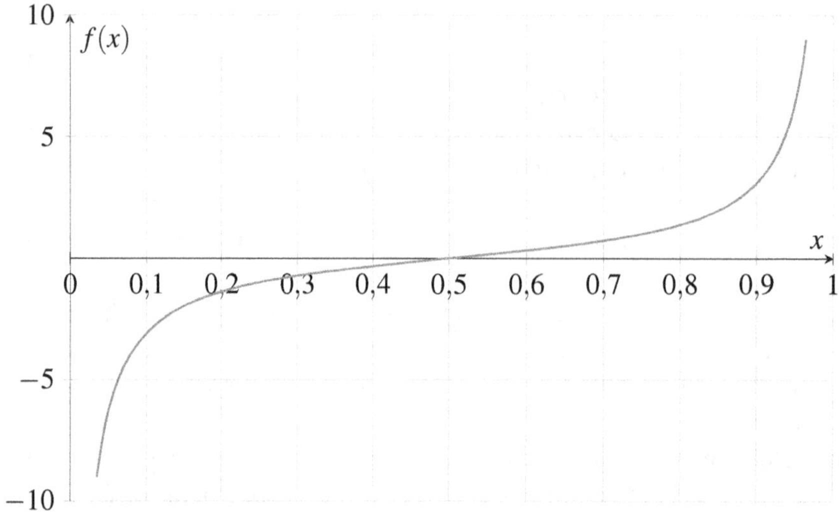

Figura 5.1.1: *Graph of the function* $f(x) = \tan\left(\pi x - \frac{\pi}{2}\right)$.

This function is continuous, strictly increasing, and surjective from $(0,1)$ to $\mathbb{R}$.
Alternatively, we can use the function $g : \mathbb{R} \to (0,1)$ defined by:

$$g(x) = \frac{1}{2} + \frac{1}{\pi}\arctan(x).$$

This function is a bijection between $\mathbb{R}$ and $(0,1)$, proving that both sets have the same cardinality. ■

R This result is surprising, as it shows that a finite and bounded interval like $(0,1)$ has as many elements as the entire set of real numbers $\mathbb{R}$.

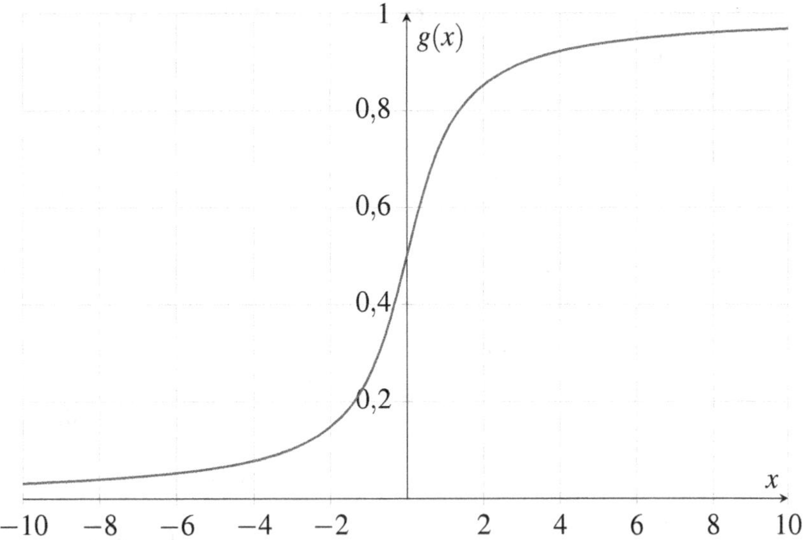

Figura 5.1.2: *Graph of the function* $g(x) = \frac{1}{2} + \frac{1}{\pi}\arctan(x)$.

Theorem 5.1.7 The Cartesian product of a finite set and an infinite set is infinite.

Demostración. Let A be a finite set with n elements, and B an infinite set. We want to show that the Cartesian product $A \times B$ is infinite.
Each element of $A \times B$ is an ordered pair (a,b), where $a \in A$ and $b \in B$. For each element $a \in A$, the set $\{a\} \times B = \{(a,b) \mid b \in B\}$ is a copy of B since it has a one-to-one correspondence with B. Since B is infinite, each set $\{a\} \times B$ is also infinite. Now, $A \times B$ is the union of n infinite sets $\{a\} \times B$ (one for each element $a \in A$). The union of a finite number of infinite sets is infinite.
Therefore, $A \times B$ is infinite. ■

Definition 5.1.4 A set A is **countably infinite** if there exists a bijection between A and $\mathbb{N}$. A set is **uncountably infinite** if it is infinite and no such bijection exists.

Example 5.3
- The set of even numbers $\{2,4,6,8,\dots\}$ is countably infinite. The function $f(n) = 2n$ establishes a bijection with $\mathbb{N}$.
- The set of irrational numbers between 0 and 1 is uncountably infinite. This is because the real numbers between 0 and 1 are uncountable, and the rational numbers are countable, so the irrationals must be uncountable.

■

Theorem 5.1.8 The countable union of countable sets is countable.

Demostración. Let $\{A_n\}_{n\in\mathbb{N}}$ be a family of countable sets. Each A_n can be enumerated as $A_n = \{a_{n1}, a_{n2}, a_{n3}, \dots\}$.
We can enumerate the elements of the union $\bigcup_{n=1}^{\infty} A_n$ using Cantor's **diagonal method**:

$$\begin{array}{lll} a_{11} & & \\ a_{12} & a_{21} & \\ a_{13} & a_{22} & a_{31} \\ a_{14} & a_{23} & a_{32} \\ \vdots & \vdots & \vdots \end{array}$$

We enumerate the elements following the diagonals, ensuring that each element is listed once. In this way, we establish a bijection between $\bigcup_{n=1}^{\infty} A_n$ and $\mathbb{N}$. ■

Exercise 5.2 Show that the set of all finite strings of characters from the English alphabet is countable. ■

Demostración. The English alphabet has 26 letters. Finite strings of these letters can be viewed as finite sequences. We can enumerate all strings of length 1, then length 2, and so on.
More formally, each string can be considered as a finite sequence of characters, and the set of all strings is the countable union of finite sets (strings of length n for each $n \in \mathbb{N}$).
Since each set of strings of fixed length is finite, and the countable union of finite sets is countable, we conclude that the set of all finite strings is countable. ■

Definition 5.1.5 A set A is **countably infinite** if its elements can be listed in an infinite sequence $a_1, a_2, a_3, \ldots$, such that each element of A appears exactly once in the sequence.

Theorem 5.1.9 The set of ordered pairs of natural numbers $\mathbb{N} \times \mathbb{N}$ is countable.

Demostración. We can enumerate the pairs (n,m) of natural numbers using the diagonal enumeration method.
We write the pairs in an infinite matrix:

$$\begin{matrix} (1,1) & (1,2) & (1,3) & \cdots \\ (2,1) & (2,2) & (2,3) & \cdots \\ (3,1) & (3,2) & (3,3) & \cdots \\ \vdots & \vdots & \vdots & \ddots \end{matrix}$$

Then, we traverse the elements along diagonals parallel to the edge of the matrix:

$$(1,1) \to (1,2) \to (2,1) \to (3,1) \to (2,2) \to (1,3) \to \ldots$$

In this way, we establish a bijection between $\mathbb{N} \times \mathbb{N}$ and $\mathbb{N}$. ■

■ **Example 5.4** The set of positive rational numbers $\mathbb{Q}^+$ is countable, since each positive rational number can be represented as a fraction $\frac{p}{q}$ with $p, q \in \mathbb{N}$. Since $\mathbb{N} \times \mathbb{N}$ is countable, and we can eliminate equivalent fractions, the resulting set remains countable. ■

(R) The distinction between finite sets, countably infinite sets, and uncountably infinite sets is essential in many areas of mathematics, including analysis, number theory, and logic. Understanding these differences allows for a deeper analysis of the structure and properties of infinite sets.

The concepts of finite and infinite sets are fundamental in mathematics. Finite sets have a defined, limited cardinality, while infinite sets can be either countably infinite or uncountably infinite, depending on whether a bijection with the natural numbers exists. The study of these sets leads to profound questions about the nature of infinity and the underlying mathematical structures.

5.1.2 Venn Diagrams for Representing Operations

Venn diagrams are graphical tools that visually represent relationships between sets and the operations performed on them. They were introduced by the British mathematician John Venn in 1880 and are widely used in logic, probability, statistics, and mathematics in general.

Definition 5.1.6 A **Venn diagram** is a graphical representation of sets using closed figures (usually circles or ellipses) within a rectangle that represents the universe of discourse U. The overlapping areas between the figures indicate the intersections between the sets.

■ **Example 5.5** Consider two sets A and B within a universe U. The Venn diagram for these sets shows two circles representing A and B, respectively. The region where the circles overlap represents the intersection $A \cap B$. ■

Venn diagrams are especially useful for visualizing basic set operations: union, intersection, difference, and complement.

- **Union** ($A \cup B$): the set of elements that belong to A, B, or both.
- **Intersection** ($A \cap B$): the set of elements that belong to both A and B.
- **Difference** ($A - B$): the set of elements that belong to A but not to B.
- **Complement** (A'): the set of elements that do not belong to A, i.e., $U - A$.

We now represent each of these operations using Venn diagrams.

Definition 5.1.7 The **union** of A and B is the set:

$$A \cup B = \{x \in U \mid x \in A \text{ or } x \in B\}.$$

The union of two sets A and B is represented by shading the areas corresponding to A, B, and their intersection.

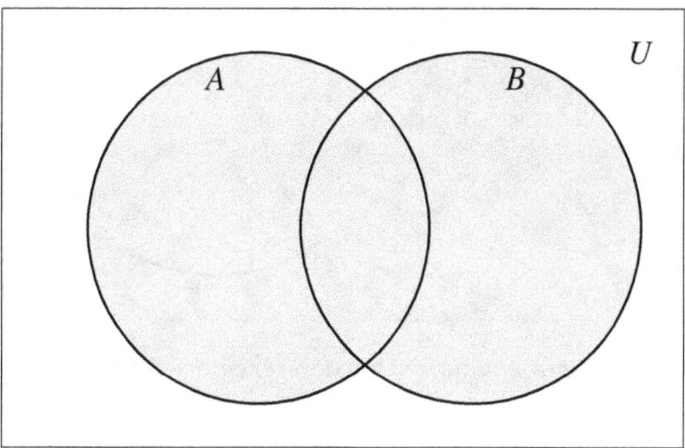

Figura 5.1.3: *Representation of the union $A \cup B$ in a Venn diagram*

Definition 5.1.8 The **intersection** of A and B is the set:

$$A \cap B = \{x \in U \mid x \in A \text{ and } x \in B\}.$$

The intersection of two sets A and B is represented by shading only the area where the circles for A and B overlap.

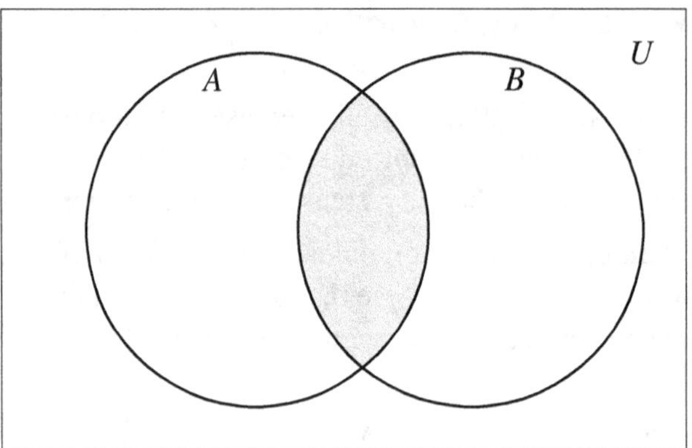

Figura 5.1.4: *Representation of the intersection $A \cap B$ in a Venn diagram*

Definition 5.1.9 The **difference** of A minus B is the set:

$$A - B = \{x \in U \mid x \in A \text{ and } x \notin B\}.$$

The difference $A - B$ is represented by shading the area of A that is not in B.

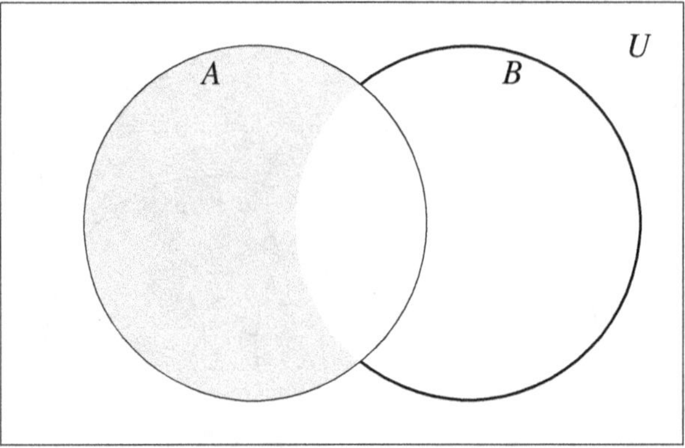

Figura 5.1.5: *Representation of the difference $A - B$ in a Venn diagram*

Definition 5.1.10 The **complement** of A is the set:

$$A' = U - A = \{x \in U \mid x \notin A\}.$$

The complement of A is represented by shading all the areas of the universe U that do not belong to A.

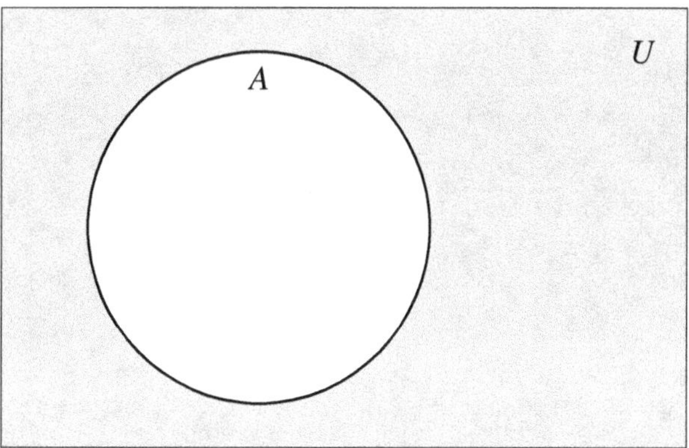

Figura 5.1.6: *Representation of the complement A' in a Venn diagram*

Definition 5.1.11 The **symmetric difference** between A and B is the set:

$$A \triangle B = (A - B) \cup (B - A).$$

The symmetric difference between A and B is represented by shading the areas that belong to A or B, but not to both.

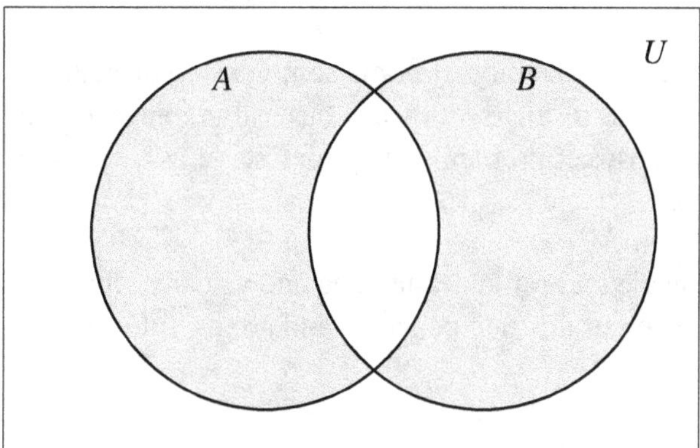

Figura 5.1.7: *Representation of the symmetric difference $A \triangle B$ in a Venn diagram*

Venn diagrams are useful for:

- Visualizing relationships between sets and operations.
- Solving probability and statistics problems.
- Simplifying expressions in set algebra.
- Understanding properties like De Morgan's Laws.

De Morgan's Laws state that:

$$(A \cup B)' = A' \cap B',$$
$$(A \cap B)' = A' \cup B'.$$

These equalities can be visualized and verified using Venn diagrams.

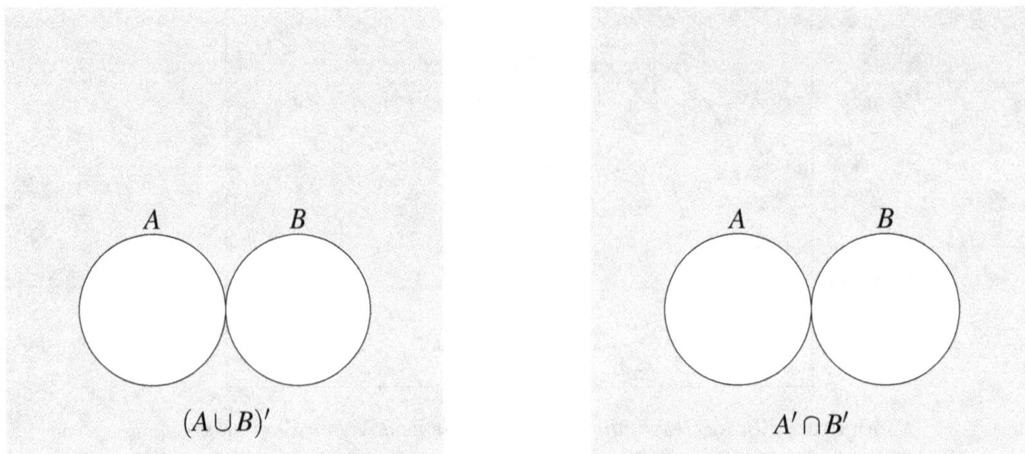

Figura 5.1.8: *Visualization of the first De Morgan's Law in Venn diagrams*

As can be observed, both graphical representations are equivalent, confirming the validity of the first De Morgan's Law.

Although Venn diagrams are powerful tools for visualizing set operations, they have limitations:

- For more than three sets, diagrams become complex and difficult to interpret.
- Not all possible patterns of intersection can be easily represented in two dimensions.
- For infinite sets or abstract problems, diagrams serve as a visual guide but do not replace mathematical rigor.

Venn diagrams are a valuable tool for understanding and visually representing set operations. They facilitate the comprehension of abstract concepts and are useful in various areas of mathematics and logic.

5.2 Venn Diagrams: Visual Representation of Sets

5.2.1 Diagrams with Three Sets

Venn diagrams can be extended to represent relationships and operations between three sets. By adding a third set, the complexity and number of regions increase, allowing for the visualization of all possible intersections and combinations between the sets.

When representing three sets A, B, and C, the Venn diagram consists of three overlapping circles arranged in such a way that eight distinct regions are created. These regions correspond to all possible combinations of an element belonging or not belonging to each of the sets.

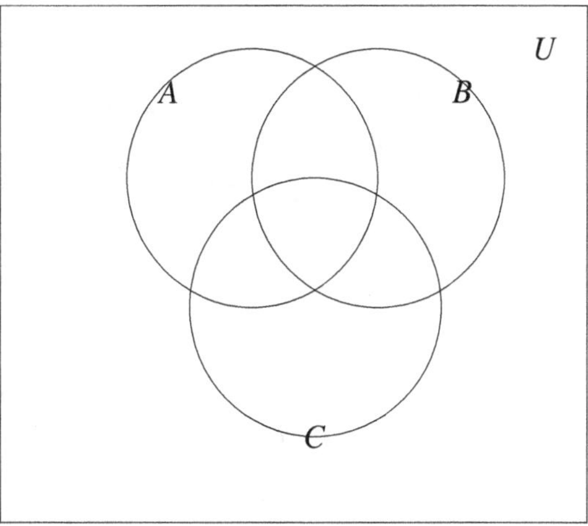

Figura 5.2.1: *Basic Venn diagram with three sets A, B, and C*

This diagram shows how the three sets overlap with each other, creating regions where two sets intersect and a central region where all three sets intersect.

When working with three sets, we can explore more complex operations, such as the union and intersection of multiple sets.

The union of three sets A, B, and C includes all elements that belong to at least one of the sets.

Definition 5.2.1 The **union** of A, B, and C is the set:

$$A \cup B \cup C = \{x \in U \mid x \in A \text{ or } x \in B \text{ or } x \in C\}.$$

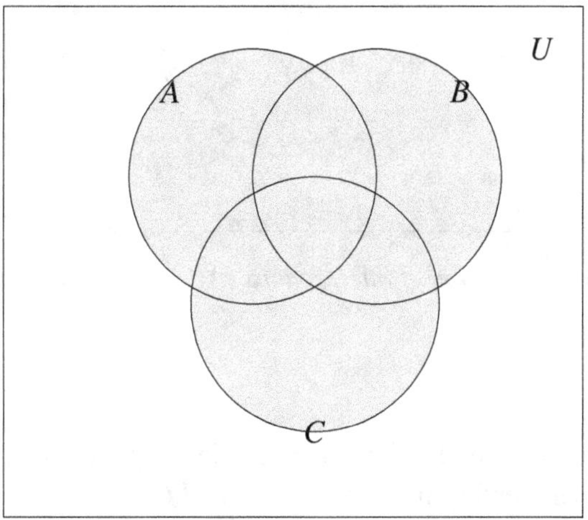

Figura 5.2.2: *Union of three sets* $A \cup B \cup C$

The intersection of three sets A, B, and C includes the elements that belong to all three sets simultaneously.

Definition 5.2.2 The **intersection** of A, B, and C is the set:

$$A \cap B \cap C = \{x \in U \mid x \in A \text{ and } x \in B \text{ and } x \in C\}.$$

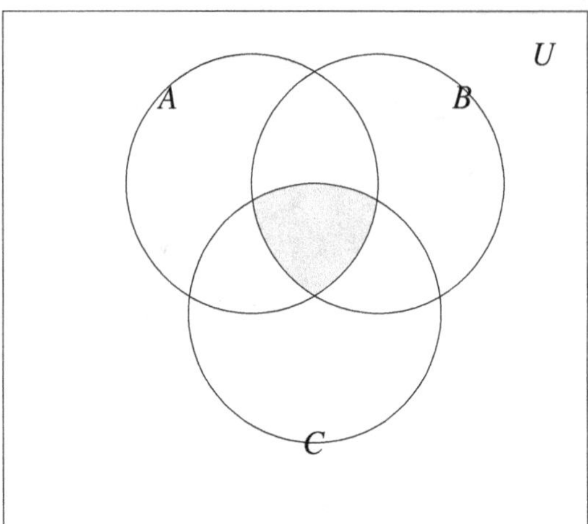

Figura 5.2.3: *Intersection of three sets* $A \cap B \cap C$

We can consider more complex operations, such as the union of intersections between sets.

■ **Example 5.6** The union of the intersections $A \cap B$, $A \cap C$, and $B \cap C$ includes all elements that simultaneously belong to two of the sets.

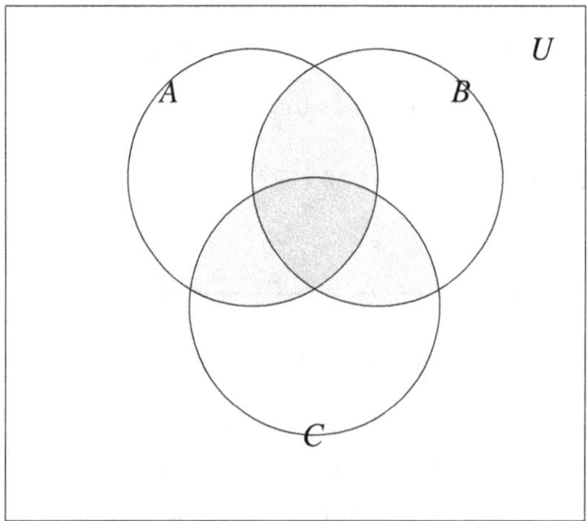

Figura 5.2.4: *Union of intersections* $(A \cap B) \cup (A \cap C) \cup (B \cap C)$

■

The Venn diagram with three sets divides the universe U into eight distinct regions, each representing a unique combination of membership in the sets A, B, and C.

The regions are:

1. Region 1: Elements that belong to none of the sets ($A' \cap B' \cap C'$).
2. Region 2: Elements that belong only to A ($A \cap B' \cap C'$).
3. Region 3: Elements that belong only to B ($A' \cap B \cap C'$).
4. Region 4: Elements that belong only to C ($A' \cap B' \cap C$).
5. Region 5: Elements that belong to A and B, but not to C ($A \cap B \cap C'$).
6. Region 6: Elements that belong to A and C, but not to B ($A \cap B' \cap C$).
7. Region 7: Elements that belong to B and C, but not to A ($A' \cap B \cap C$).
8. Region 8: Elements that belong to all three sets ($A \cap B \cap C$).

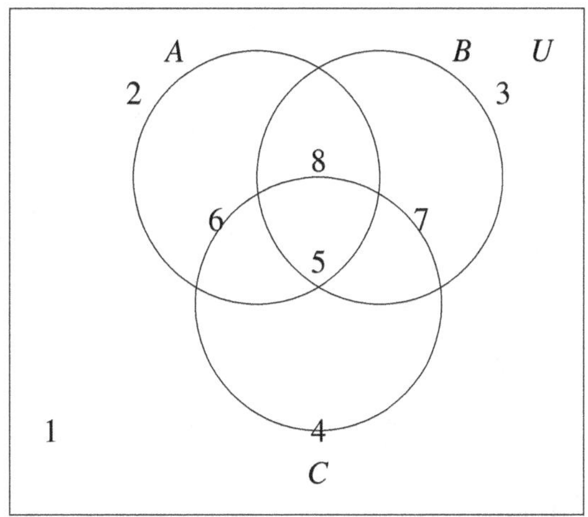

Figura 5.2.5: *Regions in a Venn diagram with three sets*

Venn diagrams with three sets allow us to visualize important laws and properties of set algebra. The distributive law states that:

$$A \cap (B \cup C) = (A \cap B) \cup (A \cap C).$$

We can graphically represent both sides of the equation and verify that they are equivalent.

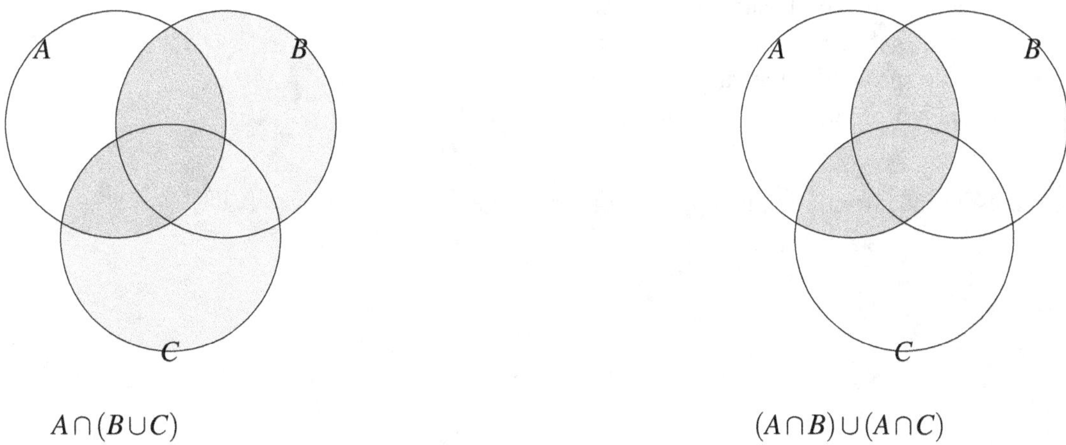

$A \cap (B \cup C)$ $\quad$ $(A \cap B) \cup (A \cap C)$

Figura 5.2.6: *Visualization of the distributive law in Venn diagrams*

Here's the translation into English, maintaining the LaTeX code as requested:
De Morgan's Laws extend to three sets:

$$(A \cup B \cup C)' = A' \cap B' \cap C',$$
$$(A \cap B \cap C)' = A' \cup B' \cup C'.$$

We can represent these equalities graphically to visualize their validity.

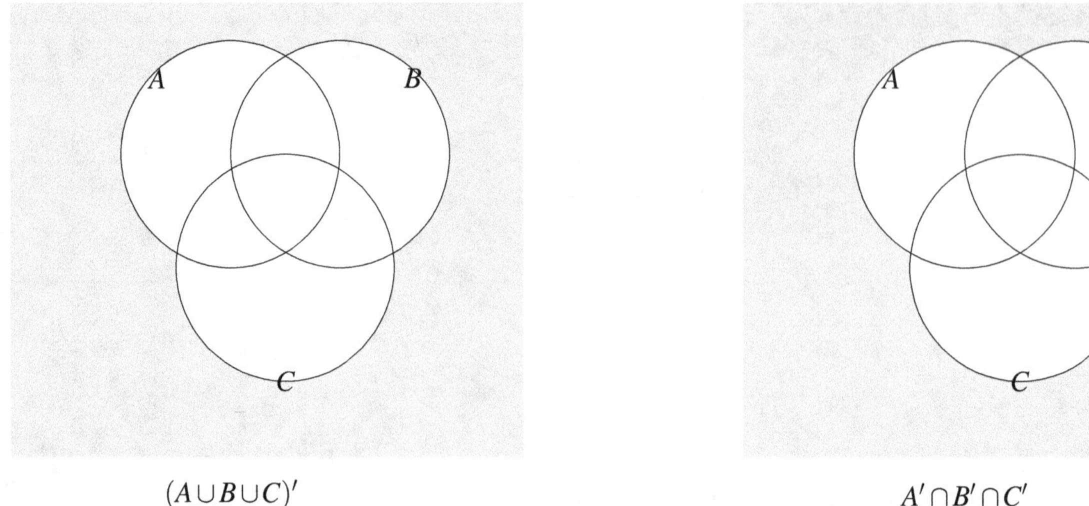

Figura 5.2.7: *Visualization of the first De Morgan's law for three sets*

The shaded areas in both diagrams are equivalent, confirming the law's validity.
Venn diagrams are particularly useful for solving problems involving counting and set analysis.

■ **Example 5.7** In a survey of 100 people, the following results were obtained:

- 60 people speak English (E).
- 30 people speak French (F).
- 20 people speak German (A).
- 25 people speak both English and French.
- 15 people speak both English and German.
- 10 people speak both French and German.
- 5 people speak all three languages.

How many people do not speak any of these languages? ■

We represent the information in a Venn diagram with three sets E, F, and A.

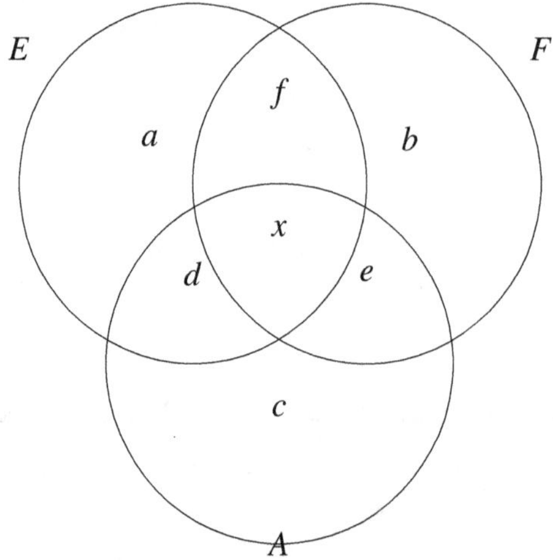

Figura 5.2.8: *Venn diagram for the language problem*

Assigning values:

- $x = 5$ (people who speak all three languages).

- $f = 25 - x = 25 - 5 = 20$ (people who speak English and French, but not German).
- $d = 15 - x = 15 - 5 = 10$ (people who speak English and German, but not French).
- $e = 10 - x = 10 - 5 = 5$ (people who speak French and German, but not English).
- $a = 60 - (f + d + x) = 60 - (20 + 10 + 5) = 25$ (people who only speak English).
- $b = 30 - (f + e + x) = 30 - (20 + 5 + 5) = 0$ (people who only speak French).
- $c = 20 - (d + e + x) = 20 - (10 + 5 + 5) = 0$ (people who only speak German).

Total number of people who speak at least one language:

$$\text{Total} = a + b + c + d + e + f + x = 25 + 0 + 0 + 10 + 5 + 20 + 5 = 65.$$

Number of people who do not speak any of the languages:

$$100 - 65 = 35.$$

When working with Venn diagrams of three sets, it is important to:

- Clearly label each region and set.
- Use different shades or patterns to distinguish between areas.
- Verify that all possible combinations of membership are represented.
- Apply principles of set algebra to simplify and solve problems.

Venn diagrams with three sets are powerful tools for visualizing and analyzing complex relationships between sets. They allow for representing all possible intersections and unions, facilitating the understanding of mathematical laws and solving practical problems in areas such as statistics, logic, and set theory.

5.2.2 Diagrams for Probability Problems

Diagrams are powerful visual tools that facilitate the understanding and solving of probability problems. By representing probabilistic situations graphically, we can visualize events, their relationships, and calculate probabilities more easily. Among the most commonly used diagrams in probability are **Venn diagrams**, **tree diagrams**, and **contingency tables**.

Venn diagrams are useful for representing events and their intersections. They help visualize the union, intersection, and complements of events, as well as calculate related probabilities.

Definition 5.2.3 In the context of probability, an **event** is a subset of the sample space S, which is the set of all possible outcomes of a random experiment.

Example 5.8 Suppose we toss a coin and roll a die. We define the following events:

- A: "Heads on the coin."
- B: "The die shows an even number."

The sample space S has $2 \times 6 = 12$ possible outcomes. We can represent A and B in a Venn diagram.

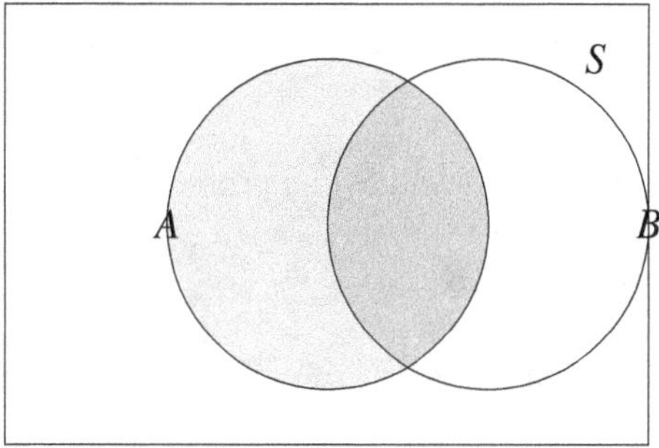

Figura 5.2.9: *Venn diagram of events A and B*

In this diagram, the darker shaded area represents $A \cap B$, that is, the outcomes where heads appear and the die shows an even number.

■

Definition 5.2.4 The **probability** of an event E, denoted by $P(E)$, is a measure of the likelihood that E occurs. For equally likely sample spaces:

$$P(E) = \frac{\text{Number of favorable outcomes for } E}{\text{Total number of outcomes in } S}$$

■ **Example 5.9** Continuing with the previous example, let's calculate:

- $P(A)$: Probability that heads appears.
- $P(B)$: Probability that the die shows an even number.
- $P(A \cap B)$: Probability that heads appears and the die shows an even number.

Demostración. The sample space S has 12 outcomes:

$$S = \{(H,1),(H,2),(H,3),(H,4),(H,5),(H,6),(T,1),(T,2),(T,3),(T,4),(T,5),(T,6)\}$$

Where H is heads and T is tails.

- A has 6 favorable outcomes (all cases where heads appears):

  $$A = \{(H,1),(H,2),(H,3),(H,4),(H,5),(H,6)\}$$

  Therefore, $P(A) = \frac{6}{12} = \frac{1}{2}$.
- B has 6 favorable outcomes (all cases where the die shows 2, 4, or 6):

  $$B = \{(H,2),(H,4),(H,6),(T,2),(T,4),(T,6)\}$$

  Therefore, $P(B) = \frac{6}{12} = \frac{1}{2}$.
- $A \cap B$ has 3 favorable outcomes (cases where heads appears and the die shows an even number):

  $$A \cap B = \{(H,2),(H,4),(H,6)\}$$

  Therefore, $P(A \cap B) = \frac{3}{12} = \frac{1}{4}$.

■

■

Theorem 5.2.1 For any pair of events A and B in a probability space:

$$P(A \cup B) = P(A) + P(B) - P(A \cap B)$$

Demostración. We want to prove that

$$P(A \cup B) = P(A) + P(B) - P(A \cap B).$$

The event $A \cup B$ represents all outcomes where at least one of the events A or B occurs. However, when adding $P(A)$ and $P(B)$, we count the outcomes where both events occur, that is, the event $A \cap B$, twice. To correct this duplication, we subtract $P(A \cap B)$.
Therefore,

$$P(A \cup B) = P(A) + P(B) - P(A \cap B),$$

as we wanted to prove. ■

■ **Example 5.10** Using the previous values, let's calculate $P(A \cup B)$.

Demostración. We have:

$$P(A \cup B) = P(A) + P(B) - P(A \cap B) = \frac{1}{2} + \frac{1}{2} - \frac{1}{4} = \frac{3}{4}$$

■

■

Tree diagrams are representations that show all possible sequences of events in compound experiments, especially useful when events occur in successive stages or steps.

Definition 5.2.5 A **tree diagram** is a graphical representation that shows all the possible paths or routes that can be followed in a random process, where each branch represents a possible outcome and is labeled with its probability.

■ **Example 5.11** Suppose there are 3 balls in a box: 1 red (R) and 2 blue (B). We draw a ball at random, record its color, and without replacing it in the box, draw a second ball. We want to determine the probabilities of obtaining different color combinations.

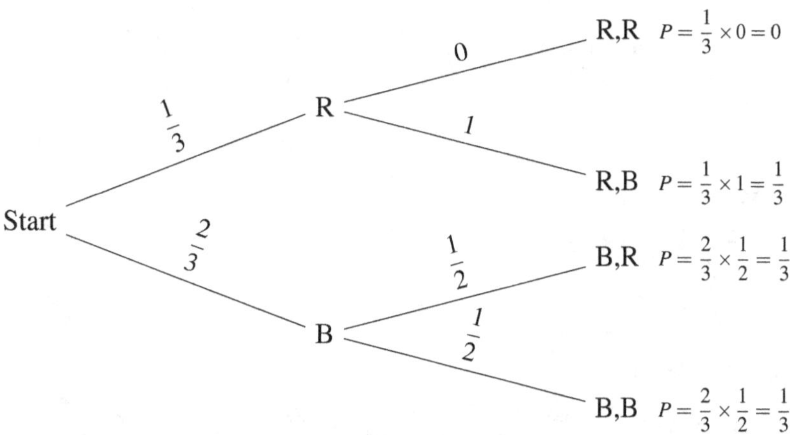

Figura 5.2.10: *Tree diagram for drawing two balls without replacement*

Demostración. We calculate the probabilities:

- Probability of drawing the red ball first (R): $P(R_1) = \frac{1}{3}$.
- Probability of drawing a red ball after drawing the red ball: $P(R_2|R_1) = 0$ (no red balls left).
- Probability of drawing a blue ball after drawing the red ball: $P(B_2|R_1) = \frac{2}{2} = 1$ (two blue balls left).

- Probability of drawing a blue ball first: $P(B_1) = \frac{2}{3}$.
- Probability of drawing a red ball after drawing a blue ball: $P(R_2|B_1) = \frac{1}{2}$ (one red and one blue left).
- Probability of drawing a blue ball after drawing a blue ball: $P(B_2|B_1) = \frac{1}{2}$ (one red and one blue left).

The probabilities of the different paths are:

- $P(R,R) = P(R_1) \times P(R_2|R_1) = \frac{1}{3} \times 0 = 0$.
- $P(R,B) = P(R_1) \times P(B_2|R_1) = \frac{1}{3} \times 1 = \frac{1}{3}$.
- $P(B,R) = P(B_1) \times P(R_2|B_1) = \frac{2}{3} \times \frac{1}{2} = \frac{1}{3}$.
- $P(B,B) = P(B_1) \times P(B_2|B_1) = \frac{2}{3} \times \frac{1}{2} = \frac{1}{3}$.

■

■

Theorem 5.2.2 The sum of the probabilities of all possible paths in a tree diagram is equal to 1.

Demostración. This is because the paths represent all mutually exclusive and exhaustive possible outcomes of the experiment. The sum of their probabilities covers the entire sample space. ■

Contingency tables are matrices that show the frequency distribution of categorical variables and help calculate joint and conditional probabilities.

■ **Example 5.12** In a class of 100 students, the following data is recorded about whether they pass an exam and whether they studied:

	Studied	Did not study	Total
Passed	60	10	70
Did not pass	20	10	30
Total	80	20	100

■

Demostración. We can calculate:

- $P(\text{Studied}) = \frac{80}{100} = 0{,}8$
- $P(\text{Passed}) = \frac{70}{100} = 0{,}7$
- $P(\text{Passed and Studied}) = \frac{60}{100} = 0{,}6$
- $P(\text{Passed} \mid \text{Studied}) = \frac{P(\text{Passed and Studied})}{P(\text{Studied})} = \frac{0{,}6}{0{,}8} = 0{,}75$
- $P(\text{Studied} \mid \text{Passed}) = \frac{P(\text{Passed and Studied})}{P(\text{Passed})} = \frac{0{,}6}{0{,}7} \approx 0{,}857$

■

Theorem 5.2.3 The conditional probability $P(A \mid B)$ is calculated as:

$$P(A \mid B) = \frac{P(A \cap B)}{P(B)}, \quad \text{if } P(B) > 0.$$

Demostración. The conditional probability $P(A \mid B)$ is defined as the probability that event A occurs given that event B has occurred. Since we know B has occurred, we restrict our space of possible outcomes to those in B.
The probability that A occurs under the condition that B has occurred is then the fraction of $P(B)$ that corresponds to $A \cap B$, that is,

$$P(A \mid B) = \frac{\text{Probability that both } A \text{ and } B \text{ occur}}{\text{Probability that } B \text{ occurs}} = \frac{P(A \cap B)}{P(B)}.$$

This is valid whenever $P(B) > 0$. Therefore,

$$P(A \mid B) = \frac{P(A \cap B)}{P(B)},$$

as we wanted to prove. ■

Tree diagrams and contingency tables are especially useful for applying the **Bayes' Theorem**, which allows updating probabilities based on new information.

Theorem 5.2.4 — Bayes' Theorem. Let $\{B_1, B_2, \ldots, B_n\}$ be a partition of the sample space S, and let A be an event such that $P(A) > 0$. Then, for each i:

$$P(B_i \mid A) = \frac{P(A \mid B_i)P(B_i)}{\sum_{k=1}^{n} P(A \mid B_k)P(B_k)}$$

Demostración. We want to prove that

$$P(B_i \mid A) = \frac{P(A \mid B_i)P(B_i)}{\sum_{k=1}^{n} P(A \mid B_k)P(B_k)}.$$

By the definition of conditional probability, we have

$$P(B_i \mid A) = \frac{P(A \cap B_i)}{P(A)}.$$

Applying the multiplication rule, $P(A \cap B_i) = P(A \mid B_i)P(B_i)$, so

$$P(B_i \mid A) = \frac{P(A \mid B_i)P(B_i)}{P(A)}.$$

To compute $P(A)$, note that $\{B_1, B_2, \ldots, B_n\}$ is a partition of the sample space S. This means A can be expressed as the disjoint union of the events $A \cap B_k$ for $k = 1, 2, \ldots, n$. Therefore,

$$P(A) = \sum_{k=1}^{n} P(A \cap B_k) = \sum_{k=1}^{n} P(A \mid B_k)P(B_k).$$

Substituting $P(A)$ into the expression for $P(B_i \mid A)$, we get

$$P(B_i \mid A) = \frac{P(A \mid B_i)P(B_i)}{\sum_{k=1}^{n} P(A \mid B_k)P(B_k)},$$

as we wanted to prove. ■

Example 5.13 A medical test has a true positive rate of 99 % and a false positive rate of 5 %. The disease has a prevalence of 1 % in the population. If a person gets a positive result, what is the probability that they actually have the disease?

Demostración. We define:
- D: The person has the disease.
- D': The person does not have the disease.
- T: The test result is positive.

We have:
- $P(D) = 0{,}01$
- $P(D') = 0{,}99$
- $P(T \mid D) = 0{,}99$
- $P(T \mid D') = 0{,}05$

Applying Bayes' Theorem:

$$P(D \mid T) = \frac{P(T \mid D)P(D)}{P(T \mid D)P(D) + P(T \mid D')P(D')}$$

We calculate:

$$P(D \mid T) = \frac{0{,}99 \times 0{,}01}{0{,}99 \times 0{,}01 + 0{,}05 \times 0{,}99} = \frac{0{,}0099}{0{,}0099 + 0{,}0495} = \frac{0{,}0099}{0{,}0594} \approx 0{,}1667$$

Therefore, the probability that the person actually has the disease is approximately 16.67 %.

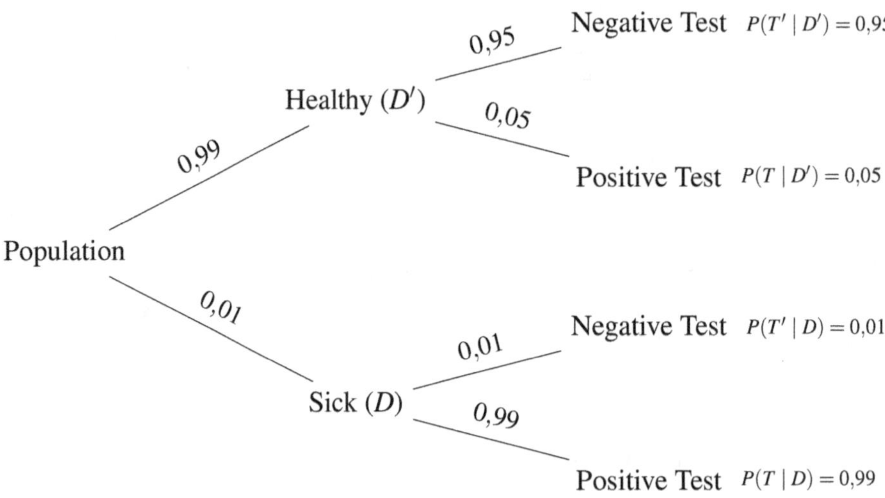

Figura 5.2.11: *Tree diagram for the medical test problem*

The tree diagram shows the probabilities on each branch, facilitating the calculation of joint and conditional probabilities.

When using diagrams to solve probability problems, it is important to:
- Clearly define the events and the sample space.
- Properly label the probabilities on each branch or region.

- Verify that the sum of probabilities is consistent (for example, branches coming from a node should sum to 1).
- Use diagrams as a visual aid, complementing with precise mathematical calculations.

Diagrams, whether Venn diagrams, tree diagrams, or contingency tables, are essential tools for solving probability problems. They facilitate understanding the relationships between events, visualizing sample spaces, and calculating complex probabilities. Proper use of these tools improves the ability to analyze and solve probabilistic situations in various contexts.

5.3 Application Problems

5.3.1 Problem Solving with Disjoint Sets

Disjoint sets are a fundamental concept in set theory and mathematics in general. Two or more sets are disjoint if they have no elements in common. Understanding and applying this concept is essential for solving problems involving set partitions, counting, probability, and other fields.

Definition 5.3.1 Two sets A and B are **disjoint** if their intersection is the empty set:

$$A \cap B = \emptyset.$$

This concept extends to collections of more than two sets:

Definition 5.3.2 A collection of sets $\{A_i\}_{i \in I}$ is **pairwise disjoint** if for all $i, j \in I$, with $i \neq j$, it holds that:

$$A_i \cap A_j = \emptyset.$$

Disjoint sets have properties that are useful when solving problems.

- The cardinality of the union of disjoint sets is equal to the sum of the cardinalities:

$$|A \cup B| = |A| + |B|, \quad \text{if } A \cap B = \emptyset.$$

- In probability, if A and B are disjoint events, then:

$$P(A \cup B) = P(A) + P(B).$$

Example 5.14 Consider the sets:

$$A = \{1,2,3\}, \quad B = \{4,5,6\}.$$

Then, A and B are disjoint since they share no elements:

$$A \cap B = \emptyset.$$

■

Example 5.15 In a sample space of a dice roll, define the events:

$$E = \{\text{rolling an even number}\} = \{2,4,6\}, O = \{\text{rolling an odd number greater than } 4\} = \{5\}.$$

Then, E and O are disjoint since:

$$E \cap O = \emptyset.$$

■

Venn diagrams are useful for visualizing disjoint sets.

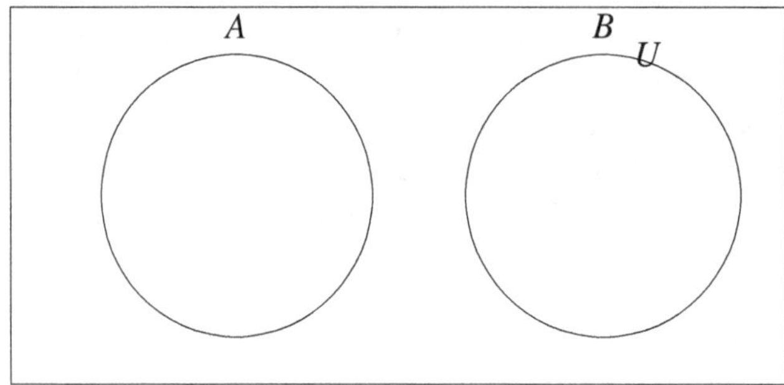

Figura 5.3.1: *Representation of disjoint sets A and B*

In this diagram, the sets A and B do not overlap, indicating that they are disjoint.
When solving problems involving disjoint sets, the properties of adding cardinalities or probabilities are often leveraged. Below are several examples illustrating how to approach these problems.

■ **Example 5.16** In a class of 40 students, 15 study French, 20 study German, and no student studies both languages. How many students study French or German?
Since no student studies both languages, the sets of students studying French (F) and German (A) are disjoint:

$$F \cap A = \emptyset.$$

Therefore, the total number of students studying French or German is:

$$|F \cup A| = |F| + |A| = 15 + 20 = 35.$$

■

■ **Example 5.17** In an experiment, two fair coins are flipped. Define the events:

$$E_1 = \{\text{The first coin lands heads}\}, \qquad E_2 = \{\text{The second coin lands heads}\}.$$

The events E_1 and E_2 are not disjoint since they can occur simultaneously. However, we can define new disjoint events.
Define:

$$A = \{\text{Both coins land heads}\}, \qquad B = \{\text{Both coins land tails}\}.$$

Now, A and B are disjoint since they cannot occur simultaneously:

$$A \cap B = \emptyset.$$

Calculate the probabilities:

$$P(A) = \frac{1}{4}, \quad P(B) = \frac{1}{4}.$$

The probability that A or B occurs is:

$$P(A \cup B) = P(A) + P(B) = \frac{1}{4} + \frac{1}{4} = \frac{1}{2}.$$

■

Disjoint sets are essential in counting problems, particularly in the principle of addition.

Theorem 5.3.1 — Addition Principle. If A and B are disjoint sets, then the number of elements in their union is the sum of the number of elements in each set:

$$|A \cup B| = |A| + |B|.$$

Demostración. We want to show that if A and B are disjoint sets, then

$$|A \cup B| = |A| + |B|.$$

Since A and B are disjoint, they have no elements in common, i.e., $A \cap B = \emptyset$. This means that each element of $A \cup B$ belongs exclusively to A or exclusively to B, but not to both.
Therefore, the number of elements in $A \cup B$ is simply the sum of the number of elements in A and the number of elements in B, since there are no duplicates when combining both sets.
Hence,

$$|A \cup B| = |A| + |B|,$$

as desired. ■

■ **Example 5.18** How many integers between 1 and 100 are multiples of 3 or multiples of 5 but not both?
Define the sets:

$$M_3 = \{\text{Multiples of 3 between 1 and 100}\},$$
$$M_5 = \{\text{Multiples of 5 between 1 and 100}\},$$
$$M_{15} = \{\text{Multiples of 15 between 1 and 100}\}.$$

The multiples of 15 are numbers that are multiples of both 3 and 5.
We calculate:

$$|M_3| = \left\lfloor \frac{100}{3} \right\rfloor = 33, \qquad |M_5| = \left\lfloor \frac{100}{5} \right\rfloor = 20, \qquad |M_{15}| = \left\lfloor \frac{100}{15} \right\rfloor = 6.$$

The numbers that are multiples of 3 or 5 but not both are:

$$(|M_3| - |M_{15}|) + (|M_5| - |M_{15}|) = (33-6) + (20-6) = 27 + 14 = 41.$$

■

Venn diagrams help visualize the relationship between sets and facilitate problem-solving.

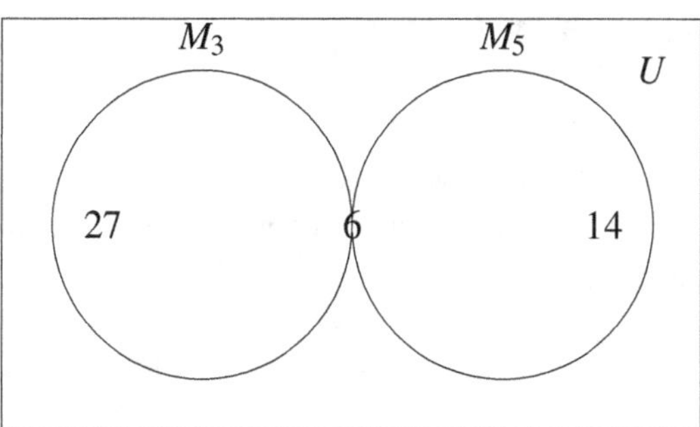

Figura 5.3.2: *Venn diagram for multiples of 3 and 5 between 1 and 100*

In this diagram, the shaded region represents the multiples of 15. The numbers in the regions indicate the count of elements in each section.
In probability, disjoint (mutually exclusive) events are those that cannot occur simultaneously.

Definition 5.3.3 Two events A and B are **mutually exclusive** if:

$$P(A \cap B) = 0.$$

■ **Example 5.19** In a standard deck of 52 cards, what is the probability of drawing an ace or a heart?
Define the events:

$$A = \{\text{Drawing an ace}\}, \quad |A| = 4, \qquad H = \{\text{Drawing a heart}\}, \quad |H| = 13.$$

The events A and H are not mutually exclusive, since the ace of hearts belongs to both.
The probability is:

$$P(A \cup H) = P(A) + P(H) - P(A \cap H) = \frac{4}{52} + \frac{13}{52} - \frac{1}{52} = \frac{16}{52} = \frac{4}{13}.$$

■

However, if we ask for the probability of drawing an ace or a king, the events are disjoint:

■ **Example 5.20** What is the probability of drawing an ace or a king from a standard deck?
The events:

$$A = \{\text{Drawing an ace}\}, \quad K = \{\text{Drawing a king}\},$$

are mutually exclusive:

$$A \cap K = \emptyset.$$

Therefore:

$$P(A \cup K) = P(A) + P(K) = \frac{4}{52} + \frac{4}{52} = \frac{8}{52} = \frac{2}{13}.$$

■

■ **Example 5.21** In a group of 100 people, it is known that:

- 45 people speak English.
- 30 people speak French.
- 20 people speak German.
- 15 people speak both English and French.
- 10 people speak both English and German.
- 5 people speak both French and German.
- 0 people speak all three languages.

How many people do not speak any of these languages?
Since no one speaks all three languages, the sets of people who speak two languages are disjoint in the intersections of two languages.
Define:

$$E = \{\text{Speak English}\}, \quad F = \{\text{Speak French}\}, \quad G = \{\text{Speak German}\}.$$

Using the inclusion-exclusion principle:

$$|E \cup F \cup G| = |E| + |F| + |G| - |E \cap F| - |E \cap G| - |F \cap G| + |E \cap F \cap G|.$$

Since $|E \cap F \cap G| = 0$, we have:

$$|E \cup F \cup G| = 45 + 30 + 20 - 15 - 10 - 5 + 0 = 65.$$

Therefore, the number of people who do not speak any of these languages is:

$$100 - 65 = 35.$$

■

Understanding disjoint sets is essential for solving problems in set theory, probability, and combinatorics. Disjoint sets simplify calculations of cardinalities and probabilities, as the intersections are empty and there is no need to adjust for common elements. Using Venn diagrams and a solid understanding of the properties of disjoint sets facilitate solving complex problems.
When addressing problems involving disjoint sets, it is crucial to correctly identify the sets and their relationships. Utilizing the properties and theorems associated with disjoint sets allows simplifying calculations and obtaining accurate solutions. Venn diagrams are valuable tools that provide a clear visual representation of the relationships between sets and support logical reasoning in problem-solving.

5.3.2 Application of Sets in Conditional Probability

Conditional probability is a fundamental tool in probability theory that allows calculating the probability of an event given that another event has occurred. Sets and their operations are essential for understanding and calculating conditional probabilities, as events can be represented as sets in a sample space.

Definition 5.3.4 — Conditional Probability. Let $(\Omega, \mathscr{F}, P)$ be a probability space, and let A and B be events with $P(B) > 0$. The **conditional probability** of A given B is defined as:

$$P(A \mid B) = \frac{P(A \cap B)}{P(B)}.$$

This definition shows how the intersection of sets and their operations are key to calculating conditional probabilities.
In terms of sets, the conditional probability $P(A \mid B)$ represents the fraction of B that is also in A. Visually, this can be represented using Venn diagrams.

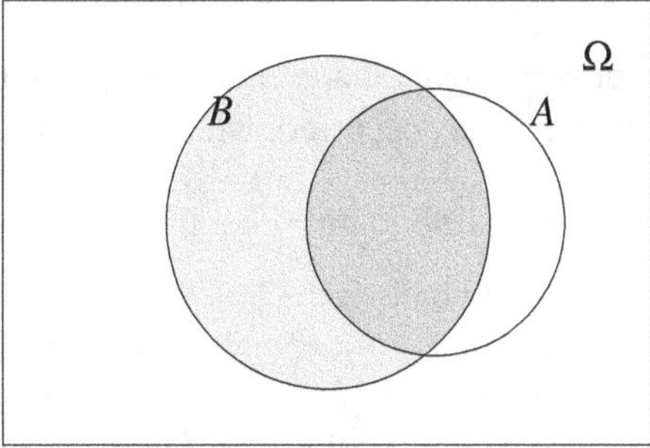

Figura 5.3.3: *Representation of the conditional probability $P(A \mid B)$ using sets*

In this diagram, the darker area represents the intersection $A \cap B$. The conditional probability $P(A \mid B)$ is the proportion of B that also belongs to A.

Theorem 5.3.2 Conditional probability satisfies the following properties:

1. $0 \leq P(A \mid B) \leq 1$.
2. If A and B are independent, then $P(A \mid B) = P(A)$.
3. If $A \subseteq B$, then $P(A \mid B) = \frac{P(A)}{P(B)}$.

Demostración. We want to prove that

$$P(A \cap B) = P(B) \cdot P(A \mid B).$$

By the definition of conditional probability, we have:

$$P(A \mid B) = \frac{P(A \cap B)}{P(B)},$$

provided that $P(B) > 0$. Rearranging this equation, we can express $P(A \cap B)$ as:

$$P(A \cap B) = P(B) \cdot P(A \mid B).$$

This shows that the probability of the intersection of A and B is the product of the probability of B and the conditional probability of A given B, which is what we wanted to prove. ■

Theorem 5.3.3 — Multiplication Rule. For any pair of events A and B with $P(B) > 0$, it holds that:

$$P(A \cap B) = P(B) \cdot P(A \mid B).$$

Demostración. We want to prove that:

$$P(A \cap B) = P(B) \cdot P(A \mid B).$$

By the definition of conditional probability, we know that:

$$P(A \mid B) = \frac{P(A \cap B)}{P(B)},$$

provided that $P(B) > 0$. Rearranging this equation, we can express $P(A \cap B)$ as:

$$P(A \cap B) = P(B) \cdot P(A \mid B).$$

This shows that the probability of both events A and B occurring is the product of the probability of B and the probability of A given that B has occurred, which is what we wanted to prove. ■

This theorem is fundamental for calculating joint probabilities using conditional probabilities.

■ **Example 5.22** In a box, there are 5 red balls and 3 blue balls. One ball is drawn at random. What is the probability that it is blue given that the drawn ball is not red?

Demostración. Let A be the event "the ball is blue.ªnd B be the event "the ball is not red."Since there are only red and blue balls, B is equivalent to A, i.e., $B = A$.
We calculate:

$$P(A \mid B) = \frac{P(A \cap B)}{P(B)} = \frac{P(A)}{P(B)}.$$

Since $B = A$, we have $P(B) = P(A)$. Therefore:

$$P(A \mid B) = \frac{P(A)}{P(A)} = 1.$$

The probability that the ball is blue given that it is not red is 1.
Alternatively, we can think that if we know the ball is not red, it must be blue. ■

■

■ **Example 5.23** In a survey of 200 people, it was found that 120 are women, 80 are men, 50 women are left-handed, and 20 men are left-handed. If a person is selected at random, what is the probability that they are left-handed given that they are a woman?

Demostración. Let Z be the event "the person is left-handed.ªnd W be the event "the person is a woman."
We have:

$$P(Z \mid W) = \frac{P(Z \cap W)}{P(W)} = \frac{\text{Number of left-handed women}}{\text{Total number of women}} = \frac{50}{120} = \frac{5}{12}.$$

Therefore, the probability that a person is left-handed given that they are a woman is $\frac{5}{12}$. ■

■

Venn diagrams are useful tools for visualizing events and their intersections, making it easier to calculate conditional probabilities.

■ **Example 5.24** At a university, 40 % of students take the Mathematics course (M), 30 % take the Physics course (F), and 20 % take both courses. If a student is selected at random and we know they are taking Mathematics, what is the probability that they are also taking Physics? ■

Demostración. We have:

$$P(M) = 0{,}4, \quad P(F) = 0{,}3, \quad P(M \cap F) = 0{,}2.$$

We want to calculate $P(F \mid M)$.
Using the definition of conditional probability:

$$P(F \mid M) = \frac{P(F \cap M)}{P(M)} = \frac{0{,}2}{0{,}4} = 0{,}5.$$

Therefore, the probability that a student is taking Physics given that they are taking Mathematics is 0.5.
We can represent this situation with a Venn diagram.

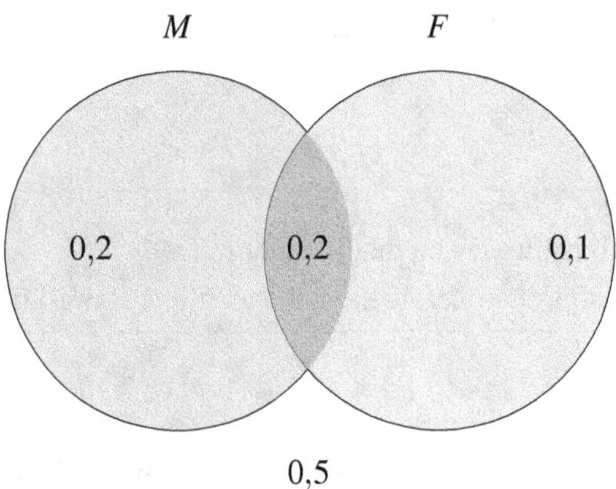

Figura 5.3.4: *Enhanced Venn diagram for Mathematics and Physics courses*

In this diagram, the areas represent the probabilities of each event and their intersections. ■

5.3.3 Bayes' Theorem with Sets

Bayes' Theorem relates conditional probabilities and is especially useful when conditions need to be inverted.

Theorem 5.3.4 — **Bayes' Theorem.** Let $\{B_1, B_2, \ldots, B_n\}$ be a partition of the sample space Ω, with $P(B_i) > 0$ for all i, and let A be an event such that $P(A) > 0$. Then, for each i,

$$P(B_i \mid A) = \frac{P(A \mid B_i)P(B_i)}{\sum_{j=1}^{n} P(A \mid B_j)P(B_j)}.$$

Demostración. We want to prove that

$$P(B_i \mid A) = \frac{P(A \mid B_i)P(B_i)}{\sum_{j=1}^{n} P(A \mid B_j)P(B_j)}.$$

By the definition of conditional probability, we have:

$$P(B_i \mid A) = \frac{P(A \cap B_i)}{P(A)}.$$

Applying the multiplication rule, we can express $P(A \cap B_i)$ as $P(A \mid B_i)P(B_i)$, which gives:

$$P(B_i \mid A) = \frac{P(A \mid B_i)P(B_i)}{P(A)}.$$

Now, to calculate $P(A)$, note that $\{B_1, B_2, \ldots, B_n\}$ is a partition of the sample space Ω. This means that A can be expressed as the disjoint union of the events $A \cap B_j$ for $j = 1, 2, \ldots, n$. Therefore,

$$P(A) = \sum_{j=1}^{n} P(A \cap B_j) = \sum_{j=1}^{n} P(A \mid B_j)P(B_j).$$

Substituting $P(A)$ into the expression for $P(B_i \mid A)$, we get:

$$P(B_i \mid A) = \frac{P(A \mid B_i)P(B_i)}{\sum_{j=1}^{n} P(A \mid B_j)P(B_j)},$$

which is what we wanted to prove. ■

■ **Example 5.25** In a factory, there are three machines (M_1, M_2, M_3) that produce 50 %, 30 %, and 20 % of the products, respectively. The defect rates are 2 % for M_1, 3 % for M_2, and 5 % for M_3. If a product is selected at random and found to be defective, what is the probability that it came from machine M_2?

Demostración. Define:

- B_1: "The product comes from M_1."
- B_2: "The product comes from M_2."
- B_3: "The product comes from M_3."
- A: "The product is defective."

We have:

$$P(B_1) = 0{,}5, \quad P(B_2) = 0{,}3, \quad P(B_3) = 0{,}2.$$

The conditional probabilities:

$$P(A \mid B_1) = 0{,}02, \quad P(A \mid B_2) = 0{,}03, \quad P(A \mid B_3) = 0{,}05.$$

Applying Bayes' Theorem:

$$P(B_2 \mid A) = \frac{P(A \mid B_2)P(B_2)}{P(A \mid B_1)P(B_1) + P(A \mid B_2)P(B_2) + P(A \mid B_3)P(B_3)}.$$

Calculating:

$$P(B_2 \mid A) = \frac{0{,}03 \times 0{,}3}{0{,}02 \times 0{,}5 + 0{,}03 \times 0{,}3 + 0{,}05 \times 0{,}2} = \frac{0{,}009}{0{,}01 + 0{,}009 + 0{,}01} = \frac{0{,}009}{0{,}029} \approx 0{,}3103.$$

Therefore, the probability that the defective product came from M_2 is approximately 31.03 %. ■

■

Conditional independence is an important concept in probability, where two events can be independent given a third event.

Definition 5.3.5 Two events A and B are **conditionally independent** given an event C with $P(C) > 0$ if:

$$P(A \cap B \mid C) = P(A \mid C) \cdot P(B \mid C).$$

■ **Example 5.26** A bag contains 3 red balls and 2 blue balls. Two balls are drawn without replacement. Let A be the event "the first ball is red,"B be the event "the second ball is blue,.ªnd C be the event "the second ball is drawn."

Since the balls are drawn without replacement, the events A and B are not independent. However, given that C is true (the second ball is drawn), we can analyze conditional independence.

We calculate:

$$P(A \mid C) = P(\text{First ball is red}) = \frac{3}{5}.$$

$$P(B \mid C) = P(\text{Second ball is blue}) = \frac{2}{5}.$$

Now, $P(A \cap B \mid C)$ is the probability that the first ball is red and the second ball is blue:

$$P(A \cap B \mid C) = P(A) \cdot P(B \mid A) = \frac{3}{5} \cdot \frac{2}{4} = \frac{3}{5} \cdot \frac{1}{2} = \frac{3}{10}.$$

However, $P(A \mid C) \cdot P(B \mid C) = \frac{3}{5} \cdot \frac{2}{5} = \frac{6}{25} \neq \frac{3}{10}$.
Therefore, A and B are not conditionally independent given C. ■

LaTeXtranslation of your content into English with the same structure:

—

Set theory provides a solid foundation for understanding and calculating conditional probabilities. Operations between sets, such as union, intersection, and complement, are directly applicable in probability calculations.

Theorem 5.3.5 — **Law of Total Probability.** Let $\{B_1, B_2, \ldots, B_n\}$ be a partition of the sample space Ω. For any event A, the following holds:

$$P(A) = \sum_{i=1}^{n} P(A \mid B_i) P(B_i).$$

Demostración. We want to prove that

$$P(A) = \sum_{i=1}^{n} P(A \mid B_i) P(B_i).$$

Since $\{B_1, B_2, \ldots, B_n\}$ is a partition of the sample space Ω, any event A can be expressed as the disjoint union of the events $A \cap B_i$ for $i = 1, 2, \ldots, n$. Thus,

$$P(A) = P\left(\bigcup_{i=1}^{n} (A \cap B_i) \right).$$

Since the events $A \cap B_i$ are mutually exclusive (disjoint), we can apply the additivity of probability:

$$P(A) = \sum_{i=1}^{n} P(A \cap B_i).$$

Using the multiplication rule, we express $P(A \cap B_i)$ as $P(A \mid B_i) P(B_i)$, so that

$$P(A) = \sum_{i=1}^{n} P(A \mid B_i) P(B_i),$$

which is what we wanted to prove. ■

This rule is fundamental in problems where it is necessary to consider all the ways an event A can occur through different events B_i.

Exercise 5.3 In a school, 60 % of the students are in primary school, and 40 % are in secondary school. 70 % of the primary school students participate in extracurricular activities, while only 50 % of the secondary school students participate in them. If a student is selected at random and we know they participate in extracurricular activities, what is the probability that they are from primary school? ■

Demostración. Let us define:

- P: The student is from primary school.
- S: The student is from secondary school.
- E: The student participates in extracurricular activities.

We have:

$$P(P) = 0{,}6, \quad P(S) = 0{,}4.$$

Conditional probabilities:

$$P(E \mid P) = 0{,}7, \quad P(E \mid S) = 0{,}5.$$

We apply Bayes' Theorem to calculate $P(P \mid E)$:

$$P(P \mid E) = \frac{P(E \mid P)P(P)}{P(E \mid P)P(P) + P(E \mid S)P(S)} = \frac{0{,}7 \times 0{,}6}{0{,}7 \times 0{,}6 + 0{,}5 \times 0{,}4} = \frac{0{,}42}{0{,}42 + 0{,}20} = \frac{0{,}42}{0{,}62} \approx 0{,}6774.$$

Therefore, the probability that the student is from primary school given that they participate in extracurricular activities is approximately 67.74 %.

■

The application of sets in conditional probability is essential for understanding and solving probabilistic problems. Set theory provides a clear and visual representation of events and their relationships, facilitating the calculation of conditional, joint, and total probabilities. Tools such as Venn diagrams and fundamental theorems like Bayes' Theorem and the Law of Total Probability are crucial in this field.

Using sets and their operations is fundamental in conditional probability. By representing events as sets and using the properties of intersection and union, we can efficiently calculate conditional probabilities and solve complex probability problems. A deep understanding of these concepts is essential for advancing in the study of statistics and probability theory.

5.4 Solved Exercises

Exercise 5.4 Prove that the sum of two even numbers is an even number. ■

Demostración. Let $a = 2k$ and $b = 2m$, where k and m are integers. The sum of a and b is:

$$a + b = 2k + 2m = 2(k + m).$$

Since $k + m$ is an integer, $a + b$ is divisible by 2 and, therefore, is an even number. ■ ■

Exercise 5.5 Find the value of x such that $3x+5=20$. ■

Demostración. Subtract 5 from both sides of the equation:

$$3x+5-5=20-5 \Rightarrow 3x=15.$$

Divide both sides by 3:

$$x=\frac{15}{3}=5.$$

Therefore, the value of x is 5. ■ ■

Exercise 5.6 Prove that the product of two rational numbers is a rational number. ■

Demostración. Let $a=\frac{p}{q}$ and $b=\frac{r}{s}$, where p,q,r,s are integers and $q,s\neq 0$. The product of a and b is:

$$a\cdot b=\frac{p}{q}\cdot\frac{r}{s}=\frac{p\cdot r}{q\cdot s}.$$

Since $p\cdot r$ and $q\cdot s$ are integers and $q\cdot s\neq 0$, $a\cdot b$ is a rational number. ■ ■

Exercise 5.7 Solve the quadratic equation $x^2-5x+6=0$. ■

Demostración. The quadratic equation $x^2-5x+6=0$ can be factored as:

$$(x-2)(x-3)=0.$$

This implies $x-2=0$ or $x-3=0$. Solving each equation, we get:

$$x=2 \quad \text{and} \quad x=3.$$

Therefore, the solutions are $x=2$ and $x=3$. ■ ■

Exercise 5.8 Prove that if n is an odd number, then n^2 is also odd. ■

Demostración. Let $n=2k+1$, where k is an integer, which means n is odd. Squaring n, we have:

$$n^2=(2k+1)^2=4k^2+4k+1=2(2k^2+2k)+1.$$

Since $2k^2+2k$ is an integer, n^2 has the form $2m+1$ (for some integer m), which implies it is odd. ■ ■

5.5 Proposed Exercises

5.5.1 Set Operations: Union, Intersection, Complement

Exercise 5.9 Given the sets $A=\{1,2,3,4\}$ and $B=\{3,4,5,6\}$, calculate $A\cup B$. ■

Exercise 5.10 Given the sets $A=\{1,2,3,4\}$ and $B=\{3,4,5,6\}$, calculate $A\cap B$. ■

Exercise 5.11 Given the universal set $U = \{1,2,3,4,5,6,7,8\}$ and the set $A = \{2,4,6\}$, find the complement of A in U. ■

Exercise 5.12 If $A = \{x \in \mathbb{Z} \mid -3 \leq x \leq 3\}$ and $B = \{x \in \mathbb{Z} \mid 1 \leq x \leq 5\}$, determine $A \cup B$. ■

Exercise 5.13 Given the sets $A = \{a,b,c,d\}$ and $B = \{c,d,e,f\}$, find $A - B$ and $B - A$. ■

5.5.2 Venn Diagrams: Visual Representation of Sets

Exercise 5.14 Draw a Venn diagram to represent two sets A and B that have a non-empty intersection. ■

Exercise 5.15 Use a Venn diagram to represent the union of three sets A, B, and C, where all have some intersection between them. ■

Exercise 5.16 Draw a Venn diagram for three sets A, B, and C, where only A and B intersect, but C does not intersect with the other sets. ■

Exercise 5.17 Represent in a Venn diagram a set A and its complement A' with respect to a universal set U. ■

Exercise 5.18 Use a Venn diagram to represent the difference $A - B$ of two sets A and B. ■

5.5.3 Application Problems

Exercise 5.19 In a survey of 100 students, 60 take mathematics and 45 take physics. If 20 take both subjects, how many students take neither subject? ■

Exercise 5.20 Out of 80 people surveyed, 50 prefer chocolate and 30 prefer vanilla. If 15 people prefer both flavors, how many people prefer neither chocolate nor vanilla? ■

Exercise 5.21 In a club, 40 people practice swimming, 30 practice cycling, and 20 practice both activities. How many people are there in total if everyone practices at least one of the activities? ■

Exercise 5.22 A class of 30 students has 18 students who speak English, 15 who speak French, and 10 who speak both languages. How many students speak neither language? ■

Exercise 5.23 In a sports survey, 70 people play soccer, 40 play basketball, and 25 play both sports. Calculate how many people play at least one of the two sports. ■

II Arithmetic Reasoning

6. Numeration System and Operations

6.1 Numeration Systems: Decimal, Binary, and Other Systems.

6.1.1 Conversion Between Numerical Systems

In this section, we explore the conversion between different numerical systems, a fundamental topic in mathematics and computer science. Understanding how to represent numbers in various bases and convert between them is essential for advanced mathematical reasoning.

Definition 6.1.1 A **positional numeral system** is a numeral system in which the value of a digit depends on its position within the number and on the base b of the system. A number in base b is represented as:

$$N = (d_n d_{n-1} \ldots d_1 d_0 . d_{-1} d_{-2} \ldots d_{-m})_b = \sum_{k=-m}^{n} d_k b^k$$

where each digit d_k satisfies $0 \leq d_k < b$.

This definition formalizes how numbers are constructed in different bases and allows us to establish methods for converting between them.

■ **Example 6.1** Consider the decimal number 45 and its conversion to base 2. We decompose 45 into powers of 2:

$$45 = 32 + 8 + 4 + 1 = 2^5 + 2^3 + 2^2 + 2^0$$

Therefore, its representation in base 2 is $(101101)_2$. ■

It is useful to establish a general algorithm for performing these conversions.

Theorem 6.1.1 The successive division algorithm allows converting a positive integer N in base

10 to any base b. It is recursively defined as:

$$d_k = N \mod b \quad \text{and} \quad N = \left\lfloor \frac{N}{b} \right\rfloor$$

repeating until $N = 0$, where d_k are the digits of the number in base b.

Demostración. We want to prove that the successive division algorithm correctly converts a positive integer N in base 10 to a base b.
Let N be the number in base 10 to be represented in base b. We start by dividing N by b and obtaining the remainder $d_0 = N \mod b$, which will be the least significant digit in the base b representation. Then, we update N as $N = \lfloor \frac{N}{b} \rfloor$ and repeat the process.
In each iteration, we calculate the new remainder $d_k = N \mod b$, representing the next digit in the base b representation, and update $N = \lfloor \frac{N}{b} \rfloor$. We continue this process until $N = 0$.
Finally, the remainders $d_0, d_1, \ldots, d_k$ correspond to the digits of the number in base b, read in reverse order from how they were calculated. This is because each remainder captures the value of the least significant position at each step, and dividing N repeatedly advances to positions of higher significance.
Thus, the successive division algorithm correctly produces the representation of N in base b. ■

This algorithm is fundamental and leads to an important corollary about the uniqueness of the representation.

Corollary 6.1.2 Every positive integer has a unique representation in any base $b \geq 2$.

Demostración. Uniqueness derives from the fact that each step of the successive division algorithm is deterministic, and the remainders obtained are unique for each division. Since the number of digits is finite and the digits belong to the set $\{0, 1, \ldots, b-1\}$, the representation is unique. ■

In addition to integers, it is important to consider fractional numbers.

Lema 6.1.1 The successive multiplication algorithm allows converting the fractional part of a decimal number to a base b using:

$$d_{-k} = \lfloor b \cdot F \rfloor \quad \text{and} \quad F = b \cdot F - d_{-k}$$

repeating until the desired precision is reached, where F is the current fractional part.

Demostración. In each iteration, we multiply the fractional part by b and extract the integer part as the next digit. The process continues with the new fractional part until the desired precision is achieved. ■

■ **Example 6.2** Convert the decimal number 0,625 to base 2:

$$0{,}625 \times 2 = 1{,}25 \Rightarrow d_{-1} = 1 \qquad 0{,}25 \times 2 = 0{,}5 \Rightarrow d_{-2} = 0 \qquad 0{,}5 \times 2 = 1{,}0 \Rightarrow d_{-3} = 1$$

Thus, 0,625 in base 2 is $(0{,}101)_2$. ■

(R) The conversion of fractional numbers may result in periodic expansions in certain bases, analogous to periodic decimal fractions in base 10.

An important aspect is understanding how divisibility properties transfer between bases.

Theorem 6.1.3 A number in base b is divisible by $b-1$ if and only if the sum of its digits is divisible by $b-1$.

Demostración. We want to prove that the successive division algorithm correctly converts a positive integer N in base 10 to a base b.
Let N be the number in base 10 to be represented in base b. We start by dividing N by b and obtaining the remainder $d_0 = N \mod b$, which will be the least significant digit in the base b representation. Then, we update N as $N = \lfloor \frac{N}{b} \rfloor$ and repeat the process.
In each iteration, we calculate the new remainder $d_k = N \mod b$, representing the next digit in the base b representation, and update $N = \lfloor \frac{N}{b} \rfloor$. We continue this process until $N = 0$.
Finally, the remainders $d_0, d_1, \ldots, d_k$ correspond to the digits of the number in base b, read in reverse order from how they were calculated. This is because each remainder captures the value of the least significant position at each step, and dividing N repeatedly advances to positions of higher significance.
Thus, the successive division algorithm correctly produces the representation of N in base b. ■

This theorem generalizes the well-known rule of divisibility by 9 in base 10.

Exercise 6.1 Convert the hexadecimal number $(2A3)_{16}$ to base 10 and then to base 8. ■

Exercise 6.2 Prove that the binary number $(1111)_2$ is divisible by 15 in base 10 using the theorem of digit sum divisibility. ■

By mastering these methods and theorems, the reader will be equipped to tackle complex problems involving different numerical systems and their interconversion, thereby strengthening advanced mathematical reasoning.

6.1.2 Applications of the Binary System in Computing

In this section, we examine how the binary system is essential in the field of computing. The binary system, based on two digits, 0 and 1, is the cornerstone of the design and operation of computers and digital devices. Understanding its properties and applications is fundamental to mathematical reasoning in computer science.

Definition 6.1.2 A **bit** (binary digit) is the basic unit of information in digital computing, which can take the value 0 or 1. A **byte** consists of a sequence of 8 bits and is a standard unit for measuring data quantity.

The binary representation allows digital systems to perform logical and arithmetic operations efficiently. Let us see an example of how decimal numbers are represented in binary.

■ **Example 6.3** The decimal number 25 can be converted to the binary system using successive divisions by 2:

$$25 \div 2 = 12 \text{ with remainder } 1$$
$$12 \div 2 = 6 \text{ with remainder } 0$$
$$6 \div 2 = 3 \text{ with remainder } 0$$
$$3 \div 2 = 1 \text{ with remainder } 1$$
$$1 \div 2 = 0 \text{ with remainder } 1$$

Reading the remainders in reverse order, we obtain $(11001)_2$. ■

Binary arithmetic operations are fundamental in computing, especially the addition and multiplication of bits.

Theorem 6.1.4 The binary addition of two bits a and b is defined as:

$$\text{Sum} = a \oplus b$$
$$\text{Carry} = a \cdot b$$

where $\oplus$ denotes the XOR (exclusive OR) operation and $\cdot$ denotes the AND (logical AND) operation.

Demostración. The truth table for binary addition is:

a	b	Sum	Carry
0	0	0	0
0	1	1	0
1	0	1	0
1	1	0	1

The operation Sum $= a \oplus b$ yields 1 when a and b are different, and 0 when they are the same. The carry Carry $= a \cdot b$ is 1 only when both a and b are 1. ■

This principle is essential in the design of **logical adders** in digital circuits.

Corollary 6.1.5 The addition of three bits a, b, and a previous carry c is calculated as:

$$\text{Sum} = a \oplus b \oplus c$$
$$\text{Outgoing Carry} = (a \cdot b) \vee (b \cdot c) \vee (a \cdot c)$$

where $\vee$ denotes the OR (logical OR) operation.

Demostración. We extend the previous theorem by considering the previous carry c and applying the properties of logical operations. The outgoing carry is 1 if at least two of the inputs are 1. ■

Logic gates implement these operations in hardware, enabling the construction of more complex circuits.

Definition 6.1.3 A **logic gate** is an electronic component that performs a logical operation on one or more input signals to produce an output signal. Basic gates include AND, OR, NOT, NAND, NOR, XOR, and XNOR.

These gates are fundamental in building **digital circuits** that perform arithmetic and logical operations.

■ **Example 6.4** A **full adder** is a circuit that adds two bits and an input carry, producing a sum and an output carry. It can be implemented using two half-adders and an OR gate. ■

Besides arithmetic operations, the binary system is crucial for data representation and manipulation.

Lema 6.1.2 Any type of data (numbers, characters, images, etc.) can be represented in binary format using appropriate encodings, allowing its processing and storage in computational systems.

Demostración. Data is converted into sequences of bits using standard encoding schemes, such as ASCII for characters or image formats like BMP, where each pixel is represented by a combination of bits. ■

(R) Efficiency in binary data representation is essential for optimizing resource usage in computational systems, impacting areas such as data compression and information transmission.

The binary system is also essential in computational logic and automata theory.

Theorem 6.1.6 **Finite automata** can be modeled using binary systems, where states and transitions are represented by bits and logical operations, enabling their implementation in hardware and software.

Demostración. Each state of a finite automaton can be encoded as a unique combination of bits. Transitions between states are defined by logical functions that depend on the inputs and the current state, allowing their representation in digital circuits. ■

This theorem connects formal language theory with practical implementation in digital systems.

■ **Example 6.5** A **shift register** is a circuit that uses flip-flops to store and shift bits in series, essential for sequential data transmission and processing. ■

Exercise 6.3 Design a logical circuit using AND, OR, and NOT gates that implements an even-parity function for a set of four input bits, and demonstrate its operation. ■

Exercise 6.4 Explain how the **RSA encryption algorithm** uses binary arithmetic operations, and analyze the importance of binary representation in the algorithm's efficiency. ■

Understanding the applications of the binary system in computing is essential for the development of technology and efficient algorithms, strengthening mathematical reasoning and its practical application in engineering and computer sciences.

6.2 Age Problems: Formulating Equations to Solve Age Problems

6.2.1 Linear Equations Applied to Age Problems

In this section, we will explore how linear equations can be applied to solve problems related to ages, combining mathematical reasoning with everyday situations. These problems are excellent for illustrating the translation of verbal information into algebraic expressions and the application of analytical methods to find solutions.

Definition 6.2.1 A **linear equation** is an algebraic equality of the form $ax+b=0$, where a and b are real constants, $a \neq 0$, and x is the unknown variable.

This definition provides the basis for formulating problems where the relationships between ages are expressed through linear equations.

■ **Example 6.6** Suppose Ana's age is twice Beatriz's age, and the sum of their ages is 36 years. We can represent this with the equations:

$$\begin{cases} A = 2B \\ A+B = 36 \end{cases}$$

Substituting A in the second equation:

$$2B+B = 36 \implies 3B = 36 \implies B = 12$$

Therefore, Ana's age is $A = 2 \times 12 = 24$ years. ■

This example demonstrates how to translate a verbal situation into a system of linear equations and solve it.

Theorem 6.2.1 Every linear system of n equations with n unknowns has a unique solution if and only if the determinant of its coefficient matrix is nonzero.

Demostración. Let $AX = B$ be a linear system of n equations with n unknowns, where A is the coefficient matrix, X is the vector of unknowns, and B is the vector of constants.
If the determinant of A, denoted $\det(A)$, is nonzero, then A is an invertible matrix. In this case, the system can be solved using the inverse of A:

$$X = A^{-1}B.$$

This expression provides a unique solution for X, as A^{-1} exists and is unique.
Conversely, if $\det(A) = 0$, then A is not invertible, and the system does not have a unique solution. In this case, the system may have infinitely many solutions or none, depending on its consistency. Thus, the system has a unique solution if and only if $\det(A) \neq 0$, as desired. ■

This result is crucial, as it ensures that properly formulated age problems will have a unique solution, reflecting the logical consistency of the posed situation.

Lema 6.2.1 In a homogeneous linear system of the form $Ax = 0$, where A is a square matrix of order n, if $\det(A) = 0$, then the system has infinitely many solutions.

Demostración. If $\det(A) = 0$, the matrix A is not invertible, implying that the solution space is a subspace of dimension at least one. Therefore, there exist infinitely many nontrivial solutions satisfying $Ax = 0$. ■

Although this lemma pertains to homogeneous systems, it is relevant in age problems when the relationships between variables lead to nonunique solutions or require additional conditions.

When solving age problems, it is essential to correctly identify the variables and establish equations that accurately reflect the temporal and arithmetic relationships described.

Consider another example involving age differences at different points in time.

■ **Example 6.7** Five years ago, Carlos's age was three times Diana's age. In five years, Carlos's age will be twice Diana's age. What are their current ages?
Let:

$$\begin{cases} C = \text{Carlos's current age} \\ D = \text{Diana's current age} \end{cases}$$

The equations are:

$$\begin{cases} C - 5 = 3(D - 5) \\ C + 5 = 2(D + 5) \end{cases}$$

Solving the system: First equation:

$$C - 5 = 3D - 15 \implies C = 3D - 10$$

Second equation:

$$C + 5 = 2D + 10 \implies C = 2D + 15$$

Equating both expressions for C:

$$3D - 10 = 2D + 15 \implies D = 25$$

Thus, $C = 2 \times 25 + 15 = 65$.
Therefore, Carlos is 65 years old, and Diana is 25 years old. ■

This example illustrates how to handle equations involving ages at different times, adding complexity to the problem.

Corollary 6.2.2 In age problems involving multiples and sums at different points in time, the number of equations required to find a unique solution equals the number of unknown variables.

Demostración. Each independent condition provides a distinct equation. For n unknown variables, n linearly independent equations are needed to ensure a unique solution, according to the fundamental theorem of linear systems. ■

(R) It is crucial to verify that the equations derived from the problem's conditions are linearly independent to ensure the uniqueness of the solution.

Now, we present an exercise to apply these concepts.

Exercise 6.5 Ten years ago, Eduardo's age was four times Fernanda's age. In ten years, Eduardo will be twice Fernanda's age. Determine their current ages. ■

This exercise is similar to the previous example and provides practice in formulating and solving systems of linear equations in the context of age problems.

Exercise 6.6 The sum of Gabriela's and Hugo's ages is 50 years. Five years ago, Gabriela was three times Hugo's age. Calculate their current ages. ■

By solving these exercises, the reader will strengthen their ability to model verbal situations with linear equations and apply algebraic methods to find solutions, essential skills in advanced mathematical reasoning.

6.3 Chronometry: Conversion of Time Units and Problem Solving

6.3.1 Conversion Between Hours, Minutes, and Seconds

In this section, we analyze the conversion between hours, minutes, and seconds from an advanced mathematical perspective. This topic is not only relevant in everyday calculations but also has implications in areas such as number theory and modular arithmetic. We will explore formal definitions, theorems, and examples to deepen the understanding of these systems of time measurement.

Definition 6.3.1 The **sexagesimal system** is a positional numeral system with base 60, historically used in various cultures to measure time and angles. In this system, one hour is divided into 60 minutes, and one minute is divided into 60 seconds.

This definition allows us to formalize operations and conversions between time units using the properties of the sexagesimal system.

Theorem 6.3.1 The conversion between hours (h), minutes (m), and seconds (s) can be expressed through the following linear relationships:

$$\begin{cases} 1 \text{ hour} = 60 \text{ minutes} \\ 1 \text{ minute} = 60 \text{ seconds} \\ 1 \text{ hour} = 3600 \text{ seconds} \end{cases}$$

Therefore, any amount of time can be represented in terms of a single unit using these relationships.

Demostración. The relationships directly follow from the definition of the sexagesimal system. Multiplying the equivalences:

$$1 \text{ hour} = 60 \text{ minutes} = 60 \times 60 \text{ seconds} = 3600 \text{ seconds}$$

■

This theorem allows precise conversions between different time units, essential for advanced temporal calculations.

Lema 6.3.1 Let $T = h + \frac{m}{60} + \frac{s}{3600}$ be a decimal representation of a time quantity in hours. Then, T can be fully converted to seconds using the relationship:

$$T_{\text{seconds}} = 3600h + 60m + s$$

Demostración. Multiplying each term by its conversion factor to seconds:

$$T_{\text{seconds}} = h \times 3600 + m \times 60 + s$$

■

This lemma is useful for normalizing different time representations into a single unit, facilitating arithmetic operations and comparisons.

■ **Example 6.8** Convert 2 hours, 45 minutes, and 30 seconds into total seconds.
Applying the previous lemma:

$$T_{\text{seconds}} = 3600 \times 2 + 60 \times 45 + 30 = 7200 + 2700 + 30 = 9930 \text{ seconds}$$

■

This example illustrates the practical application of the lemma in a concrete situation.

Theorem 6.3.2 The set of equivalences between hours, minutes, and seconds forms a vector space over the field of real numbers, where addition and scalar multiplication operations are naturally defined.

Demostración. To demonstrate that the set of equivalences between hours, minutes, and seconds forms a vector space over the field of real numbers, we verify that it satisfies the axioms of a vector space, considering hours, minutes, and seconds as vectors and defining addition and scalar multiplication naturally.
1. **Set**: We consider the set of all linear combinations of hours, minutes, and seconds, where each combination can be represented as a vector (h, m, s), with h in hours, m in minutes, and s in seconds, and the equivalences 1 hour = 60 minutes and 1 minute = 60 seconds.

2. **Addition**: We define the addition of two vectors (h_1, m_1, s_1) and (h_2, m_2, s_2) as

$$(h_1, m_1, s_1) + (h_2, m_2, s_2) = (h_1 + h_2, m_1 + m_2, s_1 + s_2).$$

This operation is closed and satisfies the axioms of commutativity and associativity, and has a neutral element $(0,0,0)$ and an additive inverse for each vector.
3. **Scalar Multiplication**: For a scalar $\alpha \in \mathbb{R}$ and a vector (h,m,s), we define

$$\alpha(h,m,s) = (\alpha h, \alpha m, \alpha s).$$

This operation is closed and satisfies the axioms of distributivity, scalar multiplication associativity, and the existence of the unit element 1.
4. **Verification of Equivalences**: Since hours, minutes, and seconds are related by constant factors (60 minutes per hour and 60 seconds per minute), any linear combination of these vectors respects these equivalences.
Thus, the set of equivalences between hours, minutes, and seconds, with the defined addition and scalar multiplication operations, satisfies all the axioms of a vector space over $\mathbb{R}$. ■

This theorem provides an algebraic structure to the set of time units, allowing the application of linear algebra techniques in its study.

Corollary 6.3.3 The operations of conversion between hours, minutes, and seconds are linear and can be represented using transformation matrices in the vector space V.

Demostración. Conversions can be expressed as matrix multiplications. For example, to convert from hours to seconds:

$$\left(T_{\text{seconds}}\right) = \left(3600\right)\left(h\right)$$

This is a linear transformation, as it satisfies the linearity property. ■

This corollary demonstrates how conversions can be integrated into a broader algebraic framework.

R The use of modular arithmetic is relevant in time measurement, especially when considering 12-hour or 24-hour clocks, where operations are performed modulo 12 or 24.

This observation connects the topic with other mathematical concepts, expanding its application.

■ **Example 6.9** If it is 22:00 hours, what time will it be in 5 hours?
Using modular arithmetic:

$$(22+5) \mod 24 = 3$$

Therefore, it will be 3:00 hours. ■

This example demonstrates the utility of modular arithmetic in temporal calculations.

Exercise 6.7 Demonstrate that adding two times expressed in hours, minutes, and seconds can result in a value requiring normalization, and formulate an algorithm to perform such normalization. ■

Exercise 6.8 If a task starts at 8:35:50 and lasts exactly 2 hours, 55 minutes, and 15 seconds, determine the exact finishing time using the studied properties. ■

By solving these exercises, the reader will apply advanced concepts of time unit conversion and manipulation, strengthening their mathematical reasoning and ability to solve complex problems.

6.3.2 Calculation of Travel Times

In this section, we analyze the calculation of travel times, considering fundamental variables such as distance, velocity, and time. This study is essential for understanding physical phenomena and optimizing processes in engineering and applied sciences.

Definition 6.3.2 Let d be the distance traveled, v the velocity, and t the time taken. The relationship between these variables in **uniform rectilinear motion** is expressed by the equation:

$$d = v \cdot t$$

This equation establishes a direct proportionality between distance and time when velocity is constant. From it, we can deduce formulas to calculate any of the three variables if the other two are known.

Theorem 6.3.4 In motion where velocity varies continuously and differentiably with respect to time, the distance traveled can be obtained by integrating the velocity:

$$d(t) = \int_{t_0}^{t} v(\tau)\, d\tau$$

Demostración. By definition, velocity is the derivative of distance with respect to time:

$$v(t) = \frac{d}{dt} d(t)$$

Integrating both sides with respect to time from t_0 to t, we obtain:

$$\int_{t_0}^{t} v(\tau)\, d\tau = \int_{t_0}^{t} \frac{d}{d\tau} d(\tau)\, d\tau = d(t) - d(t_0)$$

If we consider $d(t_0) = 0$, then:

$$d(t) = \int_{t_0}^{t} v(\tau)\, d\tau$$

■

This theorem is fundamental for analyzing motions with variable velocity, allowing distances to be calculated using integration.

Lema 6.3.2 If an object travels a distance d in two segments at constant velocities v_1 and v_2, taking times t_1 and t_2 respectively, the total distance is:

$$d = v_1 t_1 + v_2 t_2$$

Demostración. The distance traveled in each segment is $d_1 = v_1 t_1$ and $d_2 = v_2 t_2$. The total distance is the sum of both:

$$d = d_1 + d_2 = v_1 t_1 + v_2 t_2$$

■

This result is useful for breaking down complex journeys into manageable segments, simplifying the calculation of the total distance.

■ **Example 6.10** A train travels for 2 hours at a speed of 80 km/h and then for 1,5 hours at 100 km/h. Calculate the total distance traveled.
Using the previous lemma:

$$d = 80 \times 2 + 100 \times 1{,}5 = 160 + 150 = 310 \text{ km}$$

■

This example illustrates how to apply the lemma to determine the total distance in journeys with different speeds.

Corollary 6.3.5 The average velocity v_{avg} in a journey composed of several segments with constant velocities is:

$$v_{\text{avg}} = \frac{\text{Total Distance}}{\text{Total Time}} = \frac{\sum_i v_i t_i}{\sum_i t_i}$$

Demostración. The total distance is $d = \sum_i v_i t_i$ and the total time is $t_{\text{total}} = \sum_i t_i$. By the definition of average velocity:

$$v_{\text{avg}} = \frac{d}{t_{\text{total}}} = \frac{\sum_i v_i t_i}{\sum_i t_i}$$

■

This corollary allows us to calculate the average velocity without explicitly knowing the total distance, using only the speeds and times of each segment.

R It is important to distinguish between **average velocity** and **mean velocity**. The average velocity is calculated considering the distances and times traveled, while the mean velocity is simply the arithmetic mean of individual velocities and generally does not represent the effective velocity of the total journey.

Delving deeper into the analysis, let us consider the case of time-dependent velocities.

Theorem 6.3.6 If the velocity of an object is a linear function of time, i.e., $v(t) = at + b$, then the distance traveled between times t_0 and t is:

$$d(t) = \frac{a}{2}(t^2 - t_0^2) + b(t - t_0)$$

Demostración. The distance traveled $d(t)$ between times t_0 and t is obtained by integrating the velocity $v(t) = at + b$ with respect to time from t_0 to t.
The expression for the distance is

$$d(t) = \int_{t_0}^{t} v(\tau)\, d\tau = \int_{t_0}^{t} (a\tau + b)\, d\tau.$$

We decompose the integral into two terms:

$$d(t) = \int_{t_0}^{t} a\tau\, d\tau + \int_{t_0}^{t} b\, d\tau.$$

Calculating each term separately. For the first term,

$$\int_{t_0}^{t} a\tau\, d\tau = a\int_{t_0}^{t} \tau\, d\tau = a\left[\frac{\tau^2}{2}\right]_{t_0}^{t} = \frac{a}{2}(t^2 - t_0^2).$$

For the second term,

$$\int_{t_0}^{t} b\, d\tau = b\int_{t_0}^{t} 1\, d\tau = b[\tau]_{t_0}^{t} = b(t - t_0).$$

Adding both results, we obtain

$$d(t) = \frac{a}{2}(t^2 - t_0^2) + b(t - t_0),$$

as desired. ■

This theorem is useful in studying motions with constant acceleration, such as free-falling objects without air resistance.

■ **Example 6.11** A car accelerates uniformly from rest ($v_0 = 0$) with a constant acceleration of 2 m/s^2. Calculate the distance traveled in 10 seconds.
The velocity as a function of time is $v(t) = at = 2t$. Applying the theorem:

$$d(10) = \frac{2}{2}(10^2 - 0^2) = 1 \times 100 = 100 \text{ m}$$

■

This example demonstrates the application of the theorem in a constant acceleration context.

Exercise 6.9 An airplane travels along a runway, accelerating uniformly from rest with an acceleration of 3 m/s^2. If it needs to reach a velocity of 60 m/s to take off, calculate the minimum runway length required. ■

Exercise 6.10 A runner completes a 400-meter race in two stages: they accelerate uniformly from rest to a maximum velocity over 8 seconds, then maintain that velocity until the end. If their acceleration is 1,5 m/s^2, determine the total time of the race. ■

By addressing these concepts and exercises, the reader will deepen their advanced mathematical analysis of time and distance calculations in motions under different conditions, strengthening their reasoning and ability to apply mathematical principles to complex situations.

6.4 Solved Exercises

Exercise 6.11 Convert the decimal number 156 to its equivalent in base 2. ■

Demostración. To convert the decimal number 156 to base 2, we apply the method of successive divisions by 2.

$156 \div 2 = 78$, remainder 0 $\quad 78 \div 2 = 39$, remainder 0 $\quad 39 \div 2 = 19$, remainder 1

$19 \div 2 = 9$, remainder 1 $\quad 9 \div 2 = 4$, remainder 1 $\quad 4 \div 2 = 2$, remainder 0

$2 \div 2 = 1$, remainder 0 $\quad 1 \div 2 = 0$, remainder 1

Reading the remainders from bottom to top, we find that 156 in base 2 is $(10011100)_2$. ■

Exercise 6.12 Solve the system of linear equations to find the current ages of Ana and Beatriz: $A = 2B$ and $A + B = 36$. ■

Demostración. From the given system of equations:

$$A = 2B \quad \text{and} \quad A + B = 36$$

Substituting $A = 2B$ into the second equation:

$$2B + B = 36 \implies 3B = 36 \implies B = 12$$

Substituting $B = 12$ into $A = 2B$:

$$A = 2 \times 12 = 24$$

Thus, Ana is 24 years old, and Beatriz is 12 years old. ■

Exercise 6.13 Convert the hexadecimal number $(2F)_{16}$ to decimal. ■

Demostración. To convert the hexadecimal number $(2F)_{16}$ to decimal, we decompose the number into powers of 16:

$$(2F)_{16} = 2 \times 16^1 + 15 \times 16^0 = 2 \times 16 + 15 \times 1 = 32 + 15 = 47$$

Thus, the number $(2F)_{16}$ in decimal is 47. ■

Exercise 6.14 Determine if the binary number $(11101)_2$ is divisible by 3 in decimal base. ■

Demostración. First, convert $(11101)_2$ to base 10:

$$(11101)_2 = 1 \times 2^4 + 1 \times 2^3 + 1 \times 2^2 + 0 \times 2^1 + 1 \times 2^0$$
$$= 16 + 8 + 4 + 0 + 1 = 29$$

Now verify if 29 is divisible by 3. Sum the digits of the decimal number 29: $2 + 9 = 11$. Since 11 is not divisible by 3, we conclude that 29 is not divisible by 3. Therefore, $(11101)_2$ is not divisible by 3. ■

Exercise 6.15 Five years ago, Carla's age was three times Diana's age. Today, Carla is 25 years old. What is Diana's current age? ■

Demostración. Let D represent Diana's current age. Five years ago, Carla was $25 - 5 = 20$ years old, and Diana was $D - 5$ years old. According to the problem:

$$20 = 3(D - 5)$$

Solving for D:

$$20 = 3D - 15 \implies 3D = 35 \implies D = \frac{35}{3} \approx 11{,}67$$

Thus, Diana is approximately 11 years and 8 months old. ■

6.5 Proposed Exercises

6.5.1 Numeration Systems: Decimal, Binary, and Other Systems

Exercise 6.16 Convert the decimal number 45 to its equivalent in base 2. ■

Exercise 6.17 Convert the binary number $(1101)_2$ to its equivalent in base 10. ■

Exercise 6.18 Convert the hexadecimal number $(3A7)_{16}$ to decimal. ■

Exercise 6.19 Convert the octal number $(127)_8$ to base 10. ■

Exercise 6.20 Convert the decimal number 255 to its equivalent in base 16. ■

6.5.2 Age Problems: Formulating Equations to Solve Age Problems

Exercise 6.21 Five years ago, Juan's age was three times Ana's age. Currently, Juan is 40 years old. What is Ana's current age? ■

Exercise 6.22 In ten years, Pedro's age will be twice what it was five years ago. If Pedro is currently 30 years old, how old will he be in ten years? ■

Exercise 6.23 The sum of Carlos's and María's ages is 50 years. Ten years ago, Carlos's age was twice María's age. What are their current ages? ■

Exercise 6.24 Andrés's age is twice Sara's age. If the difference between their ages is 15 years, what are their ages? ■

Exercise 6.25 Eight years ago, Sofía's age was half of what it is now. What is her current age? ■

6.5.3 Chronometry: Conversion of Time Units and Problem Solving

Exercise 6.26 Convert 2 hours, 30 minutes, and 45 seconds to seconds. ■

Exercise 6.27 Convert 5000 seconds to hours, minutes, and seconds. ■

Exercise 6.28 If a task starts at $14:20:00$ and lasts exactly 3 hours, 45 minutes, and 30 seconds, what time does it end? ■

Exercise 6.29 Convert 1 day, 4 hours, and 15 minutes to minutes. ■

Exercise 6.30 How many days, hours, minutes, and seconds are in $900,000$ seconds? ■

7. Divisibility and Fractions

7.1 GCD and LCM: Methods of Prime Factorization

7.1.1 Simultaneous Factorization Method

In this section, we delve into the **simultaneous factorization method** to calculate the greatest common divisor (GCD) and least common multiple (LCM) of two or more integers. This method is based on prime factorization and leverages fundamental arithmetic properties to simplify calculations and better understand the structure of numbers.

Definition 7.1.1 The **Greatest Common Divisor** (GCD) of two integers a and b, not both zero, is the largest positive integer d that divides both numbers, i.e., $d \mid a$ and $d \mid b$. It is denoted as $\gcd(a,b)$.

Definition 7.1.2 The **Least Common Multiple** (LCM) of two integers a and b, not both zero, is the smallest positive integer m that is a multiple of both numbers, i.e., $a \mid m$ and $b \mid m$. It is denoted as $\text{lcm}(a,b)$.

To apply the simultaneous factorization method, we need to understand prime factorization.

Theorem 7.1.1 — **Fundamental Theorem of Arithmetic.** Every positive integer greater than 1 can be expressed uniquely (except for the order of the factors) as a product of prime numbers.

Demostración. We will prove the theorem in two parts: the existence and uniqueness of prime factorization.

First, consider the existence of a prime factorization. Let n be a positive integer greater than 1. If n is prime, it is already expressed as a product of a single prime number. If n is composite, then there exists at least one divisor d such that $1 < d < n$. Divide n by d and repeat the process with the resulting quotients until the divisors are prime. This process terminates because the divisors are smaller than n and we are working in a finite set of positive divisors. Thus, we have decomposed n as a product of prime numbers.

For uniqueness, suppose a number n has two distinct prime factorizations:

$$n = p_1 p_2 \dots p_k = q_1 q_2 \dots q_m,$$

where p_i and q_j are prime numbers. By Euclid's Lemma, if a prime p divides a product, it must divide at least one of the factors of that product. Applying this repeatedly, we conclude that each p_i must equal some q_j, and vice versa, until the two factorizations are identical except for the order of factors.
Thus, the factorization into primes is unique, as we intended to demonstrate. ■

With this foundation, we can proceed to the simultaneous factorization method.

■ **Example 7.1** Calculate the GCD and LCM of $a = 60$ and $b = 84$ using the simultaneous factorization method.
First, decompose each number into its prime factors:

$$60 = 2^2 \times 3 \times 5$$
$$84 = 2^2 \times 3 \times 7$$

Now, identify the common and non-common factors:
- The common factors are 2^2 and 3. - The non-common factors are 5 and 7.
The GCD is the product of the smallest exponents of the common prime factors:

$$\gcd(60, 84) = 2^2 \times 3 = 12$$

The LCM is the product of the largest exponents of all prime factors:

$$\text{lcm}(60, 84) = 2^2 \times 3 \times 5 \times 7 = 420$$

■

The simultaneous factorization method involves simultaneously factoring the numbers into their prime factors and using these factorizations to calculate the GCD and LCM.

Theorem 7.1.2 Let a and b be positive integers with prime factorizations:

$$a = p_1^{\alpha_1} p_2^{\alpha_2} \dots p_k^{\alpha_k}, \quad b = p_1^{\beta_1} p_2^{\beta_2} \dots p_k^{\beta_k},$$

where p_i are the prime numbers common and unique to the factorizations of a and b (with exponents α_i or β_i equal to zero if the prime p_i is not present in the factorization of a or b, respectively). Then,

$$\gcd(a, b) = \prod_{i=1}^{k} p_i^{\text{mín}(\alpha_i, \beta_i)}, \quad \text{lcm}(a, b) = \prod_{i=1}^{k} p_i^{\text{máx}(\alpha_i, \beta_i)}.$$

Demostración. To demonstrate the formulas for $\gcd(a, b)$ and $\text{lcm}(a, b)$ using the prime factorizations of a and b, consider the given factorizations:

$$a = \prod_i p_i^{\alpha_i}, \quad b = \prod_i p_i^{\beta_i},$$

where p_i are the prime numbers present in the factorizations of a or b, and the exponents α_i or β_i may be zero if the prime p_i is not present in the factorization of a or b, respectively.
For the greatest common divisor $\gcd(a,b)$, we seek the largest integer that divides both a and b. This number must consist of the same prime factors present in both factorizations, with each prime p_i raised to the smallest exponent between α_i and β_i, as any higher exponent would not divide both numbers. Thus,

$$\gcd(a,b) = \prod_i p_i^{\text{mín}(\alpha_i,\beta_i)}.$$

For the least common multiple $\text{lcm}(a,b)$, we seek the smallest integer divisible by both a and b. This requires each prime p_i to be raised to the largest exponent between α_i and β_i, as any smaller exponent would not ensure divisibility by a or b. Hence,

$$\text{lcm}(a,b) = \prod_i p_i^{\text{máx}(\alpha_i,\beta_i)}.$$

These formulas guarantee the greatest common divisor and the least common multiple of a and b, as we intended to prove. ■

This theorem is fundamental to the simultaneous factorization method, as it formalizes the process and provides a general formula for calculating the GCD and LCM.

> R The simultaneous factorization method is particularly useful when working with relatively small numbers or when a deeper understanding of the prime structure of the numbers involved is required. However, for larger numbers, other methods like Euclid's algorithm may be more efficient.

■ **Example 7.2** Calculate the GCD and LCM of $a = 210$ and $b = 231$.
Prime factorization:

$$210 = 2 \times 3 \times 5 \times 7$$
$$231 = 3 \times 7 \times 11$$

Common factors: 3 and 7.
GCD:

$$\gcd(210,231) = 3 \times 7 = 21$$

LCM:

$$\text{lcm}(210,231) = 2 \times 3 \times 5 \times 7 \times 11 = 2310$$

■

This example demonstrates the application of the method to numbers with more prime factors and shows how to correctly identify common and non-common factors.

Corollary 7.1.3 For any pair of positive integers a and b, the following holds:

$$a \times b = \gcd(a,b) \times \text{lcm}(a,b)$$

Demostración. Using the prime factorizations:

$$a = \prod_i p_i^{\alpha_i}, \quad b = \prod_i p_i^{\beta_i}$$

Then,

$$a \times b = \prod_i p_i^{\alpha_i+\beta_i}, \quad \gcd(a,b) \times \operatorname{lcm}(a,b) = \left(\prod_i p_i^{\min(\alpha_i,\beta_i)}\right)\left(\prod_i p_i^{\max(\alpha_i,\beta_i)}\right) = \prod_i p_i^{\alpha_i+\beta_i}$$

Thus,

$$a \times b = \gcd(a,b) \times \operatorname{lcm}(a,b)$$

■

This relationship is very useful for verifying calculations and understanding the connection between the GCD and LCM of two numbers.

Exercise 7.1 Find the GCD and LCM of $a = 252$ and $b = 105$, and verify that $a \times b = \gcd(a,b) \times \operatorname{lcm}(a,b)$. ■

Exercise 7.2 Given the numbers $a = 128$ and $b = 48$, use the simultaneous factorization method to calculate $\gcd(a,b)$ and $\operatorname{lcm}(a,b)$, and explain why this method is more convenient in this case than other methods. ■

By solving these exercises, the reader will apply the simultaneous factorization method in different contexts, strengthening their understanding and ability to handle prime factorizations, and appreciating the elegance and power of this approach in number theory.

7.1.2 Euclidean Algorithm for the GCD

The **Euclidean Algorithm** is an efficient and fundamental method in number theory for calculating the greatest common divisor (GCD) of two integers. This algorithm, over two millennia old, is essential not only for its simplicity and speed but also for its profound implications in mathematics and cryptography.

Definition 7.1.3 Given two integers a and b, not both zero, the **greatest common divisor** $\gcd(a,b)$ is the largest positive integer that divides both without leaving a remainder.

The algorithm is based on the principle that the GCD of two numbers is also the GCD of the smaller number and the remainder when the larger number is divided by the smaller.

Theorem 7.1.4 — Euclidean Algorithm. For positive integers a and b, with $a \geq b$, we have:

$$\gcd(a,b) = \gcd(b, a \mod b),$$

where $a \mod b$ is the remainder of dividing a by b. Repeating this process recursively, the last nonzero remainder is the GCD of a and b.

Demostración. We want to show that for positive integers a and b with $a \geq b$,

$$\gcd(a,b) = \gcd(b, a \mod b),$$

where $a \mod b$ is the remainder of dividing a by b.
By the division algorithm, we can write a as

$$a = bq + r,$$

where q is the quotient, $r = a \mod b$ is the remainder, and $0 \leq r < b$.
We observe that any common divisor of a and b also divides $r = a - bq$. Therefore, the common divisors of a and b are exactly the same as the common divisors of b and r. This implies that

$$\gcd(a,b) = \gcd(b,r) = \gcd(b, a \mod b).$$

Repeating this process recursively, we replace (a,b) with $(b, a \mod b)$, then with $(a \mod b, b \mod (a \mod b))$, and so on, until the remainder is zero. At this point, the last nonzero remainder is the greatest common divisor of a and b.
This completes the proof of the Euclidean Algorithm. ■

This result reduces the calculation of the GCD to a simpler problem, iterating until a trivial case is reached.

■ **Example 7.3** Calculate the GCD of $a = 252$ and $b = 198$ using the Euclidean Algorithm.
Proceed with successive divisions:

$$\begin{aligned} 252 &= 198 \times 1 + 54 && \Longrightarrow r_1 = 54 \\ 198 &= 54 \times 3 + 36 && \Longrightarrow r_2 = 36 \\ 54 &= 36 \times 1 + 18 && \Longrightarrow r_3 = 18 \\ 36 &= 18 \times 2 + 0 && \Longrightarrow r_4 = 0 \end{aligned}$$

The last nonzero remainder is 18, so $\gcd(252, 198) = 18$. ■

In addition to calculating the GCD, the Euclidean Algorithm can be extended to find integer coefficients that express the GCD as a linear combination of a and b.

Theorem 7.1.5 — **Bézout's Identity.** For integers a and b, there exist integers x and y such that:

$$\gcd(a,b) = ax + by.$$

This identity states that the GCD of a and b can be written as a linear combination of them.

Demostración. Bézout's Identity states that for integers a and b, there exist integers x and y such that

$$\gcd(a,b) = ax + by.$$

To prove this, we use the Euclidean Algorithm to calculate $\gcd(a,b)$ and express it as a linear combination of a and b.
Using the Euclidean Algorithm, we find a sequence of divisions:

$$a = bq_1 + r_1,$$

$$b = r_1 q_2 + r_2,$$
$$r_1 = r_2 q_3 + r_3,$$
$$\vdots$$
$$r_{n-2} = r_{n-1} q_n + r_n,$$
$$r_{n-1} = r_n q_{n+1} + 0,$$

where the last nonzero remainder, r_n, is $\gcd(a,b)$.
Now, we back-substitute the equations of the Euclidean Algorithm to express each remainder as a linear combination of a and b, starting with r_n. Since each remainder r_k is a linear combination of the previous remainders, successive substitution eventually expresses r_n (which is $\gcd(a,b)$) as a linear combination of a and b.
Thus, there exist integers x and y such that

$$\gcd(a,b) = ax + by,$$

which proves Bézout's Identity. ■

■ **Example 7.4** Find integers x and y such that $18 = 252x + 198y$.
Using the extended Euclidean Algorithm:

$$\begin{aligned}
252 &= 198 \times 1 + 54 && \Longrightarrow 54 = 252 - 198 \times 1 \\
198 &= 54 \times 3 + 36 && \Longrightarrow 36 = 198 - 54 \times 3 \\
54 &= 36 \times 1 + 18 && \Longrightarrow 18 = 54 - 36 \times 1 \\
36 &= 18 \times 2 + 0
\end{aligned}$$

Back-substituting:

$$\begin{aligned}
18 &= 54 - 36 \times 1 \\
&= 54 - (198 - 54 \times 3) \times 1 \\
&= 54 - 198 \times 1 + 54 \times 3 \\
&= 54 \times 4 - 198 \times 1 \\
&= (252 - 198 \times 1) \times 4 - 198 \times 1 \\
&= 252 \times 4 - 198 \times 4 - 198 \times 1 \\
&= 252 \times 4 - 198 \times 5
\end{aligned}$$

Thus, $x = 4$ and $y = -5$, and we have $18 = 252 \times 4 - 198 \times 5$. ■

(R) Bézout's Identity is essential for solving linear Diophantine equations and has applications in areas such as cryptography, particularly in the RSA algorithm.

It is important to note that the Euclidean Algorithm is efficient even for large numbers, making it relevant in computational contexts.

Lema 7.1.1 The number of steps in the Euclidean Algorithm for integers a and b with $a \geq b$ is bounded by $\log_2 b$.

Demostración. Each step of the algorithm reduces the pair (a,b) to $(b, a \mod b)$, where $a \mod b < b$. Additionally, a Fibonacci sequence shows that the number of steps is proportional to the logarithm of the smaller number. Formally, it can be shown that the maximum number of steps does not exceed $5\log_{10} b$. ■

This efficiency is key to its use in cryptographic algorithms and other applications requiring operations with large numbers.

Corollary 7.1.6 For positive integers a and b, the following holds:

$$a \times b = \gcd(a,b) \times \operatorname{lcm}(a,b),$$

where $\operatorname{lcm}(a,b)$ is the least common multiple of a and b.

Demostración. Since $\gcd(a,b)$ divides both a and b, we can write $a = \gcd(a,b) \times m$ and $b = \gcd(a,b) \times n$, where m and n are integers with no common factors. Then:

$$\operatorname{lcm}(a,b) = \gcd(a,b) \times m \times n,$$

and multiplying $\gcd(a,b)$ by $\operatorname{lcm}(a,b)$:

$$\gcd(a,b) \times \operatorname{lcm}(a,b) = \gcd(a,b)^2 \times m \times n = (\gcd(a,b) \times m) \times (\gcd(a,b) \times n) = a \times b.$$

■

This relationship is useful for calculating the LCM when the GCD is known and vice versa.

Exercise 7.3 Use the Euclidean Algorithm to find the GCD of $a = 391$ and $b = 299$, and express the GCD as a linear combination of a and b. ■

Exercise 7.4 If $\gcd(a,b) = 14$ and $a \times b = 10976$, find the LCM of a and b. ■

By understanding and applying the Euclidean Algorithm and its extensions, one strengthens skills in number theory and acquires fundamental tools for advanced mathematical reasoning and practical applications in various areas of mathematics.

7.2 Divisibility Problems: How to Identify Multiples and Divisors

7.2.1 Divisibility Problems with Prime Numbers

In this section, we will explore divisibility problems involving prime numbers, which are fundamental in number theory and have profound applications in various areas of mathematics and cryptography. Understanding the properties of prime numbers and their role in divisibility is essential for advanced mathematical reasoning.

Definition 7.2.1 A **prime number** is a positive integer greater than 1 that has exactly two distinct positive divisors: 1 and itself. That is, a number p is prime if $p > 1$ and the only positive integer solutions of $d \mid p$ are $d = 1$ and $d = p$.

This definition allows us to distinguish prime numbers from **composite numbers**, which have more than two divisors.

Theorem 7.2.1 — **Infinitude of Prime Numbers.** There are infinitely many prime numbers.

Demostración. We will prove the infinitude of prime numbers by contradiction.
Suppose there are only finitely many prime numbers, denoted $p_1, p_2, \ldots, p_n$. Consider the number

$$N = p_1 p_2 \cdots p_n + 1,$$

which is obtained by multiplying all the prime numbers and adding 1.

This number N is not divisible by any of the primes $p_1, p_2, \dots, p_n$, because dividing N by any p_i leaves a remainder of 1. Therefore, N is either a prime number or has a prime divisor not in $p_1, p_2, \dots, p_n$.
In both cases, we reach a contradiction: if N is prime, we have found a prime not in our finite list; if N is composite, it must have a prime divisor not in the list. This contradicts the assumption that there are only finitely many prime numbers.
Thus, there must be infinitely many prime numbers. ■

This fundamental theorem establishes that there are always new prime numbers to discover, which has significant implications in number theory.

Lema 7.2.1 If p is a prime number and $p \mid ab$, then $p \mid a$ or $p \mid b$.

Demostración. This is a particular case of the **Fundamental Theorem of Arithmetic**. If p divides the product ab and does not divide a, then it must divide b. This is because, if p does not divide a, a and p are coprime, and by **Euclid's Lemma**, p divides b. ■

This lemma is crucial in divisibility problems, as it allows the divisibility of products to be broken down into individual factors.

Theorem 7.2.2 — **Wilson's Theorem.** An integer $p > 1$ is a prime number if and only if:

$$(p-1)! \equiv -1 \mod p$$

Demostración. We will prove Wilson's Theorem, which states that an integer $p > 1$ is prime if and only if

$$(p-1)! \equiv -1 \pmod{p}.$$

First, consider the case where p is prime. In the set $\{1, 2, \dots, p-1\}$, each number has a distinct multiplicative inverse modulo p, except for the elements that are their own inverses, namely 1 and $p-1$. Thus, we can pair all the elements of $\{1, 2, \dots, p-1\}$ into products congruent to 1 modulo p, except for 1 and $p-1$.
Therefore,

$$(p-1)! = 1 \cdot 2 \cdot \dots \cdot (p-2) \cdot (p-1) \equiv (p-1) \pmod{p}.$$

Since $p - 1 \equiv -1 \pmod{p}$, we have

$$(p-1)! \equiv -1 \pmod{p}.$$

For the case where p is not prime, suppose $p = ab$ for integers a, b such that $1 < a, b < p$. Then, both a and b are divisors of $(p-1)!$, so p divides $(p-1)!$. This implies that

$$(p-1)! \equiv 0 \pmod{p},$$

which is not congruent to $-1 \pmod{p}$.
Thus, $(p-1)! \equiv -1 \pmod{p}$ if and only if p is prime, as desired. ■

Wilson's Theorem provides a characterization of prime numbers, though it is impractical for large values of p due to the factorial calculation.

Corollary 7.2.3 If p is a prime number greater than 2, then p is odd and $2^{p-1} \equiv 1 \mod p$.

Demostración. Since $p > 2$ is prime, it is odd. By **Fermat's Little Theorem**, for any integer a coprime to p, we have $a^{p-1} \equiv 1 \mod p$. Taking $a = 2$, we get $2^{p-1} \equiv 1 \mod p$. ■

This corollary is useful in primality tests and cryptography.

■ **Example 7.5** Determine if $p = 7$ is prime using Wilson's Theorem.
We calculate $(7-1)! = 6! = 720$. Then, check if $720 \equiv -1 \mod 7$:

$$720 \mod 7 = 720 - 7 \times 102 = 720 - 714 = 6$$

Since $6 \equiv -1 \mod 7$ (as $7 - 1 = 6$), Wilson's Theorem confirms that 7 is prime. ■

This example illustrates the application of Wilson's Theorem in a specific case.

R Although Wilson's Theorem is theoretically interesting, methods like the **Fermat Primality Test** are more efficient for large numbers.

It is also important to consider the contrapositive of Fermat's Little Theorem in divisibility problems.

Lema 7.2.2 If n is a composite integer, then there exist integers a such that $a^{n-1} \not\equiv 1 \mod n$.

Demostración. For a composite number n, not all integers a less than n are coprime to n. Furthermore, there exist values of a for which Fermat's congruence does not hold, allowing for the identification of composite numbers. ■

This lemma underpins primality tests based on Fermat's Little Theorem.

Exercise 7.5 Use Fermat's Little Theorem to demonstrate that $n = 341$ is a composite number. ■

Exercise 7.6 Prove that if p is a prime number and a is an integer not divisible by p, then p divides $(a^{p-1} - 1)$. ■

By solving these exercises, the reader will strengthen their understanding of how prime numbers interact with divisibility properties and how fundamental theorems can be used to identify primes and composites, which is essential in advanced mathematical reasoning.

7.2.2 Application in Divisibility of Composite Numbers

In this section, we explore how the properties of divisibility apply to composite numbers, delving into techniques and theorems that allow us to analyze and solve related problems. Composite numbers, being products of prime numbers, exhibit particular characteristics that we can leverage to better understand their structure and behavior in divisibility contexts.

Definition 7.2.2 A **composite number** is a positive integer greater than 1 that is not prime; that is, it has at least one positive divisor other than 1 and itself. In other words, a composite number n can be expressed as the product of two positive integers a and b, such that $1 < a \leq b < n$ and $n = a \cdot b$.

This definition allows us to identify composite numbers and analyze their factors, which is essential for studying divisibility problems associated with them.

Lema 7.2.3 If n is a composite number, then there exists a prime number p such that $p \mid n$ and $p \leq \sqrt{n}$.

Demostración. Since n is composite, it can be expressed as $n = a \cdot b$, with $1 < a \leq b < n$. If $a \leq \sqrt{n}$, then a is a divisor of n less than or equal to $\sqrt{n}$. If a is prime, then we have found a prime $p = a$ such that $p \mid n$ and $p \leq \sqrt{n}$. If a is composite, we can decompose it into prime factors, and at least one of those primes will be less than or equal to $a \leq \sqrt{n}$. Therefore, there exists a prime $p \leq \sqrt{n}$ that divides n. ■

This lemma is useful for developing factorization algorithms and for divisibility tests involving composite numbers.

Theorem 7.2.4 Let n be a composite number and a an integer such that $1 < a < n$. If $n \mid a^k$ for some positive integer k, then n divides $a^{\gcd(k,\varphi(n))}$, where $\varphi(n)$ is Euler's totient function.

Demostración. Let n be a composite number and a an integer such that $1 < a < n$. Suppose $n \mid a^k$ for some positive integer k. We aim to prove that $n \mid a^{\gcd(k,\varphi(n))}$, where $\varphi(n)$ is Euler's totient function.
Since n is composite, it can be written as $n = p_1^{e_1} p_2^{e_2} \dots p_m^{e_m}$, the product of its prime factors. By the divisibility theorem in modular arithmetic, $n \mid a^k$ implies $p_i^{e_i} \mid a^k$ for each $i = 1, 2, \dots, m$.
For each prime p_i in the factorization of n, $p_i^{e_i} \mid a^k$ implies that the minimum exponent t such that $p_i^{e_i} \mid a^t$ divides k. This exponent t is related to $\varphi(n)$, since $\varphi(n)$ is the smallest number such that $a^{\varphi(n)} \equiv 1 \pmod{p_i^{e_i}}$ for all i.
Using the Euclidean algorithm, we can express $\gcd(k, \varphi(n))$ as a linear combination of k and $\varphi(n)$, i.e., there exist integers x and y such that

$$\gcd(k, \varphi(n)) = xk + y\varphi(n).$$

Thus,

$$a^{\gcd(k,\varphi(n))} = a^{xk+y\varphi(n)} = (a^k)^x \cdot (a^{\varphi(n)})^y.$$

Since $n \mid a^k$, it follows that $(a^k)^x$ is divisible by n. Additionally, $a^{\varphi(n)} \equiv 1 \pmod{n}$ by Euler's totient property, so $(a^{\varphi(n)})^y \equiv 1 \pmod{n}$. This implies that $a^{\gcd(k,\varphi(n))}$ is divisible by n.
Hence, $n \mid a^{\gcd(k,\varphi(n))}$, as required. ■

This theorem aids in understanding how powers of numbers relate to divisibility by composite numbers, especially in the context of exponents and Euler's totient function.

■ **Example 7.6** Let $n = 8$ (a composite number) and $a = 4$. Compute the smallest k such that $n \mid a^k$.
The Euler totient function is $\varphi(8) = 4$, as the numbers less than 8 and coprime with 8 are $1, 3, 5, 7$.
We seek k such that $8 \mid 4^k$.
Calculate:

$$4^1 = 4 \mod 8 \equiv 4,$$

$$4^2 = 16 \mod 8 \equiv 0.$$

Thus, $k = 2$ is the smallest positive integer such that $8 \mid 4^k$.
Observe that $\gcd(k, \varphi(8)) = \gcd(2, 4) = 2$.
By the theorem above, n divides $a^{\gcd(k,\varphi(n))} = a^2 = 4^2 = 16$, and indeed $8 \mid 16$.

■

7.3 Operations with Fractions: Addition, Subtraction, Multiplication, and Division with Applied Problems

7.3.1 Homogeneous and Heterogeneous Fractions

In this section, we explore homogeneous and heterogeneous fractions, fundamental concepts in the study of arithmetic and algebra. Understanding these distinctions is essential for efficiently performing operations with fractions and developing solid mathematical reasoning.

Definition 7.3.1 A **homogeneous fraction** is one in which two or more fractions have the same denominator. That is, given the fractions $\frac{a}{d}$ and $\frac{b}{d}$, both are homogeneous because they share the common denominator d.

Definition 7.3.2 A **heterogeneous fraction** is one in which two or more fractions have different denominators. For example, the fractions $\frac{a}{d}$ and $\frac{b}{e}$ are heterogeneous if $d \neq e$.

Recognizing homogeneous and heterogeneous fractions is crucial when performing operations such as addition and subtraction, as the methods differ depending on whether the denominators are the same or different.

Theorem 7.3.1 The sum of homogeneous fractions $\frac{a}{d}$ and $\frac{b}{d}$ is another homogeneous fraction whose numerator is the sum of the numerators and whose denominator is the common denominator:

$$\frac{a}{d}+\frac{b}{d}=\frac{a+b}{d}$$

Demostración. We aim to prove that the sum of the homogeneous fractions $\frac{a}{d}$ and $\frac{b}{d}$ is another homogeneous fraction with numerator $a+b$ and denominator d.
To add $\frac{a}{d}$ and $\frac{b}{d}$, we write

$$\frac{a}{d}+\frac{b}{d}=\frac{a+b}{d}.$$

Since both fractions have the same denominator d, we directly add the numerators and retain the common denominator. Therefore,

$$\frac{a}{d}+\frac{b}{d}=\frac{a+b}{d},$$

which is the desired form for the sum of two homogeneous fractions, as we wanted to prove. ■

This theorem simplifies the addition process when fractions are homogeneous. However, when dealing with heterogeneous fractions, it is necessary to find a common denominator.

Lema 7.3.1 To add or subtract heterogeneous fractions, it is necessary to convert them into homogeneous fractions by determining a common denominator, preferably the least common multiple (LCM) of the denominators.

Demostración. By finding the LCM of the denominators, we ensure that the resulting fractions have the smallest possible common denominator, simplifying calculations and avoiding unnecessarily large denominators. ■

Let us look at an example illustrating this process.

■ **Example 7.7** Add the heterogeneous fractions $\frac{2}{3}$ and $\frac{5}{4}$.
First, find the LCM of the denominators 3 and 4, which is 12. Then, convert the fractions into equivalent fractions with denominator 12:

$$\frac{2}{3}=\frac{2\times 4}{3\times 4}=\frac{8}{12},\quad \frac{5}{4}=\frac{5\times 3}{4\times 3}=\frac{15}{12}$$

Now that the fractions are homogeneous, add the numerators:

$$\frac{8}{12}+\frac{15}{12}=\frac{23}{12}$$

■

This example demonstrates the importance of converting heterogeneous fractions into homogeneous ones to perform basic arithmetic operations.

Theorem 7.3.2 The product of two fractions $\frac{a}{b}$ and $\frac{c}{d}$ is another fraction whose numerator is the product of the numerators and whose denominator is the product of the denominators:

$$\frac{a}{b}\times\frac{c}{d}=\frac{a\times c}{b\times d}$$

Demostración. We aim to prove that the product of the fractions $\frac{a}{b}$ and $\frac{c}{d}$ is another fraction whose numerator is the product of the numerators and whose denominator is the product of the denominators.
By definition of fraction multiplication, we have

$$\frac{a}{b}\times\frac{c}{d}=\frac{a\cdot c}{b\cdot d}.$$

When multiplying fractions, we multiply the numerators and denominators directly. Thus, the numerator of the resulting fraction is $a\cdot c$, and the denominator is $b\cdot d$.
Therefore,

$$\frac{a}{b}\times\frac{c}{d}=\frac{a\cdot c}{b\cdot d},$$

as we wanted to prove. ■

Multiplication and division of fractions do not require the fractions to be homogeneous, simplifying these operations compared to addition and subtraction.

Corollary 7.3.3 The division of two fractions $\frac{a}{b}$ and $\frac{c}{d}$ (with $c\neq 0$) is equivalent to multiplying the first fraction by the reciprocal of the second:

$$\frac{a}{b}\div\frac{c}{d}=\frac{a}{b}\times\frac{d}{c}$$

Demostración. We aim to prove that the division of fractions $\frac{a}{b}$ and $\frac{c}{d}$ (with $c\neq 0$) is equivalent to multiplying the first fraction by the reciprocal of the second.

By definition of fraction division, we have

$$\frac{a}{b} \div \frac{c}{d} = \frac{a}{b} \times \frac{d}{c}.$$

Dividing by $\frac{c}{d}$ is equivalent to multiplying by its reciprocal $\frac{d}{c}$, given that $c \neq 0$. Upon multiplication, we obtain

$$\frac{a}{b} \times \frac{d}{c} = \frac{a \cdot d}{b \cdot c}.$$

Therefore,

$$\frac{a}{b} \div \frac{c}{d} = \frac{a}{b} \times \frac{d}{c},$$

as we wanted to prove. ■

R Simplifying fractions before performing operations can reduce calculation complexity and avoid arithmetic errors.

■ **Example 7.8** Calculate $\frac{7}{9} \div \frac{14}{27}$.
First, find the reciprocal of $\frac{14}{27}$, which is $\frac{27}{14}$.
Then:

$$\frac{7}{9} \times \frac{27}{14} = \frac{7 \times 27}{9 \times 14}$$

Simplify before multiplying: - 27 and 9 share a common factor of 9:

$$\frac{7 \times 3}{1 \times 14} = \frac{21}{14}$$

Simplify $\frac{21}{14}$ by dividing numerator and denominator by 7:

$$\frac{21 \div 7}{14 \div 7} = \frac{3}{2}$$

■

This example illustrates how simplification can facilitate calculations and yield a more manageable result.

Exercise 7.7 Add the fractions $\frac{5}{6}$ and $\frac{7}{8}$ and simplify the result. ■

Exercise 7.8 Multiply the fractions $\frac{3}{4}$ and $\frac{16}{9}$ and express the result in its simplest form. ■

By understanding and applying the properties of homogeneous and heterogeneous fractions, one strengthens arithmetic skills and builds a solid foundation for more advanced studies in algebra and mathematical analysis.

7.3.2 Application in Financial Problems

The concepts of fractions and operations with them are fundamental in the financial field. In this section, we will explore how fractions and advanced arithmetic operations are applied to financial problems such as calculating interest, amortizations, and investment analysis.

Definition 7.3.3 Simple interest is interest calculated only on the initial principal, without considering accrued interest from previous periods. It is defined by the formula:

$$I = C \cdot r \cdot t$$

where I is the interest, C is the initial principal, r is the interest rate (expressed as a fraction or percentage), and t is the time.

This basic concept of simple interest can be extended and analyzed using operations with fractions to better understand how interest varies concerning changes in the rate or time.

■ **Example 7.9** Suppose an investor deposits $C = \$10,000$ in an account that pays an annual simple interest rate of $r = \frac{5}{100} = 0{,}05$. If the money is invested for $t = 3$ years, the interest generated is:

$$I = \$10,000 \times 0{,}05 \times 3 = \$1,500$$

The total amount at the end of the period is $C + I = \$11,500$. ■

However, in finance, **compound interest** is commonly used, where the interest generated is added to the principal to calculate future interest.

Definition 7.3.4 Compound interest is interest calculated on the initial principal and the accumulated interest from previous periods. The general formula is:

$$C_t = C_0 \left(1 + \frac{r}{n}\right)^{nt}$$

where C_t is the capital after t periods, C_0 is the initial principal, r is the nominal annual interest rate, n is the number of compounding periods per year, and t is the time in years.

Compound interest can be analyzed using geometric series and exponential properties, requiring advanced handling of fractions and exponents.

Theorem 7.3.4 For an interest rate r, as the number of compounding periods n approaches infinity, the accumulated amount in continuous compound interest approximates:

$$C_t = C_0 e^{rt}$$

where e is the base of the natural logarithm.

Demostración. Consider an initial investment C_0 and an interest rate r, with compounding n times per year. The accumulated amount C_t after t years in compound interest is

$$C_t = C_0 \left(1 + \frac{r}{n}\right)^{nt}.$$

We want to find the limit of C_t as n approaches infinity, i.e.,

$$\lim_{n\to\infty} C_0 \left(1 + \frac{r}{n}\right)^{nt}.$$

Observe that

$$\lim_{n\to\infty} \left(1 + \frac{r}{n}\right)^{n} = e^r,$$

by the definition of e as the limit of $\left(1 + \frac{1}{n}\right)^n$ as $n \to \infty$. Therefore,

$$\lim_{n\to\infty} \left(1 + \frac{r}{n}\right)^{nt} = e^{rt}.$$

Substituting into the expression for C_t, we obtain

$$C_t = C_0 \lim_{n\to\infty} \left(1+\frac{r}{n}\right)^{nt} = C_0 e^{rt}.$$

Thus, the accumulated amount in continuous compound interest approximates

$$C_t = C_0 e^{rt},$$

as desired. ■

This theorem is fundamental in finance to understand how continuous compound interest works, which is a useful idealization in advanced financial calculations.

Corollary 7.3.5 The growth of capital in continuous compound interest is exponential, and the time required for the capital to double can be calculated by:

$$t = \frac{\ln 2}{r}$$

Demostración. Since the accumulated amount in continuous compound interest is given by

$$C_t = C_0 e^{rt},$$

we want to find the time t required for the initial capital C_0 to double, i.e., for $C_t = 2C_0$. Substituting $C_t = 2C_0$ into the formula, we have

$$2C_0 = C_0 e^{rt}.$$

Divide both sides by C_0:

$$2 = e^{rt}.$$

Take the natural logarithm of both sides:

$$\ln 2 = rt.$$

Solving for t, we obtain

$$t = \frac{\ln 2}{r}.$$

Therefore, the time required for the capital to double in continuous compound interest is

$$t = \frac{\ln 2}{r},$$

as desired. ■

This result allows us to calculate the time required to double an investment at a given interest rate, which is a practical application in financial planning.

■ **Example 7.10** If an investment grows at a continuous compound interest rate of 6 % per year ($r = 0{,}06$), the time required to double the capital is:

$$t = \frac{\ln 2}{0{,}06} \approx \frac{0{,}6931}{0{,}06} \approx 11{,}55 \text{ years}$$

■

The concept of **Net Present Value** (NPV) also uses operations with fractions and summations to evaluate the feasibility of investment projects, considering future cash flows discounted to the present.

Definition 7.3.5 The **Net Present Value** of a series of cash flows C_t is:

$$NPV = \sum_{t=0}^{n} \frac{C_t}{(1+r)^t}$$

where r is the discount rate.

This calculation involves operations with fractions and powers and is essential for decision-making in investments.

Exercise 7.9 An investment project requires an initial outlay of \$50,000 and promises to generate \$15,000 at the end of each of the next 5 years. If the discount rate is 8 %, determine the NPV of the project and assess whether it is a profitable investment. ■

Exercise 7.10 Calculate the accumulated amount after 10 years for an investment of \$20,000 at an annual compound interest rate of 5 %, compounded semi-annually. ■

These exercises allow the application of the concepts of fractions and advanced arithmetic operations in real financial contexts, strengthening understanding and mathematical reasoning skills in the financial domain.

7.4 Solved Exercises

Exercise 7.11 Find the GCD and LCM of the numbers 48 and 180 using the prime factorization method. ■

Demostración. First, we decompose each number into prime factors:

$$48 = 2^4 \times 3, \quad 180 = 2^2 \times 3^2 \times 5.$$

For the GCD, take the smallest exponent of the common factors:

$$\text{GCD}(48, 180) = 2^2 \times 3 = 12.$$

For the LCM, take the largest exponent of all the factors present:

$$\text{LCM}(48, 180) = 2^4 \times 3^2 \times 5 = 720.$$

Thus, the GCD is 12, and the LCM is 720. ■

Exercise 7.12 Convert the decimal number 125 to base 2. ■

Demostración. We divide 125 by 2 successively, recording the remainders:

$125 \div 2 = 62$, remainder $162 \div 2 = 31$, remainder $031 \div 2 = 15$, remainder $115 \div 2 = 7$, remainder $17 \div 2 = 3$, remainder $13 \div 2 = 1$, remainder $11 \div 2 = 0$, remainder 1

Reading the remainders from bottom to top, we obtain: $125_{10} = 1111101_2$. ■

Exercise 7.13 Solve the following age problem: Ten years ago, Carlos was twice as old as Ana. If their current ages sum to 50, what are their current ages? ■

Demostración. Let C be Carlos's current age and A be Ana's current age. The equations are:

$$C - 10 = 2(A - 10)$$

$$C + A = 50.$$

Solving the first equation:

$$C - 10 = 2A - 20 \Rightarrow C = 2A - 10.$$

Substituting into the second equation:

$$2A - 10 + A = 50 \Rightarrow 3A = 60 \Rightarrow A = 20.$$

Thus, $C = 50 - 20 = 30$. Therefore, Carlos is 30 years old, and Ana is 20 years old. ■

Exercise 7.14 A principal of \$2000 is invested at a compound interest rate of 5 % per year for 3 years. Calculate the total amount at the end of the period. ■

Demostración. The formula for compound interest is:

$$C_t = C_0 \left(1 + \frac{r}{n}\right)^{nt}.$$

Here, $C_0 = 2000$, $r = 0{,}05$, $n = 1$, and $t = 3$. Substituting the values:

$$C_t = 2000\,(1 + 0{,}05)^3 = 2000 \times 1{,}157625 = 2315{,}25.$$

Therefore, the total amount at the end of the period is \$2315.25. ■

Exercise 7.15 Prove that if p is a prime number and $p \mid ab$, then $p \mid a$ or $p \mid b$. ■

Demostración. Assume that p divides the product ab but does not divide a. Since p is prime and does not divide a, a and p are coprime. By the **Euclid's Lemma**, if p divides ab and p does not divide a, then p must divide b. Therefore, $p \mid a$ or $p \mid b$, as required. ■

7.5 Proposed Exercises

7.5.1 GCD and LCM: Prime Factorization Methods

Exercise 7.16 Find the GCD of the numbers 96 and 144 using the prime factorization method. ■

Exercise 7.17 Calculate the LCM of the numbers 45 and 60 using prime factorization. ■

Exercise 7.18 Determine the GCD and LCM of the numbers 84 and 120, and verify that the product of both numbers equals the product of their GCD and LCM. ■

Exercise 7.19 Using the prime factorization method, find the GCD of the numbers 210 and 315. ■

Exercise 7.20 Calculate the LCM of the numbers 18, 24, and 36 using prime factorization. ■

7.5.2 Divisibility Problems: Identifying Multiples and Divisors

Exercise 7.21 Determine whether the number 135 is divisible by 3, 5, and 9. ■

Exercise 7.22 Find all divisors of the number 72 and determine whether it is a perfect number (the sum of its proper divisors equals the number). ■

Exercise 7.23 Given the number 210, find all multiples of 7 less than 210. ■

Exercise 7.24 Verify whether 385 is divisible by 5, 7, and 11. Justify your answer using divisibility rules. ■

Exercise 7.25 Find the greatest common divisor of 56 and 98 using the Euclidean algorithm. ■

7.5.3 Operations with Fractions: Addition, Subtraction, Multiplication, and Division with Applied Problems

Exercise 7.26 Simplify the fraction $\frac{84}{126}$ to its irreducible form. ■

Exercise 7.27 Add the fractions $\frac{3}{4}$ and $\frac{5}{6}$, and express the result in its simplest form. ■

Exercise 7.28 Multiply the fractions $\frac{7}{8}$ and $\frac{4}{9}$ and simplify the result. ■

Exercise 7.29 Calculate the result of dividing $\frac{5}{12}$ by $\frac{3}{8}$. ■

Exercise 7.30 A cake is divided into $\frac{3}{4}$ for Ana and $\frac{2}{5}$ for Beatriz. How much cake remains undistributed? ■

8. Ratios, Proportions, and Percentages

8.1 Ratios and Proportions: Simplification and Problem Solving

8.1.1 Direct and Inverse Proportions

In this section, we will study direct and inverse proportions, fundamental concepts in the analysis of functional relationships between variables in mathematics. These concepts are essential for understanding how quantities change in relation to one another and have applications in various fields such as physics, economics, and engineering.

Definition 8.1.1 Two quantities x and y are in **direct proportion** if there exists a constant $k \neq 0$ such that:

$$y = kx$$

The constant k is called the **constant of proportionality**.

This definition implies that as x increases, y increases proportionally, and vice versa. Let us see an example to illustrate this concept.

■ **Example 8.1** If a car travels at a constant speed, the distance d traveled is directly proportional to the time t of travel. If the speed is v, then:

$$d = vt$$

Here, the constant of proportionality is the speed v. ■

In addition to direct proportions, it is important to understand inverse proportions.

Definition 8.1.2 Two quantities x and y are in **inverse proportion** if there exists a constant $k \neq 0$ such that:

$$y = \frac{k}{x}$$

In this case, as x increases, y decreases proportionally, and vice versa. An example will help clarify this relationship.

■ **Example 8.2** The intensity I of illumination at a point is inversely proportional to the square of the distance r from the light source. This is expressed as:

$$I = \frac{k}{r^2}$$

where k is a constant depending on the light source. ■

It is crucial to recognize the mathematical properties governing these proportions.

Theorem 8.1.1 If x and y are in direct proportion, then the ratio $\frac{y}{x}$ is constant and equal to k for all values of x and y.

Demostración. If x and y are in direct proportion, this means y is directly proportional to x, which can be expressed as

$$y = kx,$$

where k is the constant of proportionality.
Dividing both sides of this equation by x (assuming $x \neq 0$), we obtain

$$\frac{y}{x} = k.$$

This shows that the ratio $\frac{y}{x}$ is equal to k and is constant for all values of x and y that satisfy the direct proportionality relationship.
Thus, if x and y are in direct proportion, then $\frac{y}{x}$ is constant and equal to k, as we intended to prove. ■

Similarly, we can establish an analogous property for inverse proportions.

Theorem 8.1.2 If x and y are in inverse proportion, then the product xy is constant and equal to k for all values of x and y.

Demostración. If x and y are in inverse proportion, this means y is inversely proportional to x, which can be expressed as

$$y = \frac{k}{x},$$

where k is the constant of proportionality.
Multiplying both sides of this equation by x (assuming $x \neq 0$), we obtain

$$xy = k.$$

This shows that the product xy is equal to k and is constant for all values of x and y that satisfy the inverse proportionality relationship.
Thus, if x and y are in inverse proportion, then xy is constant and equal to k, as we intended to prove. ■

These properties allow us to solve practical problems where it is necessary to determine one variable as a function of another.

 Direct and inverse proportions are particular cases of linear and rational variation functions, respectively. Understanding their behavior is key to studying more complex functions.

Let us see another example illustrating the application of inverse proportions.

■ **Example 8.3** The time t it takes to fill a tank is inversely proportional to the flow rate q of filling. If the tank fills in $t_1 = 4$ hours with a flow rate of $q_1 = 50$ liters per hour, how long t_2 will it take to fill with a flow rate of $q_2 = 100$ liters per hour?
We know that $tq = k$, so:

$$t_1 q_1 = t_2 q_2 \implies 4 \times 50 = t_2 \times 100 \implies t_2 = \frac{4 \times 50}{100} = 2 \text{ hours.}$$

■

It is interesting to analyze how these relationships are represented graphically.

Theorem 8.1.3 The graph of a direct proportion $y = kx$ is a straight line passing through the origin, while the graph of an inverse proportion $y = \frac{k}{x}$ is a rectangular hyperbola in the quadrants where $x \neq 0$.

Demostración. For the direct proportion, $y = kx$ is the equation of a straight line with slope k and y-intercept 0.
For the inverse proportion, $y = \frac{k}{x}$ is a hyperbolic function. The points (x, y) such that $xy = k$ form a rectangular hyperbola centered at the origin. ■

This graphical understanding is useful for visualizing the behavior of the variables and anticipating how changes in one affect the other.

Exercise 8.1 If y is directly proportional to x and $y = 15$ when $x = 5$, determine the constant of proportionality and find the value of y when $x = 9$. ■

Exercise 8.2 If x is inversely proportional to z and $x = 8$ when $z = 6$, calculate the value of z when $x = 12$. ■

By solving these exercises, the reader will apply the concepts of direct and inverse proportions, strengthening their understanding and ability to model mathematical situations using these fundamental relationships.

8.1.2 Scale and Map Problems

In this section, we will study scales and their application in interpreting and creating maps. Scales are fundamental tools that allow real-world distances and areas to be represented in manageable dimensions, facilitating geographic and architectural analysis. Understanding the mathematics behind scales is essential for solving problems involving proportionality and geometric similarity.

Definition 8.1.3 A **scale** is the mathematical relationship that indicates how many times a real-world measurement has been reduced or enlarged to represent it on a map or model. It is expressed as a ratio of the form $1 : n$, where 1 unit on the map corresponds to n units in reality.

This definition allows us to establish a precise correspondence between map measurements and real-world measurements. Let us see how this concept applies in a practical context.

■ **Example 8.4** If a map has a scale of 1 : 50,000, this means that 1 centimeter on the map represents 50,000 centimeters in reality, or 500 meters. Therefore, a distance of 7 cm on the map corresponds to:

$$7 \text{ cm} \times 50{,}000 = 350{,}000 \text{ cm} = 3{,}5 \text{ km}.$$

■

This example illustrates how to use the scale to convert map measurements into real-world measurements. Now, let us delve into the mathematical properties that support this process.

Theorem 8.1.4 On a map with a scale of $1 : n$, the linear distances on the map (d_m) and the real-world distances (d_r) are related by a direct proportion:

$$\frac{d_m}{d_r} = \frac{1}{n}.$$

Demostración. By the definition of scale, 1 unit on the map corresponds to n units in reality. Therefore, for any distance d_m on the map, the real-world distance is $d_r = n \cdot d_m$. Rearranging, we obtain the proportion:

$$\frac{d_m}{d_r} = \frac{1}{n}.$$

■

This theorem confirms that the ratio between map distances and real-world distances is constant and determined by the scale. Additionally, it is important to consider how areas and volumes relate in maps.

Lema 8.1.1 The areas on a map (A_m) and the real-world areas (A_r) are related by the square of the scale:

$$\frac{A_m}{A_r} = \left(\frac{1}{n}\right)^2.$$

Demostración. If the linear dimensions are reduced by a factor of $\frac{1}{n}$, then the areas, being two-dimensional, are reduced by the square of that factor. Therefore:

$$A_m = A_r \times \left(\frac{1}{n}\right)^2 \implies \frac{A_m}{A_r} = \left(\frac{1}{n}\right)^2.$$

■

R This relationship extends to volumes, where the proportion is cubic:

$$\frac{V_m}{V_r} = \left(\frac{1}{n}\right)^3.$$

This is crucial in three-dimensional models and applications such as architectural models.

Continuing with practical applications, let us consider the following example.

■ **Example 8.5** An architect designs a model at a scale of 1 : 100 of a building whose real height is 50 meters. The height of the model will be:

$$h_m = \frac{h_r}{n} = \frac{50\text{ m}}{100} = 0{,}5\text{ m}.$$

If we want to calculate the volume of the building in the model, knowing that the real volume is $V_r = 10{,}000\text{ m}^3$, then:

$$V_m = V_r \times \left(\frac{1}{n}\right)^3 = 10{,}000\text{ m}^3 \times \left(\frac{1}{100}\right)^3 = 0{,}01\text{ m}^3.$$

■

This example demonstrates how to apply scale relationships in three-dimensional contexts.

Theorem 8.1.5 When changing the scale of a map or model from $1 : n_1$ to $1 : n_2$, the new measurements can be obtained by multiplying the original measurements by the relative scale factor $k = \frac{n_1}{n_2}$.

Demostración. Let d_{m1} be a measurement on the original map. The corresponding real-world measurement is $d_r = n_1 \cdot d_{m1}$. On the new scale, the measurement on the map will be:

$$d_{m2} = \frac{d_r}{n_2} = \frac{n_1 \cdot d_{m1}}{n_2} = d_{m1} \cdot \frac{n_1}{n_2}.$$

■

Corollary 8.1.6 If $n_2 < n_1$, the map or model is enlarged; if $n_2 > n_1$, it is reduced. The relative scale factor determines the degree of enlargement or reduction.

This knowledge is essential when modifying maps to fit different formats or purposes.

Exercise 8.3 On a map with a scale of 1 : 25,000, the surface area of a lake is 8 cm^2. Calculate the real surface area of the lake in square kilometers. ■

Exercise 8.4 An engineer needs to create a detailed plan at a scale of 1 : 500 of a rectangular plot measuring 200 m by 150 m. Determine the dimensions of the plan and calculate the area of the plot on the plan in square centimeters. ■

By understanding and applying the mathematical principles of scales, we are better equipped to interpret and create accurate representations on maps and models, which is fundamental in disciplines such as cartography, architecture, and engineering.

8.2 Rule of Three: Direct and Inverse.

8.2.1 Applications in Mixture Problems

In this section, we explore mixture problems, which are common scenarios in chemistry, physics, and other applied fields, where substances or components are combined in different proportions. Solving these problems requires solid mathematical reasoning and the use of algebraic equations that model the relationships between the quantities involved.

Definition 8.2.1 A **mixture problem** involves combining two or more substances with different concentrations or properties to determine the final composition or the necessary amounts of each component to achieve a mixture with specific characteristics.

Mixture problems often involve proportions, percentages, and linear equations. It is essential to understand how to set up equations that represent the relationships between the quantities and concentrations of the components.

Example 8.6 Suppose we want to obtain 100 liters of a saline solution with a 30 % concentration. We have solutions with 20 % and 50 % concentrations. How many liters of each should be mixed to achieve the desired solution?
Let x be the amount in liters of the 20 % solution, and y be the amount in liters of the 50 % solution. We have the system of equations:

$$\begin{cases} x+y=100 \\ 0{,}20x+0{,}50y=0{,}30\times 100 \end{cases}$$

Solving the system, we obtain:

$$\begin{aligned} x+y&=100 \\ 0{,}20x+0{,}50y&=30 \end{aligned}$$

Subtracting 0,20 times the first equation from the second:

$$0{,}20x+0{,}50y-0{,}20x-0{,}20y=30-20 \implies 0{,}30y=10 \implies y=\frac{10}{0{,}30}=33.\overline{3}$$

Thus, $x=100-y=66.\overline{6}$. Therefore, approximately $66.\overline{6}$ liters of the 20 % solution and $33.\overline{3}$ liters of the 50 % solution are needed.

■

This example illustrates how to formulate and solve linear equations in the context of mixture problems. To generalize these methods, consider the following theorem.

Theorem 8.2.1 To obtain a quantity Q of a mixture with concentration C_m, combining two solutions with concentrations C_1 and C_2, where $C_1 < C_m < C_2$, the quantities Q_1 and Q_2 of each solution needed to achieve the desired mixture are given by:

$$\begin{aligned} Q_1 &= Q\times\frac{C_2-C_m}{C_2-C_1} \\ Q_2 &= Q-Q_1 \end{aligned}$$

Demostración. The total amount of solute in the mixture is C_mQ. This solute comes from the quantities Q_1 and Q_2 of the initial solutions, i.e.,

$$C_1Q_1+C_2Q_2=C_mQ$$

Additionally, since $Q=Q_1+Q_2$, we can substitute $Q_2=Q-Q_1$. Substituting into the previous equation:

$$C_1Q_1 + C_2(Q - Q_1) = C_mQ \implies C_1Q_1 + C_2Q - C_2Q_1 = C_mQ$$

Rearranging terms:

$$(C_1 - C_2)Q_1 = C_mQ - C_2Q \implies Q_1 = Q\frac{C_2 - C_m}{C_2 - C_1}$$

And $Q_2 = Q - Q_1$.

■

This theorem provides a direct formula to calculate the required quantities of each component in a mixture, simplifying the resolution of these problems.

R It is important that $C_1 \neq C_2$ and that C_m lies between C_1 and C_2. Otherwise, the problem would not have a physical solution, as a concentration outside the range of the initial concentrations cannot be achieved by simple mixing.

Let us see another example applying the theorem.

■ **Example 8.7** We want to prepare 200 grams of a metal alloy with a purity of 70%. If we have two alloys, one with a purity of 60% and the other with 90%, how many grams of each should be mixed?
Applying the theorem:

$$Q_1 = 200 \times \frac{90\% - 70\%}{90\% - 60\%} = 200 \times \frac{20\%}{30\%} = 200 \times \frac{2}{3} \approx 133.\overline{3} \text{ grams}$$
$$Q_2 = 200 - 133.\overline{3} \approx 66.\overline{6} \text{ grams}$$

Therefore, approximately $133.\overline{3}$ grams of the 60% alloy and $66.\overline{6}$ grams of the 90% alloy should be mixed.

■

Mixture problems can also involve differential equations when considering dynamic processes, such as continuous mixing in tanks.

Lema 8.2.1 In a system where a solution flows into a tank at a constant rate and the mixture is kept uniform, the amount of solute $A(t)$ in the tank as a function of time satisfies the first-order linear differential equation:

$$\frac{dA}{dt} = \text{Rate of solute inflow} - \text{Rate of solute outflow}$$

Demostración. The rate of change of the solute amount in the tank equals the difference between the rate at which the solute enters and the rate at which it exits. If r_{in} is the inflow rate and C_{in} the inflow concentration, and r_{out} and C_{out} the outflow rate and concentration, then:

$$\frac{dA}{dt} = r_{\text{in}}C_{\text{in}} - r_{\text{out}}C_{\text{out}}$$

If the tank volume remains constant, and the mixture is uniform, then $C_{\text{out}} = \frac{A(t)}{V}$, where V is the tank volume.

■

This lemma is fundamental in dynamic mixture models and is the basis for solving problems in chemical and environmental engineering.

Exercise 8.5 A 100-liter tank initially contains pure water. A 20 % saline solution is added at a rate of 5 liters per minute, and the mixture leaves the tank at the same rate, maintaining a constant volume. Find the amount of salt in the tank after 10 minutes. ■

Exercise 8.6 A 15 % acid solution needs to be diluted to prepare 500 ml of a 5 % acid solution. How much water should be added to the original solution? ■

By mastering these concepts and methods, the reader will be well-prepared to tackle mixture problems in various contexts, applying advanced mathematical reasoning to find precise and efficient solutions.

8.2.2 Compound Rule of Three

In this section, we address the **compound rule of three**, a fundamental tool for solving problems involving multiple magnitudes related proportionally. Understanding this method is essential in advanced mathematical reasoning, as it enables the modeling and solving of complex situations of direct and inverse proportionality.

Definition 8.2.2 The **compound rule of three** is a procedure that allows calculating the unknown value of a magnitude when two or more magnitudes are proportionally related, either directly or inversely. It is based on establishing a proportion between the known and unknown magnitudes, taking into account the type of proportionality between them.

To apply the compound rule of three, it is crucial to correctly identify the proportional relationships between the magnitudes involved and establish an equation that reflects these relationships.

■ **Example 8.8** Suppose 6 identical machines can produce 180 units of a product in 8 hours. How many units will 9 machines produce working for 6 hours?
First, we identify the magnitudes involved:

- Number of machines (M)
- Working time in hours (T)
- Units produced (U)

We analyze the proportional relationships:

1. The number of units produced is directly proportional to the number of machines ($U \propto M$).
2. The number of units produced is directly proportional to the working time ($U \propto T$).

We establish the relationship:

$$\frac{U_1}{U_2} = \frac{M_1}{M_2} \times \frac{T_1}{T_2}$$

Substituting the known values:

$$\frac{180}{U_2} = \frac{6}{9} \times \frac{8}{6}$$

We calculate:

$$\frac{180}{U_2} = \frac{6 \times 8}{9 \times 6} = \frac{48}{54} = \frac{8}{9}$$

Solving for U_2:

$$U_2 = 180 \times \frac{9}{8} = 202{,}5$$

Therefore, 9 machines working 6 hours will produce 202,5 units. If only whole units can be produced, they will produce 202 complete units.

■

Theorem 8.2.2 In the compound rule of three, if all magnitudes are directly proportional, the unknown value is calculated as:

$$\text{Unknown value} = \text{Known value} \times \prod_{i=1}^{n} \frac{\text{Known magnitude}_i}{\text{Unknown magnitude}_i}$$

Demostración. In a compound rule of three where all magnitudes are directly proportional, increasing or decreasing one magnitude results in the same proportional change in the others.
Suppose we have a known value V_k related to several magnitudes $M_{i,\text{known}}$ and we want to calculate an unknown value V_d corresponding to other magnitudes $M_{i,\text{unknown}}$, maintaining direct proportionality.
Since each magnitude is directly proportional, the unknown value V_d relates to the known value V_k multiplied by the product of the ratios between each known magnitude and its corresponding unknown magnitude:

$$V_d = V_k \times \prod_{i=1}^{n} \frac{M_{i,\text{known}}}{M_{i,\text{unknown}}}.$$

This formula maintains direct proportionality among the magnitudes and allows calculating the unknown value based on the known value and the relationships among the magnitudes.
Thus, we have demonstrated that

$$\text{Unknown value} = \text{Known value} \times \prod_{i=1}^{n} \frac{\text{Known magnitude}_i}{\text{Unknown magnitude}_i},$$

■

When some magnitudes are inversely proportional, the corresponding ratios are inverted.

Corollary 8.2.3 If any magnitude is inversely proportional, its ratio is inverted in the formula:

$$V = V_0 \times \prod_{\text{direct}} \frac{M_i}{M_{0i}} \times \prod_{\text{inverse}} \frac{M_{0j}}{M_j}$$

Demostración. In the compound rule of three, when magnitudes are directly proportional, the ratio of each magnitude is included in the formula as is. However, if any magnitude is inversely proportional, an increase in this magnitude implies a decrease in the final value, and vice versa.
To incorporate this inverse relationship, we invert the ratio of each inversely proportional magnitude in the formula. Thus, if V_0 is the initial value and we want to calculate V after changing several magnitudes M_i in direct proportionality and M_j in inverse proportionality, the formula is expressed as:

$$V = V_0 \times \prod_{\text{direct}} \frac{M_i}{M_{0i}} \times \prod_{\text{inverse}} \frac{M_{0j}}{M_j}.$$

This ensures that the effect of inversely proportional magnitudes is correctly adjusted in the calculation of the final value V.
Thus, if any magnitude is inversely proportional, its ratio is inverted in the formula, as we wanted to demonstrate. ■

■ **Example 8.9** A team of 5 workers can complete a project in 12 days working 8 hours per day. How many days will a team of 8 workers need if they work 6 hours per day to complete the same project?
Magnitudes:

- Number of workers (O)
- Daily working hours (H)
- Total time in days (D)

Proportional relationships:

1. Time D is inversely proportional to the number of workers ($D \propto \frac{1}{O}$).
2. Time D is inversely proportional to daily working hours ($D \propto \frac{1}{H}$).

We apply the compound rule of three, inverting the inverse magnitudes:

$$\frac{D}{D_0} = \frac{O_0}{O} \times \frac{H_0}{H}$$

Substituting:

$$\frac{D}{12} = \frac{5}{8} \times \frac{8}{6}$$

We calculate:

$$\frac{D}{12} = \frac{5 \times 8}{8 \times 6} = \frac{5}{6}$$

Solving for D:

$$D = 12 \times \frac{5}{6} = 10 \text{ days}$$

Therefore, the team of 8 workers will take 10 days working 6 hours per day.

■

This example illustrates how to handle inverse proportionalities when applying the compound rule of three.

(R) It is essential to correctly identify the type of proportionality between the magnitudes to properly apply the compound rule of three. An incorrect analysis can lead to inaccurate results.

Exercise 8.7 A vehicle travels 360 km in 6 hours at a constant speed. If its speed increases by 20 %, how long will it take to travel 480 km? ■

Exercise 8.8 A recipe for 4 people requires 600 grams of flour. If the same recipe is to be prepared for 10 people with a 15 % increase to compensate for losses, how much flour is needed? ■

By mastering the compound rule of three, the reader will be equipped to solve complex problems involving multiple related magnitudes, applying rigorous and structured mathematical reasoning.

8.3 Percentages: Problems of Increase and Discount.

8.3.1 Application in Financial Problems.

Mathematical reasoning is a fundamental tool in the analysis and solution of financial problems. Through mathematical concepts, we can model real-life situations and make informed decisions in the economic field.

Definition 8.3.1 An *annuity* is a series of equal payments or receipts made at regular intervals of time.

Annuities are common in loans, mortgages, and savings plans, where payments or deposits are made periodically.

Theorem 8.3.1 The formula to calculate the future value FV of an ordinary annuity (payments at the end of each period) is:

$$FV = P\left(\frac{(1+r)^n - 1}{r}\right),$$

where P is the periodic payment, r is the interest rate per period, and n is the total number of payments.

Demostración. To calculate the future value FV of an ordinary annuity, consider n periodic payments of amount P, made at the end of each period, with an interest rate per period r.
The future value of each individual payment P depends on how many periods it remains invested until the end of the annuity. The first payment accumulates for $n-1$ periods, the second for $n-2$ periods, and so on, until the last payment, which does not accumulate interest.
The future value FV of the annuity is the sum of the future value of all payments:

$$FV = P(1+r)^{n-1} + P(1+r)^{n-2} + \cdots + P(1+r)^0.$$

Factorizing P from the expression:

$$FV = P\left((1+r)^{n-1} + (1+r)^{n-2} + \cdots + 1\right).$$

The expression inside the parentheses is a geometric sum with n terms, a ratio of $1+r$, and the first term equal to 1. The sum of a geometric series is

$$\sum_{k=0}^{n-1}(1+r)^k = \frac{(1+r)^n - 1}{r}.$$

Thus,

$$FV = P\left(\frac{(1+r)^n - 1}{r}\right),$$

as required. ■

This theorem allows us to determine how much will accumulate at the end of a given period when making periodic payments.

■ **Example 8.10** If we deposit \$500 at the end of each month into an account paying a monthly interest rate of 0.5 %, how much will we have after 5 years?

$$FV = 500\left(\frac{(1+0{,}005)^{60}-1}{0{,}005}\right) \approx 500 \times 69{,}7617 = \$34{,}880{,}85.$$

■

In addition to calculating the future value, it is important to understand how to determine the periodic payment required to reach a financial goal.

Corollary 8.3.2 The formula to calculate the periodic payment P required to reach a future value FV in an ordinary annuity is:

$$P = FV\left(\frac{r}{(1+r)^n-1}\right).$$

Demostración. Starting from the formula for the future value FV of an ordinary annuity:

$$FV = P\left(\frac{(1+r)^n-1}{r}\right),$$

where P is the periodic payment, r is the interest rate per period, and n is the total number of payments.
To isolate P, divide both sides by $\frac{(1+r)^n-1}{r}$:

$$P = FV \cdot \frac{r}{(1+r)^n-1}.$$

Thus, we obtain:

$$P = FV\left(\frac{r}{(1+r)^n-1}\right),$$

as required. ■

R This formula is useful for planning savings or investments with a specific financial goal in mind.

■ **Example 8.11** We want to have \$100,000 in 10 years for a child's college education. If the account pays an annual interest rate of 6 %, how much should we deposit at the end of each year?

$$P = 100{,}000\left(\frac{0{,}06}{(1+0{,}06)^{10}-1}\right) \approx 100{,}000 \times 0{,}0609 = \$6{,}090{,}24.$$

■

In the context of loans, mathematical reasoning helps us understand amortization.

Lema 8.3.1 The periodic payment P to amortize a loan of amount L with an interest rate per period r over n periods is:

$$P = L\left(\frac{r(1+r)^n}{(1+r)^n-1}\right).$$

Demostración. The periodic payment is calculated such that the present value of the future payments equals the loan amount, using the present value formula for an annuity. ■

Exercise 8.9 A mortgage loan of \$200,000 is to be paid over 30 years with monthly payments and an annual interest rate of 4.5 %. Calculate the required monthly payment. ■

Exercise 8.10 You plan to withdraw \$40,000 annually for 20 years from your retirement fund. If the fund earns an annual interest rate of 5 %, how much should you have saved at the start of retirement? ■

These concepts and exercises enable the application of mathematical reasoning to make sound financial decisions and effectively plan for economic goals.

8.3.2 Progressive Discounts in Commerce.

Progressive discounts are a common practice in commerce, where multiple discounts are applied successively to a product or service. Understanding how to calculate the total discount is essential for proper financial management and mathematical reasoning in commercial situations.

Definition 8.3.2 A *progressive discount* is a series of percentage discounts applied sequentially to the price of a product or service. Each discount is calculated on the resulting price after the previous discount has been applied.

It is important to note that progressive discounts are not additive in percentage terms; the sum of the discount percentages does not equal the total percentage discount applied to the original price.

Theorem 8.3.3 Let P be the original price of a product, and n progressive discounts are applied with percentages $d_1, d_2, \dots, d_n$, expressed as decimal fractions. The final price P_f after all discounts is:

$$P_f = P\prod_{k=1}^{n}(1-d_k).$$

Demostración. Let P be the original price of a product, and assume n successive discounts with percentages $d_1, d_2, \dots, d_n$, expressed as decimal fractions.
After the first discount of d_1, the price reduces to:

$$P_1 = P(1-d_1).$$

Applying the second discount of d_2 to the new price P_1, we get:

$$P_2 = P_1(1-d_2) = P(1-d_1)(1-d_2).$$

Continuing this way, after applying all the discounts, the final price P_f becomes:

$$P_f = P\prod_{k=1}^{n}(1-d_k).$$

Thus, the final price after applying n successive discounts is:

$$P_f = P\prod_{k=1}^{n}(1-d_k),$$

as required. ■

This theorem allows for efficient calculation of the final price without needing to apply each discount individually in sequence.

Example 8.12 A store offers a 20 % discount followed by an additional 10 % discount on the reduced price. If the original price is \$100, what is the final price?
Applying the theorem:

$$P_f = 100 \times (1 - 0{,}20) \times (1 - 0{,}10) = 100 \times 0{,}80 \times 0{,}90 = \$72.$$

■

We observe that the total discount is not 30 %, but the final price is \$72, representing a total discount of 28 %.

Corollary 8.3.4 The total discount percentage D_t resulting from applying n progressive discounts with percentages $d_1, d_2, \ldots, d_n$ is:

$$D_t = 1 - \prod_{k=1}^{n} (1 - d_k).$$

Demostración. Let P be the original price of a product and P_f the final price after applying n successive discounts with percentages $d_1, d_2, \ldots, d_n$, expressed as decimal fractions. From the previous theorem, the final price P_f is:

$$P_f = P \prod_{k=1}^{n} (1 - d_k).$$

The total discount percentage D_t is the fraction of the original price that has been reduced, given by:

$$D_t = 1 - \frac{P_f}{P}.$$

Substituting P_f into the expression, we have:

$$D_t = 1 - \frac{P \prod_{k=1}^{n} (1 - d_k)}{P}.$$

Simplifying, we get:

$$D_t = 1 - \prod_{k=1}^{n} (1 - d_k).$$

Thus, the total discount percentage is:

$$D_t = 1 - \prod_{k=1}^{n} (1 - d_k),$$

as required. ■

R It is common to think that adding the individual discount percentages gives the total discount, but as we have seen, this is incorrect due to the multiplicative nature of progressive discounts.

Understanding these formulas allows businesses and consumers to make informed decisions when offering or taking advantage of multiple discounts.

■ **Example 8.13** Calculate the total discount percentage when three progressive discounts of 15 %, 10 %, and 5 % are applied.
Applying the corollary:

$$D_t = 1-(1-0{,}15)(1-0{,}10)(1-0{,}05) = 1-(0{,}85\times 0{,}90\times 0{,}95) = 1-0{,}72675 = 0{,}27325.$$

The total discount is 27.325 %. ■

Lema 8.3.2 If all the progressive discounts are equal, that is, $d_k = d$ for all k, then the final price is:

$$P_f = P(1-d)^n.$$

Demostración. Substituting $d_k = d$ into the theorem:

$$P_f = P\prod_{k=1}^{n}(1-d_k) = P(1-d)^n.$$

■

This particular case is useful when equal discounts are applied in special promotions.

Exercise 8.11 A store applies two equal progressive discounts on a product whose original price is \$150. If the final price is \$108, determine the percentage of each discount. ■

Exercise 8.12 A customer has a 25 % discount coupon and is offered an additional 15 % discount at the checkout. What is the total percentage discount applied to the original price? ■

These concepts are fundamental for analyzing pricing strategies and promotions in commerce and demonstrate how mathematical reasoning applies to real economic contexts.

8.4 Solved Exercises

Exercise 8.13 Two magnitudes are directly proportional. If $x = 6$ when $y = 24$, find the value of y when $x = 9$. ■

Demostración. Since the magnitudes are directly proportional, we can write:

$$\frac{y}{x} = \frac{24}{6} = 4$$

Thus, when $x = 9$, we have:

$$y = 4\times 9 = 36$$

Therefore, the value of y is 36. ■

Exercise 8.14 On a map with a scale of $1:50,000$, the distance between two points is 3 cm. What is the real distance between these points in kilometers? ■

Demostración. The scale $1:50,000$ means that 1 cm on the map represents $50,000$ cm in reality. Thus, the real distance is:

$$3\times 50,000 = 150,000 \text{ cm}$$

We convert this distance to kilometers:

$$150,000 \text{ cm} = 1{,}5 \text{ km}$$

Therefore, the real distance between the points is 1,5 km. ■

Exercise 8.15 It is desired to prepare 100 liters of a solution with 25 % concentration. If solutions with 10 % and 50 % concentrations are available, how many liters of each should be mixed? ■

Demostración. Let x be the amount of 10 % solution, and y be the amount of 50 % solution. We have the system of equations:

$$x+y=100$$

$$0{,}10x+0{,}50y=0{,}25\times 100=25$$

Substituting $y=100-x$ into the second equation:

$$0{,}10x+0{,}50(100-x)=25$$

$$0{,}10x+50-0{,}50x=25$$

$$-0{,}40x=-25$$

$$x=62{,}5$$

Thus, $y=100-62{,}5=37{,}5$. We need 62,5 liters of the 10 % solution and 37,5 liters of the 50 % solution. ■

Exercise 8.16 An item costs \$200, and two successive discounts of 10 % and 20 % are applied. What is the final price of the item? ■

Demostración. Apply the first discount of 10 %:

$$200\times(1-0{,}10)=200\times 0{,}90=180$$

Then apply the second discount of 20 % to the reduced price:

$$180\times(1-0{,}20)=180\times 0{,}80=144$$

Therefore, the final price of the item is \$144. ■

Exercise 8.17 An initial investment of \$1,000 is placed in an account that pays an annual compound interest of 5 %. What will the value of the investment be after 3 years? ■

Demostración. The compound interest formula is:

$$A=P(1+r)^t$$

where P is the initial amount, r is the interest rate, and t is the time in years. Substituting the values:

$$A=1000\times(1+0{,}05)^3$$

$$A=1000\times(1{,}05)^3$$

$$A\approx 1000\times 1{,}157625=1157{,}63$$

Thus, the value of the investment after 3 years is approximately \$1,157.63. ■

8.5 Proposed Exercises

8.5.1 Ratios and Proportions: Simplification and Problem Solving

Exercise 8.18 Simplify the ratio $24 : 36$. ■

Exercise 8.19 If a and b are directly proportional, and when $a = 3$, $b = 12$, find the value of b when $a = 8$. ■

Exercise 8.20 In a recipe, the ratio of sugar to flour is $2 : 5$. If 500 grams of flour are used, how many grams of sugar are needed? ■

Exercise 8.21 A map has a scale of $1 : 200,000$. If the distance between two points on the map is 7 cm, what is the actual distance in kilometers? ■

Exercise 8.22 If three numbers are in the ratio $2 : 3 : 5$ and their sum is 50, find each of the numbers. ■

8.5.2 Rule of Three: Direct and Inverse

Exercise 8.23 If 4 workers can complete a task in 10 days, how many days will 8 workers take to complete the same task, assuming all work at the same rate? ■

Exercise 8.24 A machine produces 200 items in 5 hours. How many items will it produce in 8 hours at the same production rate? ■

Exercise 8.25 If 6 meters of fabric cost $90, how much will 10 meters of the same fabric cost? ■

Exercise 8.26 A vehicle traveling at 60 km/h takes 3 hours to cover a certain distance. How long will it take to cover the same distance at a speed of 80 km/h? ■

Exercise 8.27 If 5 painters can paint a wall in 12 hours, how many painters are needed to paint the same wall in 8 hours? ■

8.5.3 Percentages: Problems of Increase and Discount

Exercise 8.28 An item is priced at $120 and a 15 % discount is applied. What is the final price after the discount? ■

Exercise 8.29 A product costs $80. If its price increases by 20 %, what will its new price be? ■

Exercise 8.30 A store offers a 10 % discount on the first item and an additional 5 % discount on the second item. If the first item costs $150 and the second $100, what is the total price after applying the discounts? ■

Exercise 8.31 An investor earns an 8 % annual return on an investment of $10,000. How much will they have at the end of one year? ■

9. Sequences and Series

9.1 Arithmetic Sequences: General Term Formula and Sum of Terms.

9.1.1 Applications in Simple Interest Problems.

Simple interest is a fundamental concept in financial mathematics that analyzes how investments or debts grow over time without considering interest capitalization. Unlike compound interest, simple interest is calculated only on the initial principal, simplifying certain calculations and being applicable in specific financial scenarios.

Definition 9.1.1 Let P be the *principal amount* or initial invested or loaned amount, r the *simple interest rate* per period (expressed as a decimal fraction), and t the number of periods. The *simple interest* I accumulated is defined as:

$$I = P \cdot r \cdot t.$$

This definition establishes the basis for calculating interest earned or owed based on time, rate, and principal amount.

Theorem 9.1.1 The *total amount* A after t periods under simple interest is:

$$A = P(1 + rt).$$

Demostración. The total amount is the sum of the principal and the accumulated interest:

$$A = P + I = P + Prt = P(1 + rt).$$

■

This result is crucial for determining the future value of an investment or loan when simple interest is applied.

■ **Example 9.1** If we invest \$5,000 at an annual simple interest rate of 6 % for 4 years, the accumulated interest and total amount will be:

$$I = 5000 \times 0{,}06 \times 4 = \$1,200,$$

$$A = 5000(1 + 0{,}06 \times 4) = 5000 \times 1{,}24 = \$6,200.$$

■

This example illustrates how to apply simple interest formulas to calculate earnings on fixed-term investments.

Lema 9.1.1 If the total amount A, principal P, and time t are known, the simple interest rate r can be determined by:

$$r = \frac{A - P}{Pt}.$$

Demostración. Rearranging the total amount formula to isolate r:

$$A = P(1 + rt) \implies \frac{A}{P} = 1 + rt \implies r = \frac{\frac{A}{P} - 1}{t} = \frac{A - P}{Pt}.$$

■

This lemma is useful for finding the interest rate applied to an investment or loan when other parameters are known.

> R It is important to note that simple interest does not consider capitalization; that is, the generated interest is not reinvested to generate additional interest.

Understanding this difference is essential when comparing with compound interest and choosing the appropriate type of investment.

Corollary 9.1.2 The time t required to reach a total amount A from a principal P with a simple interest rate r is:

$$t = \frac{A - P}{Pr}.$$

Demostración. Rearranging the total amount formula to isolate t:

$$A = P(1 + rt) \implies \frac{A}{P} = 1 + rt \implies t = \frac{\frac{A}{P} - 1}{r} = \frac{A - P}{Pr}.$$

■

This corollary is useful for determining the required term to achieve specific financial goals.

■ **Example 9.2** We want an investment of \$2,000 to grow to \$2,500 at an annual simple interest rate of 5 %. How long will it take?

$$t = \frac{2500 - 2000}{2000 \times 0{,}05} = \frac{500}{100} = 5 \text{ years}.$$

■

This calculation helps plan future investments based on specific financial objectives.

Definition 9.1.2 The *commercial discount* D is the simple interest deducted from the face value N of a document due at a future date when discounted before maturity. It is calculated as:

$$D = Nrt,$$

where r is the discount rate and t is the time until maturity.

Commercial discount is essential in financial operations such as the discounting of promissory notes or bills of exchange.

Lema 9.1.2 The *present value* V of a commercially discounted document is:

$$V = N(1 - rt).$$

Demostración. The present value is the nominal value minus the discount:

$$V = N - D = N - Nrt = N(1 - rt).$$

■

This result allows calculating the effective amount received by someone discounting the document before its maturity.

■ **Example 9.3** A promissory note with a nominal value of \$8,000 is due in 9 months and is discounted at an annual simple interest rate of 7 %. The present value is:

$$D = 8000 \times 0{,}07 \times \frac{9}{12} = 8000 \times 0{,}07 \times 0{,}75 = \$420,$$

$$V = 8000 - 420 = \$7,580.$$

■

This example demonstrates how to calculate the present value of a document discounted before its maturity date.

(R) In commercial discount, the interest is calculated on the nominal value, not on the present value, distinguishing this method from *rational discount*.

Understanding this distinction is vital for analyzing and comparing different financing options.

Exercise 9.1 An investor wants to earn \$900 in interest over 3 years using simple interest. If the annual interest rate is 4 %, how much should they invest initially? ■

Exercise 9.2 A document with a nominal value of \$15,000 is due in 1 year. If it is discounted today at an annual simple interest rate of 5 %, what is the present value of the document? ■

These concepts and results enable the application of mathematical reasoning in financial contexts, facilitating informed decision-making and the analysis of investments and loans under the simple interest scheme.

9.1.2 Problem Solving with Arithmetic Progressions.

Arithmetic progressions are numerical sequences in which the difference between consecutive terms is constant. These progressions appear in multiple areas of mathematics and are essential tools for solving various problems.

Definition 9.1.3 An *arithmetic progression* is a sequence of numbers $(a_n)_{n=1}^{\infty}$ such that for all $n \in \mathbb{N}$, the following holds:

$$a_{n+1} = a_n + d,$$

where d is the *common difference* of the progression.

Understanding the structure of arithmetic progressions allows us to analyze patterns and solve problems involving sums and specific terms.

Theorem 9.1.3 The general term a_n of an arithmetic progression is expressed as:

$$a_n = a_1 + (n-1)d,$$

where a_1 is the first term and d is the common difference.

Demostración. We proceed by mathematical induction. For $n = 1$, we have $a_1 = a_1 + (1-1)d = a_1$, which is true. Suppose the formula is valid for some $n = k$, i.e., $a_k = a_1 + (k-1)d$. Then:

$$a_{k+1} = a_k + d = [a_1 + (k-1)d] + d = a_1 + kd = a_1 + [(k+1)-1]d,$$

which shows that the formula is valid for $n = k+1$. ■

This formula allows us to find any term of the progression without listing all the previous terms.

▪ **Example 9.4** Consider an arithmetic progression where the first term is 3 and the common difference is 5. The fifth term is:

$$a_5 = 3 + (5-1) \times 5 = 3 + 20 = 23.$$

▪

In addition to finding individual terms, it is often necessary to calculate the sum of the terms of an arithmetic progression.

Theorem 9.1.4 The sum S_n of the first n terms of an arithmetic progression is:

$$S_n = \frac{n}{2}(a_1 + a_n) = \frac{n}{2}[2a_1 + (n-1)d].$$

Demostración. The sum of the first n terms is:

$$S_n = a_1 + a_2 + a_3 + \cdots + a_n.$$

Rewriting the sum in reverse order:

$$S_n = a_n + a_{n-1} + a_{n-2} + \cdots + a_1.$$

Adding both expressions term by term:

$$2S_n = (a_1 + a_n) + (a_2 + a_{n-1}) + \cdots + (a_n + a_1).$$

Since each pair sums to $a_1 + a_n$, and there are n terms, we have:

$$2S_n = n(a_1 + a_n) \implies S_n = \frac{n}{2}(a_1 + a_n).$$

Substituting $a_n = a_1 + (n-1)d$, we get:

$$S_n = \frac{n}{2}[a_1 + a_1 + (n-1)d] = \frac{n}{2}[2a_1 + (n-1)d].$$

■

■ **Example 9.5** Calculate the sum of the first 20 terms of an arithmetic progression where $a_1 = 7$ and $d = 3$.

$$S_{20} = \frac{20}{2}[2\times 7 + (20-1)\times 3] = 10[14+57] = 10\times 71 = 710.$$

■

Arithmetic progressions are also useful in solving problems involving numerical patterns and finite series.

Lema 9.1.3 In an arithmetic progression, the average of the terms equidistant from the extremes is equal to the average of the first and last terms:

$$\frac{a_k + a_{n-k+1}}{2} = \frac{a_1 + a_n}{2}.$$

Demostración. Consider the terms a_k and a_{n-k+1}. Using the general term formula:

$$a_k = a_1 + (k-1)d, \quad a_{n-k+1} = a_1 + [n-(k-1)-1]d = a_1 + (n-k)d.$$

The sum of these terms is:

$$a_k + a_{n-k+1} = [a_1 + (k-1)d] + [a_1 + (n-k)d] = 2a_1 + (n-1)d = a_1 + a_n.$$

Thus, their average is:

$$\frac{a_k + a_{n-k+1}}{2} = \frac{a_1 + a_n}{2}.$$

■

R This result indicates that in an arithmetic progression, the terms symmetric with respect to the center have the same average as the extremes, which is useful in various calculations and proofs.

Corollary 9.1.5 If an arithmetic progression has an odd number of terms, the middle term is equal to the average of the extremes:

$$a_{\frac{n+1}{2}} = \frac{a_1 + a_n}{2}.$$

Demostración. For odd n, $k = \frac{n+1}{2}$. Applying the previous lemma:

$$\frac{a_k + a_{n-k+1}}{2} = \frac{a_1 + a_n}{2}.$$

But $n - k + 1 = n - \frac{n+1}{2} + 1 = \frac{n+1}{2}$, so $a_k = a_{n-k+1} = a_{\frac{n+1}{2}}$, hence:

$$\frac{a_{\frac{n+1}{2}} + a_{\frac{n+1}{2}}}{2} = \frac{a_1 + a_n}{2} \implies a_{\frac{n+1}{2}} = \frac{a_1 + a_n}{2}.$$

■

This corollary is especially useful when working with arithmetic progressions of odd length.

Example 9.6 In an arithmetic progression of 9 terms where $a_1 = 4$ and $a_9 = 20$, the fifth term is:

$$a_5 = \frac{a_1 + a_n}{2} = \frac{4+20}{2} = 12.$$

■

Arithmetic progressions are also applied to solve everyday and advanced mathematical problems.

Exercise 9.3 Find the number of terms in an arithmetic progression where $a_1 = 5$, $d = 3$, and the last term is 62. ■

Exercise 9.4 The sum of a certain number of consecutive terms of an arithmetic progression is 210. If the first term is 7 and the common difference is 3, how many terms were added? ■

These exercises deepen the understanding and application of arithmetic progressions in various mathematical contexts.

9.2 Geometric Sequences: Common Ratio and Sum of the Series.

9.2.1 Applications in Exponential Growth Problems.

Exponential growth is a phenomenon that appears in various fields such as biology, economics, and physics. It is characterized by a rate of change proportional to the current value of the function, leading to accelerated growth. Mathematical reasoning is essential for modeling and understanding such processes.

Definition 9.2.1 A function $f(t)$ exhibits *exponential growth* if it satisfies the differential equation:

$$\frac{df}{dt} = kf(t),$$

where $k > 0$ is a constant called the *growth rate*.

This definition formalizes the concept that the rate of change of $f(t)$ is proportional to its current value.

Theorem 9.2.1 The general solution to the exponential growth differential equation is:

$$f(t) = f_0 e^{kt},$$

where f_0 is the initial value of the function at $t = 0$.

Demostración. Consider the differential equation $\frac{df}{dt} = kf(t)$. Separating variables:

$$\frac{df}{f} = kdt.$$

Integrating both sides:

$$\int \frac{df}{f} = \int kdt \implies \ln|f| = kt + C,$$

where C is the constant of integration. Exponentiating both sides:

$$f(t) = e^{kt+C} = e^C e^{kt}.$$

Let $f_0 = e^C$, then:

$$f(t) = f_0 e^{kt}.$$

■

This result is fundamental for modeling processes that follow an exponential growth pattern.

■ **Example 9.7** The population of a colony of bacteria doubles every 3 hours. If there are initially 100 bacteria, what will the population be after t hours?
Using the exponential growth formula, the doubling time is $T = 3$ hours, so the growth rate k satisfies:

$$2 = e^{kT} \implies k = \frac{\ln 2}{T} = \frac{\ln 2}{3}.$$

Thus, the population as a function of time is:

$$f(t) = 100e^{(\ln 2/3)t} = 100 \times 2^{t/3}.$$

■

This example shows how to use the exponential model to predict population growth.

Lema 9.2.1 For any time t, the growth rate of the exponential function is constant and equal to the rate k:

$$\frac{f'(t)}{f(t)} = k.$$

Demostración. Given $f(t) = f_0 e^{kt}$, we differentiate $f(t)$ with respect to t:

$$f'(t) = f_0 k e^{kt} = k f(t).$$

Thus:

$$\frac{f'(t)}{f(t)} = \frac{k f(t)}{f(t)} = k.$$

■

This lemma confirms that in exponential growth, the relative rate of change is constant.

Corollary 9.2.2 The doubling time T of a quantity growing exponentially with rate k is:

$$T = \frac{\ln 2}{k}.$$

Demostración. We want to find T such that $f(T) = 2f_0$. Using the formula for $f(t)$:

$$2f_0 = f_0 e^{kT} \implies 2 = e^{kT} \implies \ln 2 = kT \implies T = \frac{\ln 2}{k}.$$

■

This corollary is useful for determining how long it takes for a quantity to double in exponential growth processes.

R Similarly, the tripling time and other multiplications can be calculated by replacing $\ln 2$ with $\ln 3$, $\ln 4$, etc.

Besides biological applications, exponential growth appears in finance, especially in compound interest.

Definition 9.2.2 *Continuous compound interest* is calculated using the formula:

$$A = Pe^{rt},$$

where A is the final amount, P is the initial principal, r is the annual interest rate, and t is the time in years.

This formula is a specific case of exponential growth applied to finance.

■ **Example 9.8** If we invest \$1,000 at an annual interest rate of 5 % compounded continuously, how much will we have after 10 years?
Applying the continuous compound interest formula:

$$A = 1000 \times e^{0,05 \times 10} = 1000 \times e^{0,5} \approx 1000 \times 1,64872 = \$1,648,72.$$

■

This example demonstrates the effect of continuous compound interest on long-term investments.

Theorem 9.2.3 The present value P needed to reach a future amount A after time t under exponential growth with rate k is:

$$P = Ae^{-kt}.$$

Demostración. Solving for P from the exponential growth formula:

$$A = Pe^{kt} \implies P = Ae^{-kt}.$$

■

This theorem is essential for calculating the initial investments required for future goals.

Lema 9.2.2 The *half-life* $T_{1/2}$ of a radioactive substance decaying exponentially with a decay rate k is:

$$T_{1/2} = \frac{\ln 2}{k}.$$

Demostración. Exponential decay is modeled by $N(t) = N_0 e^{-kt}$. The half-life is the time $T_{1/2}$ such that $N(T_{1/2}) = \frac{N_0}{2}$. Thus:

$$\frac{N_0}{2} = N_0 e^{-kT_{1/2}} \implies \frac{1}{2} = e^{-kT_{1/2}} \implies \ln\left(\frac{1}{2}\right) = -kT_{1/2} \implies T_{1/2} = \frac{\ln 2}{k}.$$

■

This result is fundamental in nuclear physics and chemistry for understanding radioactive decay processes.

Exercise 9.5 A population of cells increases exponentially with a growth rate of 8 % per hour. If there are initially 1,000 cells, how many will there be after 15 hours? ■

Exercise 9.6 A radioactive isotope has a half-life of 5 years. What percentage of the original substance will remain after 20 years? ■

These exercises allow the application of exponential growth and decay concepts in practical contexts.

Exponential growth can lead to extremely large quantities in relatively short periods of time, emphasizing the importance of understanding and managing such processes.

Through these results, we can appreciate how mathematical reasoning and analytical tools are essential for modeling and solving problems involving exponential growth in various disciplines.

9.2.2 Resolution of Infinite Series.

Infinite series are fundamental in mathematical analysis and appear in various fields such as physics, engineering, and economics. Understanding how to solve and analyze these series is essential for advanced mathematical reasoning.

Definition 9.2.3 An *infinite series* is an expression of the form

$$\sum_{n=1}^{\infty} a_n,$$

where $\{a_n\}_{n=1}^{\infty}$ is a sequence of real or complex numbers.

The study of infinite series focuses on determining whether this sum makes sense, that is, whether it converges to a finite value or diverges.

Definition 9.2.4 An infinite series $\sum_{n=1}^{\infty} a_n$ *converges* if the sequence of partial sums $S_N = \sum_{n=1}^{N} a_n$ has a finite limit as $N \to \infty$. Otherwise, the series *diverges*.

To analyze the convergence of infinite series, there are various criteria and tests that allow us to determine the behavior of a series without explicitly calculating the partial sums.

Theorem 9.2.4 — **Comparison Test.** Let $\sum_{n=1}^{\infty} a_n$ be a series with positive terms, and let $\sum_{n=1}^{\infty} b_n$ be a known convergent series. If there exists N such that for all $n \geq N$, $0 \leq a_n \leq b_n$, then the series $\sum_{n=1}^{\infty} a_n$ also converges.

To prove the Comparison Test, let $\sum_{n=1}^{\infty} a_n$ be a series with positive terms, and suppose that $\sum_{n=1}^{\infty} b_n$ is a convergent series, also with positive terms. Moreover, there exists an integer N such that for all $n \geq N$, $0 \leq a_n \leq b_n$.
Since $\sum_{n=1}^{\infty} b_n$ converges, its sequence of partial sums $\{S_m\}_{m=1}^{\infty}$, where $S_m = \sum_{n=1}^{m} b_n$, is bounded. That is, there exists a constant $M > 0$ such that for all m, $S_m \leq M$.
Now, consider the partial sums of the series $\sum_{n=1}^{\infty} a_n$, denoted by $T_m = \sum_{n=1}^{m} a_n$. For $m \geq N$, we have:

$$T_m = \sum_{n=1}^{m} a_n = \sum_{n=1}^{N-1} a_n + \sum_{n=N}^{m} a_n.$$

Since $0 \leq a_n \leq b_n$ for $n \geq N$, it follows that:

$$\sum_{n=N}^{m} a_n \leq \sum_{n=N}^{m} b_n \leq M - \sum_{n=1}^{N-1} b_n.$$

Thus, T_m is bounded for $m \geq N$, which implies that the sequence $\{T_m\}$ is bounded. By the convergence test for series with positive terms, this implies that $\sum_{n=1}^{\infty} a_n$ converges.
Therefore, we have proved the Comparison Test.
This theorem is useful when we can compare the given series with another series whose behavior is already known.

■ **Example 9.9** Consider the series

$$\sum_{n=1}^{\infty} \frac{1}{n^2}.$$

We know that the series $\sum_{n=1}^{\infty} \frac{1}{n^p}$ converges if $p > 1$. Since $p = 2 > 1$, the series converges. ■

In addition to the Comparison Test, another powerful method is the Ratio Test.

Theorem 9.2.5 — D'Alembert's Ratio Test. Let $\sum_{n=1}^{\infty} a_n$ be a series with positive terms. If there exists

$$L = \lim_{n\to\infty} \frac{a_{n+1}}{a_n},$$

then:

- If $L < 1$, the series converges.
- If $L > 1$, the series diverges.
- If $L = 1$, the test is inconclusive.

Let $\sum_{n=1}^{\infty} a_n$ be a series with positive terms, and suppose the limit

$$L = \lim_{n\to\infty} \frac{a_{n+1}}{a_n}$$

exists. Consider the following cases:
1. **If $L < 1$:**
In this case, there exists a number r such that $L < r < 1$. Then, there exists an integer N such that for all $n \geq N$,

$$\frac{a_{n+1}}{a_n} < r.$$

This implies $a_{n+1} < ra_n$ for $n \geq N$, so the terms of the series decrease geometrically from a certain index onward. Since a geometric series with ratio $r < 1$ converges, the series $\sum_{n=1}^{\infty} a_n$ also converges.
2. **If $L > 1$:**
In this case, for some r such that $r < L$, there exists an integer N such that for all $n \geq N$,

$$\frac{a_{n+1}}{a_n} > r > 1.$$

This means the terms a_n grow without bound from a certain point onward, implying the series $\sum_{n=1}^{\infty} a_n$ diverges, since its terms do not approach zero.
3. **If $L = 1$:**
In this case, the Ratio Test does not provide conclusive information about the convergence or divergence of the series.
This concludes the proof of D'Alembert's Ratio Test.
This test is particularly useful for series involving factorials or powers.

■ **Example 9.10** Analyze the convergence of the series

$$\sum_{n=1}^{\infty} \frac{n!}{n^n}.$$

We calculate

$$L = \lim_{n\to\infty} \frac{a_{n+1}}{a_n} = \lim_{n\to\infty} \frac{(n+1)!}{(n+1)^{n+1}} \cdot \frac{n^n}{n!} = \lim_{n\to\infty} \left(\frac{n}{n+1}\right)^n = \frac{1}{e} < 1.$$

Therefore, the series converges. ■

Another essential criterion is the Integral Test.

Theorem 9.2.6 — Cauchy's Integral Test. Let $f : [1,\infty) \to \mathbb{R}$ be a continuous, positive, and decreasing function. Then, the series $\sum_{n=1}^{\infty} f(n)$ and the integral $\int_1^{\infty} f(x)dx$ either both converge or both diverge.

To prove Cauchy's Integral Test, let $f : [1,\infty) \to \mathbb{R}$ be a continuous, positive, and decreasing function. Consider the series $\sum_{n=1}^{\infty} f(n)$ and the improper integral $\int_1^{\infty} f(x)\,dx$.
We express the improper integral as a sum of partial integrals:

$$\int_1^{\infty} f(x)\,dx = \lim_{b\to\infty} \int_1^b f(x)\,dx.$$

Since f is decreasing, for each integer $n \geq 1$, we have:

$$\int_n^{n+1} f(x)\,dx \leq f(n) \quad \text{and} \quad f(n+1) \leq \int_n^{n+1} f(x)\,dx.$$

Summing these inequalities for $n = 1,2,\ldots,N$, we get:

$$\int_1^{N+1} f(x)\,dx \leq \sum_{n=1}^{N} f(n) \leq f(1) + \int_1^N f(x)\,dx.$$

Taking the limit as $N \to \infty$, we obtain:

$$\int_1^{\infty} f(x)\,dx \leq \sum_{n=1}^{\infty} f(n) \leq f(1) + \int_1^{\infty} f(x)\,dx.$$

Thus, if the integral $\int_1^{\infty} f(x)\,dx$ converges, then the series $\sum_{n=1}^{\infty} f(n)$ also converges. Similarly, if the integral diverges, then the series also diverges.
This concludes the proof of Cauchy's Integral Test.
This criterion is useful when we can integrate the function associated with the terms of the series.

■ **Example 9.11** Consider the harmonic series

$$\sum_{n=1}^{\infty} \frac{1}{n}.$$

The function $f(x) = \frac{1}{x}$ is continuous, positive, and decreasing on $[1,\infty)$. We calculate the integral:

$$\int_1^{\infty} \frac{1}{x} dx = \lim_{b\to\infty} \ln b = \infty.$$

Since the integral diverges, the harmonic series also diverges. ■

In some cases, it is useful to consider alternating series.

Theorem 9.2.7 — Leibniz's Test. Let $\{a_n\}$ be a sequence of positive real numbers that decreases monotonically to zero. Then, the alternating series

$$\sum_{n=1}^{\infty}(-1)^{n+1}a_n$$

converges.

To prove Leibniz's Test, let $\{a_n\}$ be a sequence of positive real numbers that decreases monotonically to zero, i.e., $a_{n+1} \leq a_n$ for all n and $\lim_{n\to\infty} a_n = 0$.
Consider the alternating series

$$\sum_{n=1}^{\infty}(-1)^{n+1}a_n = a_1 - a_2 + a_3 - a_4 + \dots$$

and its partial sums $S_N = \sum_{n=1}^{N}(-1)^{n+1}a_n$.
We observe that the partial sums S_N form an alternating sequence. Moreover, since $\{a_n\}$ is decreasing and tends to zero, the sequence of partial sums $\{S_N\}$ is bounded and converges monotonically. This implies that the sequence $\{S_N\}$ has a finite limit.
Therefore, the series $\sum_{n=1}^{\infty}(-1)^{n+1}a_n$ converges.
This concludes the proof of Leibniz's Test.
This criterion is particularly useful for series where the terms alternate in sign.

■ **Example 9.12** The alternating series

$$\sum_{n=1}^{\infty}\frac{(-1)^{n+1}}{n}$$

converges because $\frac{1}{n}$ is decreasing and tends to zero. ■

(R) Although the harmonic series diverges, its alternating version converges, highlighting the importance of the sign of the terms in the convergence of a series.

For more complex series, we can use Raabe's Test or Cauchy's Test.

Theorem 9.2.8 — Raabe's Test. Let $\sum_{n=1}^{\infty} a_n$ be a series with positive terms. If the limit

$$\lim_{n\to\infty} n\left(\frac{a_n}{a_{n+1}} - 1\right) = R$$

exists, then:

- If $R > 1$, the series converges.
- If $R < 1$, the series diverges.
- If $R = 1$, the test is inconclusive.

To prove Raabe's Test, consider the series $\sum_{n=1}^{\infty} a_n$ with positive terms and suppose the limit

$$\lim_{n\to\infty} n\left(\frac{a_n}{a_{n+1}} - 1\right) = R$$

exists. We analyze the following three cases based on the value of R:

1. **If $R > 1$:**
In this case, Raabe's Test implies that the series $\sum_{n=1}^{\infty} a_n$ converges. This is because, when $R > 1$, the term a_n decreases "sufficiently fast"such that the series converges, similar to the behavior of a p-series with $p > 1$.
2. **If $R < 1$:**
In this case, the series $\sum_{n=1}^{\infty} a_n$ diverges. This occurs because, when $R < 1$, the term a_n does not decrease fast enough, resembling a p-series with $p \leq 1$, which diverges.
3. **If $R = 1$:**
In this case, the test is inconclusive. The convergence or divergence of the series cannot be determined solely from this criterion.
This concludes the proof of Raabe's Test.

■ **Example 9.13** Consider the series

$$\sum_{n=1}^{\infty} \frac{1}{n^p},$$

with $p > 0$. We compute

$$n\left(\frac{a_n}{a_{n+1}} - 1\right) = n\left(\frac{n^p}{(n+1)^p} - 1\right).$$

For large n,

$$\frac{n^p}{(n+1)^p} \approx \left(1 - \frac{1}{n}\right)^p \approx 1 - \frac{p}{n}.$$

Thus,

$$n\left(\left(1 - \frac{p}{n}\right) - 1\right) = -p.$$

Therefore, the limit is $R = p$. If $p > 1$, $R > 1$ and the series converges; if $p \leq 1$, $R \leq 1$ and the series diverges. ■

Exercise 9.7 Determine the convergence of the series

$$\sum_{n=2}^{\infty} \frac{1}{n(\ln n)^2}.$$

■

Exercise 9.8 Study the convergence of the series

$$\sum_{n=1}^{\infty} \frac{(-1)^{n+1}}{\sqrt{n}}.$$

■

These exercises allow the application of the studied criteria and deepen the understanding of infinite series.

Ⓡ Solving infinite series requires not only knowing the convergence tests but also selecting the most appropriate one for each particular series.

Through these results and examples, the importance of mathematical reasoning in analyzing and solving infinite series is demonstrated, an essential tool in various fields of science and engineering.

9.3 Patterns and Generalization: Identification and Analysis of Patterns.

9.3.1 Identifying Patterns in Geometric Figures.

Identifying patterns in geometric figures is essential for developing advanced mathematical reasoning skills. Analyzing these patterns allows us to predict behaviors, formulate conjectures, and establish general properties.

Definition 9.3.1 A **geometric pattern** is an ordered sequence of geometric figures where each term follows a specific rule or relationship with respect to the previous one.

Understanding these patterns involves recognizing numerical and spatial relationships between the figures.

■ **Example 9.14** Consider a sequence of figures formed by equilateral triangles where, at each step, triangles are added to form a larger figure:
1. Step 1: One equilateral triangle. 2. Step 2: A larger triangle composed of 4 smaller equilateral triangles. 3. Step 3: An even larger triangle composed of 9 smaller equilateral triangles.
We observe that the number of small triangles at each step forms the perfect squares 1^2, 2^2, 3^2, etc. ■

This example leads us to establish mathematical properties about the relationship between the number of steps and the number of triangles.

Theorem 9.3.1 In the described construction, the total number of small equilateral triangles T at step n is $T = n^2$.

Demostración. We observe that at each step n, the figure consists of a triangular matrix with side length n, where each level adds n smaller triangles. Therefore, the total is the sum of the first n natural numbers:

$$T = \sum_{k=1}^{n} k = \frac{n(n+1)}{2}.$$

However, since the structure forms a square of $n \times n$ triangles when rearranged, we conclude that $T = n^2$. ■

This result simplifies the calculation of the number of triangles at any step of the construction.

Corollary 9.3.2 At step n, the perimeter P of the figure is proportional to n, given by $P = 3n$ units of length.

Demostración. Each side of the large triangle has a length of n times the side length of the small triangle. Since an equilateral triangle has three sides, the perimeter is $P = 3n$. ■

It is important to recognize how numerical patterns are reflected in geometric properties.

Lema 9.3.1 In a sequence of regular polygons inscribed in circles of constant radius, the perimeter of the polygon P_n tends to the circumference perimeter as the number of sides n approaches infinity.

Demostración. The perimeter of a regular polygon inscribed in a circle of radius r is:

$$P_n = 2nr\sin\left(\frac{\pi}{n}\right).$$

As $n \to \infty$, $\sin\left(\frac{\pi}{n}\right) \approx \frac{\pi}{n}$, so:

$$P_n \approx 2nr\left(\frac{\pi}{n}\right) = 2r\pi = C,$$

where C is the circumference perimeter. ■

This result shows how polygonal shapes can approximate curved shapes by increasing the number of sides, a fundamental concept in calculus and geometry.

Let's see an example illustrating this concept.

Example 9.15 Calculate the perimeter of a regular hexagon inscribed in a circle with radius $r = 1$ unit. ■

Using the formula:

$$P_6 = 2 \times 6 \times 1 \times \sin\left(\frac{\pi}{6}\right) = 12 \times \frac{1}{2} = 6 \text{ units.}$$

The perimeter of the hexagon is 6 units.
To consolidate these concepts, we present the following exercises.

Exercise 9.9 Consider a sequence of figures where each figure is a circle surrounded by six tangent circles of the same radius, forming a hexagonal pattern. If each circle has radius r, find the relationship between the radius of the central circle and the total perimeter of the figure formed. ■

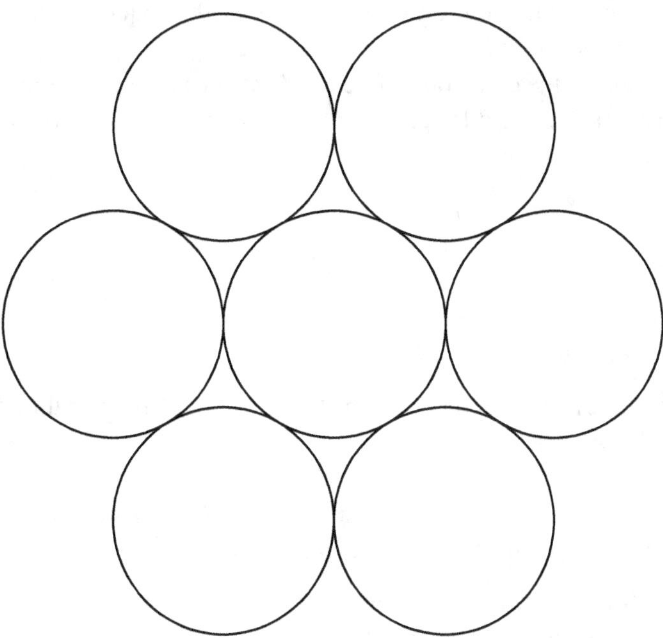

Figura 9.3.1: *Pattern of tangent circles forming a hexagonal structure.*

Exercise 9.10 A square spiral is constructed by adding consecutive squares around an initial square with side a. Each new square has a side length that is a constant multiple of the previous one. If the constant is $k > 0$, find the expression for the total perimeter after n squares have been added. ■

These exercises allow the application of geometric pattern analysis to solve complex problems and develop deeper mathematical thinking.

9.3.2 Generalization of Numerical Patterns.

The generalization of numerical patterns is a fundamental skill in mathematics that allows identifying regularities and formulating conjectures that can be rigorously proven. Through mathematical reasoning, it is possible to extend patterns observed in particular cases to general formulations that apply in broader contexts.

Definition 9.3.2 A *numerical pattern* is a sequence or arrangement of numbers that follows a specific rule or relationship. The *generalization* of a numerical pattern involves finding a mathematical expression that describes that rule for any term in the sequence.

To illustrate how to generalize numerical patterns, let us consider sequences defined by recursive relations or explicit formulas.

■ **Example 9.16** Consider the sequence of odd natural numbers: $1,3,5,7,9,\ldots$. The pattern is that each number is two units greater than the previous one. The generalization of this pattern is $a_n = 2n-1$, where n is a natural number. ■

The ability to generalize numerical patterns leads to the study of sequences and series and the development of general formulas that describe their behavior.

Theorem 9.3.3 Let $\{a_n\}$ be a sequence defined by $a_n = kn+b$, where k and b are real constants. Then, $\{a_n\}$ is an *arithmetic progression* with common difference $d = k$.

To prove that $\{a_n\}$ is an arithmetic progression, consider the sequence defined by $a_n = kn+b$, where k and b are real constants.
In an arithmetic progression, the common difference d between consecutive terms must be constant. Let's verify this by calculating the difference between consecutive terms of $\{a_n\}$:

$$a_{n+1} - a_n = (k(n+1)+b) - (kn+b).$$

Simplifying, we get:

$$a_{n+1} - a_n = kn+k+b-kn-b = k.$$

This difference is constant and equal to k, which shows that $\{a_n\}$ is an arithmetic progression with common difference $d = k$.
This concludes the proof.
This theorem shows how a general formula can describe a linear numerical pattern, which is a powerful tool for analyzing and predicting behaviors in sequences.

Lema 9.3.2 Let $\{a_n\}$ be a sequence defined by $a_n = r^{n-1}a_1$, where $r \neq 0$ is a real constant. Then, $\{a_n\}$ is a *geometric progression* with common ratio r.

Demostración. In a geometric progression, the ratio between consecutive terms is constant. We compute $\frac{a_{n+1}}{a_n} = \frac{r^n a_1}{r^{n-1} a_1} = r$, which is constant for all n. Therefore, $\{a_n\}$ is a geometric progression with common ratio r. ■

The generalization of numerical patterns also involves more complex sequences, such as those defined by recursive relations.

■ **Example 9.17** The Fibonacci sequence is defined by the recurrence relation $F_n = F_{n-1} + F_{n-2}$, with $F_1 = 1$ and $F_2 = 1$. The generalization of this numerical pattern is the explicit formula known as *Binet's Formula*:

$$F_n = \frac{\varphi^n - \psi^n}{\varphi - \psi},$$

where $\varphi = \frac{1+\sqrt{5}}{2}$ and $\psi = \frac{1-\sqrt{5}}{2}$. ■

This generalization allows calculating any term in the sequence without referring to previous terms.

Theorem 9.3.4 Let $\{a_n\}$ be a sequence defined by a linear recurrence relation of order k:

$$a_n = c_1 a_{n-1} + c_2 a_{n-2} + \cdots + c_k a_{n-k},$$

where c_i are constants and $n > k$. Then, the general solution of the sequence is a linear combination of exponential functions of the form $a_n = \sum_{i=1}^{k} \lambda_i r_i^n$, where r_i are the roots of the associated characteristic equation.

To prove this theorem, consider the sequence $\{a_n\}$ defined by a linear recurrence relation of order k:

$$a_n = c_1 a_{n-1} + c_2 a_{n-2} + \cdots + c_k a_{n-k},$$

where c_i are constants and $n > k$.
To solve this recurrence, we propose a solution of the form $a_n = r^n$, where r is a constant to be determined. Substituting into the recurrence relation, we get

$$r^n = c_1 r^{n-1} + c_2 r^{n-2} + \cdots + c_k r^{n-k}.$$

Dividing both sides by r^{n-k} (assuming $r \neq 0$), we obtain the **characteristic equation**:

$$r^k = c_1 r^{k-1} + c_2 r^{k-2} + \cdots + c_k.$$

This is a polynomial equation of degree k in r. Let $r_1, r_2, \ldots, r_k$ be the roots of this characteristic equation, which may be real or complex and may have multiplicity.
The general solution of the recurrence relation depends on these roots. If the roots are distinct, the general solution is a linear combination of powers of the roots:

$$a_n = \lambda_1 r_1^n + \lambda_2 r_2^n + \cdots + \lambda_k r_k^n,$$

where $\lambda_1, \lambda_2, \ldots, \lambda_k$ are constants determined by the initial conditions of the sequence.
If a root has multiplicity $m > 1$, then the corresponding terms in the general solution are multiplied by increasing powers of n up to n^{m-1}.
Thus, the general solution of the sequence is a linear combination of terms of the form $a_n = \sum_{i=1}^{k} \lambda_i r_i^n$, where r_i are the roots of the associated characteristic equation. This concludes the proof.

This theorem generalizes the numerical pattern defined by a linear recurrence relation, allowing explicit formulas for sequences.

(R) The identification and generalization of numerical patterns can extend to multidimensional patterns and more complex structures, such as power series and generating functions.

9.4 Solved Exercises

Exercise 9.11 Find the general term of an arithmetic sequence where the first term is $a_1 = 4$ and the common difference is $d = 3$. Calculate the tenth term of the sequence.

Demostración. The general term of an arithmetic sequence is defined as:

$$a_n = a_1 + (n-1)d.$$

Substituting the given values:

$$a_{10} = 4 + (10-1)\cdot 3 = 4 + 9\cdot 3 = 4 + 27 = 31.$$

Therefore, the tenth term is $a_{10} = 31$. ■

Exercise 9.12 Calculate the sum of the first 15 terms of an arithmetic progression where the first term is $a_1 = 7$ and the common difference is $d = 5$.

Demostración. The formula for the sum of the first n terms of an arithmetic progression is:

$$S_n = \frac{n}{2}(2a_1 + (n-1)d).$$

Substituting the given values:

$$S_{15} = \frac{15}{2}(2\cdot 7 + (15-1)\cdot 5) = \frac{15}{2}(14+70) = \frac{15}{2}\cdot 84 = 15\cdot 42 = 630.$$

Therefore, the sum of the first 15 terms is $S_{15} = 630$. ■

Exercise 9.13 Determine the value of the infinite sum of a geometric series whose first term is $a = 8$ and the common ratio is $r = \frac{1}{2}$.

Demostración. The formula for the sum of an infinite geometric series is:

$$S = \frac{a}{1-r},$$

where $|r| < 1$. Substituting the given values:

$$S = \frac{8}{1-\frac{1}{2}} = \frac{8}{\frac{1}{2}} = 8\cdot 2 = 16.$$

Therefore, the infinite sum of the series is $S = 16$. ■

Exercise 9.14 In a geometric sequence, the second term is 12 and the fourth term is 48. Find the first term and the common ratio of the sequence.

Demostración. Let the first term be a and the common ratio be r. We know that:

$$a_2 = a \cdot r = 12 \quad \text{and} \quad a_4 = a \cdot r^3 = 48.$$

Dividing the second equation by the first:

$$\frac{a \cdot r^3}{a \cdot r} = \frac{48}{12} \Rightarrow r^2 = 4 \Rightarrow r = 2.$$

Substituting $r = 2$ into the first equation:

$$a \cdot 2 = 12 \Rightarrow a = 6.$$

Therefore, the first term is $a = 6$ and the common ratio is $r = 2$. ■

Exercise 9.15 In a sequence of odd natural numbers $1, 3, 5, 7, 9, \ldots$, find the general term of the sequence and calculate the twentieth term.

Demostración. We observe that each term in the sequence is an odd number and increases by 2 relative to the previous term. The first term is 1, and the common difference is $d = 2$. The formula for the general term is:

$$a_n = a_1 + (n-1)d.$$

Substituting $a_1 = 1$ and $d = 2$:

$$a_n = 1 + (n-1) \cdot 2 = 1 + 2n - 2 = 2n - 1.$$

For the twentieth term:

$$a_{20} = 2 \cdot 20 - 1 = 40 - 1 = 39.$$

Therefore, the twentieth term is $a_{20} = 39$. ■

9.5 Proposed Exercises

9.5.1 Arithmetic Sequences: General Term Formula and Sum of Terms

Exercise 9.16 Find the general term of an arithmetic sequence where the first term is 3 and the common difference is 5.

Exercise 9.17 In an arithmetic sequence, the fifth term is 18 and the tenth term is 33. Find the first term and the common difference.

Exercise 9.18 Calculate the sum of the first 15 terms of an arithmetic progression with the first term 7 and a common difference of 4.

Exercise 9.19 If in an arithmetic sequence the sum of the first n terms is 150 and the common difference is 2, find the value of n if the first term is 3. ■

Exercise 9.20 An arithmetic progression has 20 terms, the first term is 6, and the last term is 56. Calculate the sum of all the terms. ■

9.5.2 Geometric Sequences: Common Ratio and Sum of the Series

Exercise 9.21 Determine the general term of a geometric sequence where the first term is 5 and the common ratio is 3. ■

Exercise 9.22 In a geometric sequence, the second term is 12 and the fourth term is 48. Find the first term and the common ratio. ■

Exercise 9.23 Calculate the sum of the first 8 terms of a geometric sequence with the first term 2 and a common ratio of 3. ■

Exercise 9.24 Find the value of the infinite sum of a geometric series whose first term is 4 and the common ratio is $\frac{1}{2}$. ■

Exercise 9.25 If in a geometric sequence the third term is 16 and the sixth term is 128, find the common ratio and the first term. ■

9.5.3 Patterns and Generalization: Identification and Analysis of Patterns

Exercise 9.26 Identify the pattern and find the general term of the sequence: $2, 5, 10, 17, 26, \ldots$ ■

Exercise 9.27 Observe the sequence of figures formed by equilateral triangles where the first step has 1 triangle, the second step has 4 triangles, the third step has 9 triangles, and so on. Find the number of triangles in the tenth step. ■

Exercise 9.28 Generalize the pattern of the sequence $3, 7, 15, 31, 63, \ldots$ and find the general term. ■

Exercise 9.29 In a spiral of squares where each square has a side length that is double the previous one, find the area of the tenth square if the first square has a side length of 1 unit. ■

Exercise 9.30 Determine a general formula for the sum of the first n terms of an arithmetic progression whose first term is a and whose common difference is d. ■

III Algebraic Reasoning

10. Mathematical Modeling

10.1 Linear Equations: Problem Formulation and Modeling

10.1.1 Problems of Motion and Speed.

Problems of motion and speed are fundamental in the study of linear equations, as they allow us to model situations where objects move at constant speeds. Through algebraic reasoning, we can establish relationships between distance, speed, and time, facilitating the solution of both practical and theoretical problems.

Definition 10.1.1 *Constant speed* is the ratio between the distance traveled and the time taken when that ratio remains unchanged. Mathematically, it is expressed as:

$$v = \frac{d}{t},$$

where v is the speed, d is the distance, and t is the time.

This basic relationship allows us to set up linear equations to describe the motion of objects traveling at constant speed.

■ **Example 10.1** A car travels a distance of 180 km in 3 hours at constant speed. What is its speed? We apply the formula:

$$v = \frac{d}{t} = \frac{180 \text{ km}}{3 \text{ h}} = 60 \text{ km/h}.$$

■

When two objects move simultaneously, we can model their motions with linear equations and analyze their interactions.

Theorem 10.1.1 If two objects move in opposite directions with constant speeds v_1 and v_2, the total distance D between them after a time t is:

$$D = (v_1 + v_2)t.$$

To prove this statement, consider two objects moving in opposite directions with constant speeds v_1 and v_2, starting from an initial position where they are zero distance apart.
After a time t:
- The first object travels a distance $d_1 = v_1 t$. - The second object travels a distance $d_2 = v_2 t$.
Since they move in opposite directions, the total distance D between them is the sum of the distances traveled by each:

$$D = d_1 + d_2 = v_1 t + v_2 t = (v_1 + v_2)t.$$

This concludes the proof.
This theorem is useful for calculating distances or times when two objects are moving away from or towards each other.

■ **Example 10.2** Two trains leave a station at the same time in opposite directions. One travels at 80 km/h, and the other at 100 km/h. What distance will be between them after 2 hours?

$$D = (80\text{ km/h} + 100\text{ km/h}) \times 2\text{ h} = 180\text{ km/h} \times 2\text{ h} = 360\text{ km}.$$

■

When objects move in the same direction, the distance between them changes based on the difference in their speeds.

Corollary 10.1.2 If an object A chases an object B that has an initial lead of d kilometers and both move in the same direction with constant speeds $v_A > v_B$, the time t it takes for A to catch B is:

$$t = \frac{d}{v_A - v_B}.$$

To prove this corollary, consider that object B has an initial lead of d kilometers over object A, and both move in the same direction with constant speeds v_A and v_B, where $v_A > v_B$.
Let t be the time it takes for object A to catch up to object B. During this time, both objects will have traveled the same distance from the point where A started its pursuit to the meeting point. To catch B, the distance traveled by A must overcome the initial lead d.
The distance traveled by A in time t is $v_A t$, and the distance traveled by B in the same time is $v_B t$.
When A catches B, the additional distance that A travels relative to B is precisely d. This can be expressed as:

$$v_A t - v_B t = d.$$

Factoring t, we get:

$$t(v_A - v_B) = d.$$

Solving for t, we obtain:

$$t = \frac{d}{v_A - v_B}.$$

This concludes the proof.

■ **Example 10.3** A runner who can run at 12 km/h tries to catch another runner who runs at 10 km/h and who started running 15 minutes earlier. How long will it take to catch up?
First, convert 15 minutes to hours: $t_0 = \frac{15}{60} = 0{,}25$ h.
The lead in distance is:

$$d = v_B t_0 = 10 \text{ km/h} \times 0{,}25 \text{ h} = 2{,}5 \text{ km}.$$

The time to catch up is:

$$t = \frac{d}{v_A - v_B} = \frac{2{,}5 \text{ km}}{12 \text{ km/h} - 10 \text{ km/h}} = \frac{2{,}5}{2} = 1{,}25 \text{ h}.$$

■

(R) When solving motion and speed problems, it is crucial to maintain consistency in units of measurement and clearly define the variables and equations involved.

Understanding these concepts allows modeling more complex situations and developing skills in formulating and solving linear equations.

Exercise 10.1 A cyclist leaves point A toward point B at a constant speed of 20 km/h. Two hours later, another cyclist leaves A in the same direction at 30 km/h. At what distance from A will the second cyclist catch the first? ■

Exercise 10.2 Two ships leave the same port at the same time. One sails north at 15 km/h, and the other sails east at 20 km/h. How far apart will they be after 3 hours? ■

These exercises provide additional practice in applying linear equations to motion and speed problems, strengthening mathematical reasoning and the ability to model real-world situations.

10.1.2 Problems Involving Costs and Prices.

Problems related to costs and prices are fundamental in applied mathematics, especially in economics and finance. By setting up linear equations, we can model situations involving production costs, selling prices, profits, and losses, thereby enabling deeper understanding and informed decision-making.

Definition 10.1.2 The *total cost* C of production is the sum of fixed costs C_f and variable costs C_v. Mathematically, it is expressed as:

$$C = C_f + C_v.$$

Fixed costs are those that do not depend on the level of production, such as rent for premises or administrative salaries. Variable costs depend directly on the quantity produced, such as raw materials or direct labor.

Definition 10.1.3 The *total revenue* I is the product of the selling price p and the number of units sold q:

$$I = p \times q.$$

Analyzing total revenue is crucial for determining the profitability of a product or service.

Theorem 10.1.3 The *break-even point* is reached when the total revenue equals the total cost. That is:

$$I = C \implies p \times q = C_f + C_v.$$

To prove this statement, consider the concept of the **break-even point**, which occurs when the total revenue I equals the total cost C.
Let:
- p be the selling price per unit, - q be the number of units sold, - C_f be the total fixed cost (costs that do not depend on the quantity produced), - C_v be the total variable cost (costs that depend on the quantity produced).
The total revenue I is given by:

$$I = p \times q.$$

The total cost C is the sum of fixed and variable costs:

$$C = C_f + C_v.$$

At the break-even point, we require $I = C$. This implies:

$$p \times q = C_f + C_v.$$

Thus, the break-even point is reached when the total revenue equals the total cost, confirming the statement.
This concludes the proof.
This theorem is fundamental for businesses, as knowing the break-even point helps set sales targets and pricing strategies.

■ **Example 10.4** A company has fixed costs of $10,000 and variable costs of $50 per unit produced. If the selling price is $100 per unit, how many units must be sold to reach the break-even point?
We apply the break-even theorem:

$$p \times q = C_f + C_v \implies 100q = 10,000 + 50q.$$

Solving for q:

$$100q - 50q = 10,000 \implies 50q = 10,000 \implies q = 200.$$

Therefore, the company must sell 200 units to reach the break-even point. ■

Lema 10.1.1 The *profit* B is the difference between the total revenue and the total cost:

$$B = I - C = pq - (C_f + C_v).$$

Demostración. Profit represents the net earnings after covering all costs associated with producing and selling goods or services. ■

Corollary 10.1.4 To maximize profit, it is necessary to maximize the difference $pq-(C_f+C_v)$. If fixed and variable costs are constant, the maximum profit is achieved by increasing q and/or p, considering market and demand constraints.

(R) It is important to consider the price elasticity of demand when adjusting prices, as an increase in price could reduce the quantity sold, negatively affecting total revenue and, consequently, profit.

■ **Example 10.5** Suppose a company can sell up to 500 units of its product at the current price of \$80. Fixed costs are \$8,000, and variable costs are \$40 per unit. What is the profit if all possible units are sold?
Calculate the total revenue:

$$I = p\times q = 80\times 500 = \$40,000.$$

Calculate the total cost:

$$C = C_f + C_v = 8,000 + (40\times 500) = 8,000 + 20,000 = \$28,000.$$

The profit is:

$$B = I - C = 40,000 - 28,000 = \$12,000.$$

■

This example shows how to apply the formulas to determine profit based on costs and selling price.

Theorem 10.1.5 If a target profit B_0 is desired, the minimum number of units q that must be sold is:

$$q = \frac{C_f + B_0}{p - c_v},$$

where c_v is the variable cost per unit.

To prove this statement, consider that we want to achieve a target profit B_0. This means that the total revenue must exceed the total cost by an amount equal to B_0.
Let:
- p be the selling price per unit, - q be the number of units sold, - C_f be the total fixed cost, - c_v be the variable cost per unit.
The total revenue I from selling q units is:

$$I = p\times q.$$

The total cost C is the sum of fixed and variable costs:

$$C = C_f + c_v\times q.$$

To achieve a target profit B_0, the total revenue minus the total cost must equal B_0:

$$I - C = B_0.$$

Substituting the expressions for I and C, we get:

$$p \times q - (C_f + c_v \times q) = B_0.$$

Rearranging terms, we have:

$$(p - c_v)q = C_f + B_0.$$

Solving for q, we get:

$$q = \frac{C_f + B_0}{p - c_v}.$$

This concludes the proof.

> **Exercise 10.3** A company wants to achieve a profit of \$15,000. Its fixed costs are \$5,000, the variable cost per unit is \$25, and the selling price is \$75 per unit. How many units must be sold to achieve this profit? ■

> **Exercise 10.4** A store offers a 10% discount on a product originally priced at \$200. If the variable cost per unit is \$120 and fixed costs are \$4,000, how many units must be sold to reach the break-even point? ■

These exercises provide practice in applying linear equations in cost and price contexts, reinforcing mathematical reasoning and the ability to model and solve real-world problems.

> (R) Understanding how costs and prices interact is essential not only in mathematics but also in economics and business. Using linear equations to model these relationships provides a powerful tool for analysis and strategic decision-making.

Through these concepts and examples, the importance of algebraic reasoning in modeling practical situations becomes evident, enabling resource optimization and profit maximization in various business contexts.

10.2 Graphical and Algebraic Solutions: Representation in the Cartesian Plane.

10.2.1 Graphs of First-Degree Equations.

First-degree equations, also known as linear equations, are fundamental in mathematics and have graphical representations that are straight lines in the Cartesian plane. Understanding how to graph these equations and analyze their properties is essential for mathematical reasoning and applications in various fields.

> **Definition 10.2.1** A *linear equation* in two variables x and y is an equation of the form:
>
> $$ax + by + c = 0,$$
>
> where a, b, and c are real constants, and at least one of a or b is different from zero.

This general form can be manipulated to obtain different representations that facilitate graphing and analysis.

> **Theorem 10.2.1** The graph of a linear equation in two variables is a straight line in the Cartesian plane.

Demostración. Consider the equation $ax+by+c=0$. For any value of x, we can solve for y (if $b \neq 0$) and obtain an ordered pair (x,y) that satisfies the equation. The collection of all such ordered pairs forms a straight line because the relationship between x and y is linear. Similarly, if $a \neq 0$, we can solve for x in terms of y. ■

To graph a linear equation, it is common to use the slope-intercept form.

Definition 10.2.2 The *slope-intercept form* of a linear equation is:

$$y = mx + b,$$

where m is the *slope* of the line and b is the *y-intercept*.

This form is especially useful for quickly graphing a line, as b indicates where the line crosses the y-axis, and m indicates the inclination of the line.

■ **Example 10.6** Let's graph the linear equation $y = 2x+1$.
The slope is $m=2$, and the intercept is $b=1$. This means the line crosses the y-axis at the point $(0,1)$ and has a slope that indicates for each unit x increases, y increases by 2 units.

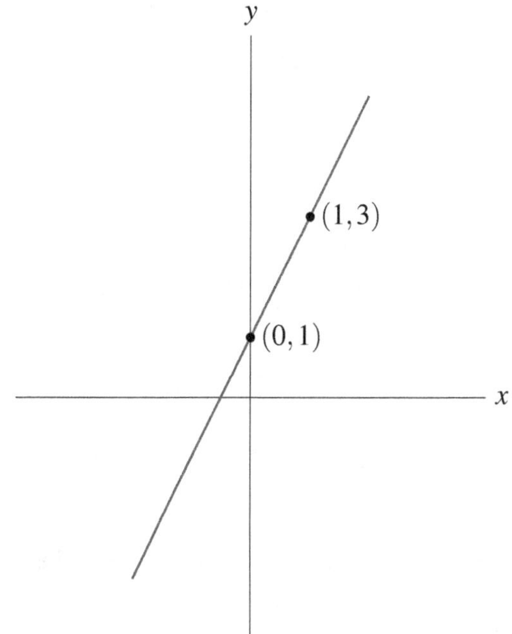

Figura 10.2.1: *Graph of the equation* $y = 2x+1$.

■

Definition 10.2.3 The *slope* m of a line passing through two points (x_1,y_1) and (x_2,y_2) is:

$$m = \frac{y_2 - y_1}{x_2 - x_1},$$

if $x_2 \neq x_1$.

The slope indicates the inclination and direction of the line: if $m > 0$, the line is ascending; if $m < 0$, it is descending; if $m = 0$, it is horizontal; and if m is undefined (when $x_2 = x_1$), the line is vertical.

Theorem 10.2.2 Two lines are parallel if and only if they have the same slope, that is, $m_1 = m_2$.

Demostración. To prove this statement, consider two lines with slopes m_1 and m_2.

1. If the lines are parallel, then $m_1 = m_2$:

Two lines are parallel if they do not intersect at any point. If both lines have the same slope, their inclinations are equal, which means they extend in the same direction without ever crossing. Therefore, if the lines are parallel, necessarily $m_1 = m_2$.

2. If $m_1 = m_2$, then the lines are parallel:

If two lines have the same slope, they have the same inclination and direction. This implies that the lines are parallel since they will never intersect.

Therefore, two lines are parallel if and only if $m_1 = m_2$, which concludes the proof. ■

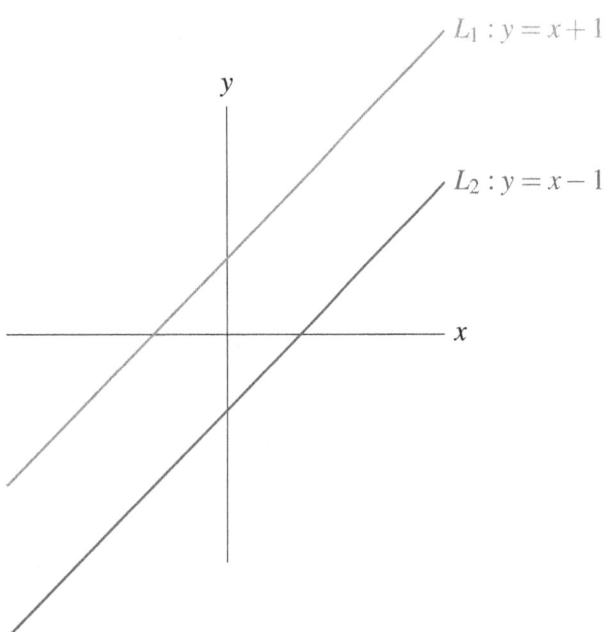

Theorem 10.2.3 Two lines are perpendicular if and only if the product of their slopes is -1, that is, $m_1 \cdot m_2 = -1$.

Demostración. To prove this statement, consider two lines with slopes m_1 and m_2.

1. If the lines are perpendicular, then $m_1 \cdot m_2 = -1$:

Two lines are perpendicular if the angle between them is $90°$. If one line has a slope m_1, then a line perpendicular to it must have a slope m_2 such that the product $m_1 \cdot m_2$ is -1. This is due to the geometric relationship of the slopes of perpendicular lines in the Cartesian plane.

2. If $m_1 \cdot m_2 = -1$, then the lines are perpendicular:

If $m_1 \cdot m_2 = -1$, the lines have slopes that are negative reciprocals, implying that the angle between them is $90°$, i.e., the lines are perpendicular.

Therefore, two lines are perpendicular if and only if $m_1 \cdot m_2 = -1$, which concludes the proof. ■

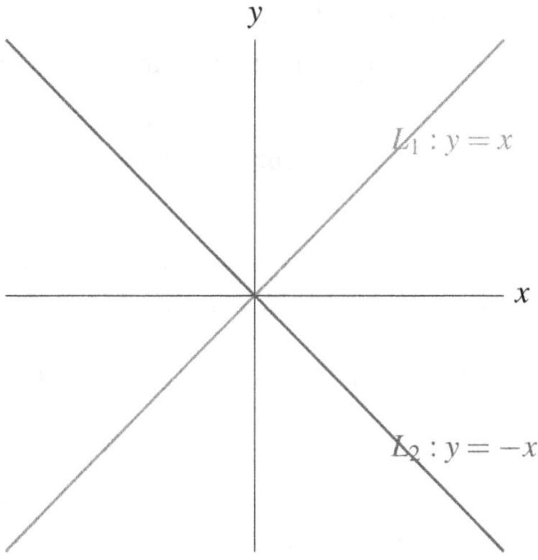

beginfigure[h]

■ **Example 10.7** Determine if the lines $y = 3x + 2$ and $y = -\frac{1}{3}x + 5$ are perpendicular. The slopes are $m_1 = 3$ and $m_2 = -\frac{1}{3}$. We calculate the product:

$$m_1 m_2 = 3\left(-\frac{1}{3}\right) = -1.$$

Therefore, the lines are perpendicular.

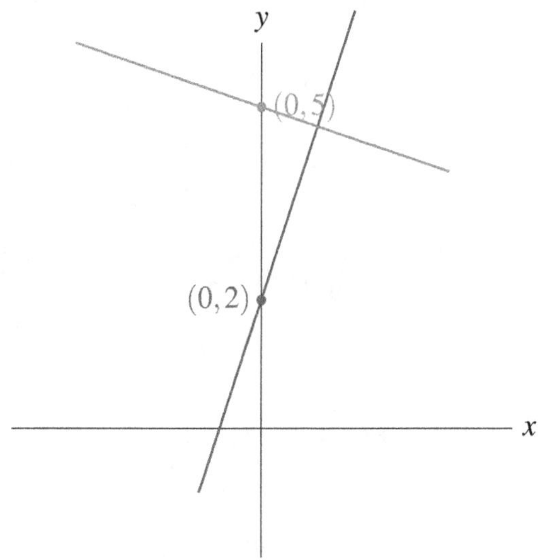

Figura 10.2.2: *Graphs of the lines* $y = 3x + 2$ *and* $y = -\frac{1}{3}x + 5$.

■

> R Understanding the properties of slopes and their relationship to line orientation is crucial for graphically analyzing linear equations and solving geometric problems.

In addition to the slope-intercept form, another useful representation is the general form.

Definition 10.2.4 The *general form* of a line equation is:

$$Ax + By + C = 0,$$

where A, B, and C are constants, and A and B are not both zero.

This form is convenient for certain algebraic calculations and for determining intersections with the axes.

Lema 10.2.1 The intersection of a line with the x-axis occurs when $y = 0$, and with the y-axis when $x = 0$.

Demostración. To find the intersection with the x-axis, we set $y = 0$ in the line equation and solve for x. Similarly, for the intersection with the y-axis, we set $x = 0$ and solve for y. ■

■ **Example 10.8** Find the intersections with the axes of the line $3x - 2y + 6 = 0$.
For the x-axis ($y = 0$):

$$3x + 6 = 0 \implies x = -2.$$

Intersection at $(-2, 0)$.
For the y-axis ($x = 0$):

$$-2y + 6 = 0 \implies y = 3.$$

Intersection at $(0, 3)$.

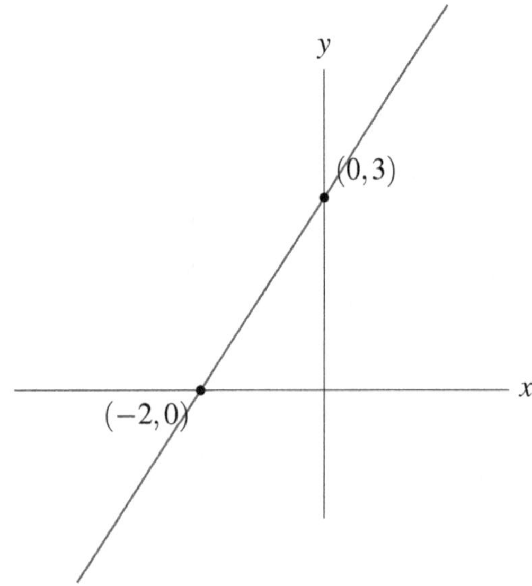

Figura 10.2.3: *Graph of the line* $3x - 2y + 6 = 0$ *and its intersections with the axes.*

■

Exercise 10.5 Determine the equation of the line passing through the points $(1, 4)$ and $(3, 0)$. Graph it and find its intersection with the x-axis. ■

Exercise 10.6 Verify if the lines given by the equations $2x + 3y - 6 = 0$ and $4x + 6y - 12 = 0$ are coincident, parallel, or intersecting. Justify your answer and graph them. ■

The graphical representation of first-degree equations is an essential tool that allows visualizing linear relationships and solving geometric and algebraic problems more intuitively.

 The ability to transform between different forms of a line equation and interpret these equations graphically is fundamental in mathematical reasoning and a deeper understanding of linear algebra.

With these tools, we are equipped to explore systems of equations and analyze more complex situations involving multiple lines and their intersections.

10.2.2 Intersections and Slopes in Graphs.

Understanding intersections and slopes in equation graphs is fundamental for analyzing and solving mathematical problems. These concepts help identify key points on graphs and understand how linear and nonlinear functions change.

Definition 10.2.5 The *slope* m of a line in the Cartesian plane is a measure of its inclination and is defined as the rate of change of y with respect to x between two distinct points (x_1, y_1) and (x_2, y_2) on the line:

$$m = \frac{y_2 - y_1}{x_2 - x_1},$$

if $x_2 \neq x_1$.

The slope tells us how y changes for each unit increase in x. A positive slope indicates a rising line, while a negative slope indicates a falling line.

Definition 10.2.6 The *intersection* of a graph with an axis is the point where the graph crosses that axis. The intersection with the y-axis occurs when $x = 0$, and the intersection with the x-axis occurs when $y = 0$.

Knowing the intersections is essential for accurately drawing graphs and solving equations.

Theorem 10.2.4 The equation of a line in the Cartesian plane can be expressed in the slope-intercept form:

$$y = mx + b,$$

where m is the slope of the line and b is the y-intercept.

Demostración. Consider a line with slope m that intersects the y-axis at the point $(0, b)$. For any point (x, y) on the line, the change in y with respect to x must satisfy:

$$m = \frac{y - b}{x - 0} = \frac{y - b}{x}.$$

Solving for y, we get:

$$y = mx + b.$$

■

This theorem provides a direct way to graph lines by knowing their slope and y-intercept.

■ **Example 10.9** Draw the graph of the line $y = -\frac{2}{3}x + 4$ and identify its slope and intersections with the axes.

- **Slope**: $m = -\frac{2}{3}$.
- **Intersection with the y-axis**: When $x = 0$, $y = 4$.

- **Intersection with the x-axis**: When $y=0$, solve $0=-\frac{2}{3}x+4$:

$$\frac{2}{3}x=4 \implies x=\frac{4\times 3}{2}=6.$$

Therefore, the intersection with the x-axis is at $(6,0)$.

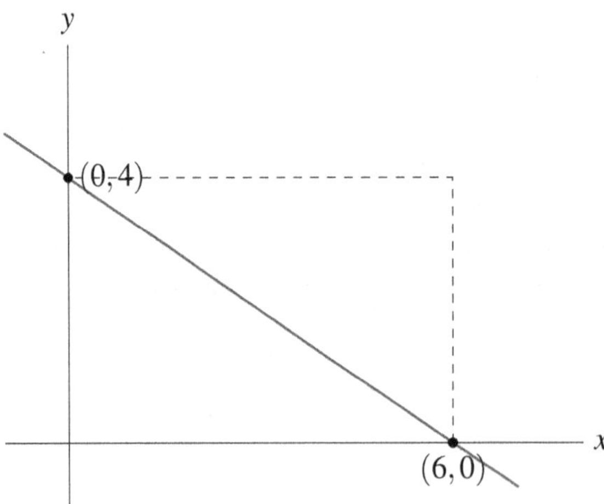

Figura 10.2.4: *Graph of the line* $y=-\frac{2}{3}x+4$ *with its intersections with the axes.*

We observe that the line descends from left to right due to the negative slope, and the intersections with the axes help us draw it accurately.

■

The relationship between the slope and intersections is key to understanding how lines behave in the plane.

Lema 10.2.2 Two non-vertical lines are parallel if and only if they have the same slope.

Demostración. If two lines have slopes m_1 and m_2, and are parallel, then they never cross and have the same inclination, so $m_1=m_2$. Conversely, if $m_1=m_2$, their slopes are equal, and therefore the lines are parallel. ■

This result is useful for determining if two lines are parallel by analyzing their equations.

Theorem 10.2.5 Two non-vertical lines are perpendicular if and only if the product of their slopes is -1:

$$m_1 \cdot m_2 = -1.$$

Demostración. If two lines are perpendicular, the angle between them is $90°$. The slope of one is the negative reciprocal of the other:

$$m_2=-\frac{1}{m_1} \implies m_1 m_2=-1.$$

Conversely, if $m_1 m_2=-1$, then $m_2=-\frac{1}{m_1}$, indicating that the lines are perpendicular. ■

■ **Example 10.10** Determine if the lines $L_1: y=\frac{1}{2}x-3$ and $L_2: y=-2x+5$ are perpendicular. The slopes are $m_1=\frac{1}{2}$ and $m_2=-2$. We calculate the product:

$$m_1 m_2=\left(\frac{1}{2}\right)(-2)=-1.$$

Since the product is -1, the lines L_1 and L_2 are perpendicular.

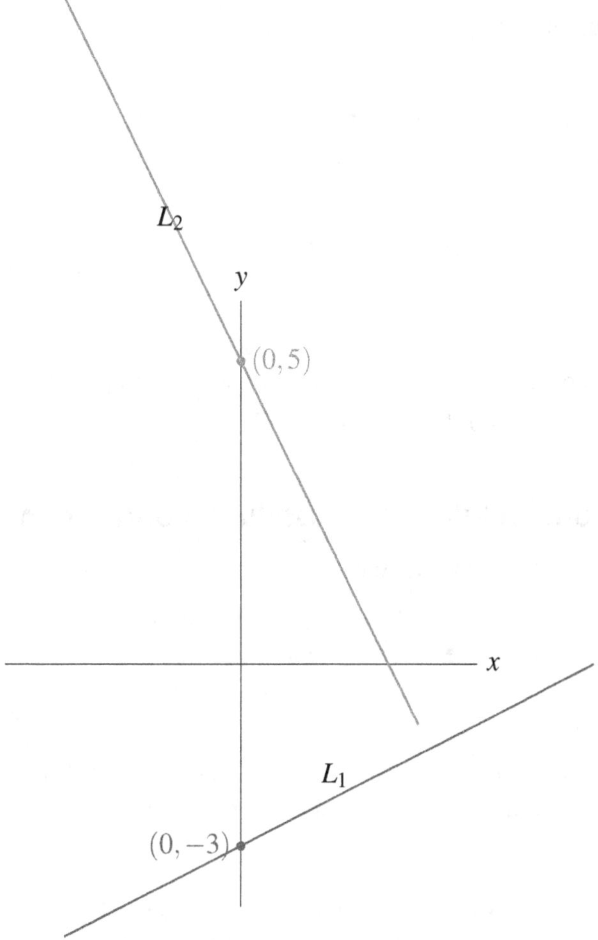

Figura 10.2.5: *Graphs of lines L_1 and L_2, which are perpendicular.*

■

This example illustrates how to use slopes to determine the relationship between two lines.

Corollary 10.2.6 If a line L has slope m, then any line perpendicular to L will have a slope $m' = -\frac{1}{m}$, provided $m \neq 0$.

Demostración. This result follows directly from the previous theorem, since if $mm' = -1$, then $m' = -\frac{1}{m}$. ■

(R) When $m = 0$, meaning the line is horizontal, the perpendicular line is vertical, whose slope is undefined. Similarly, a vertical line has an undefined slope, and its perpendicular is horizontal with a slope of $m = 0$.

Exercise 10.7 Find the equation of the line that passes through the point $(2,-1)$ and is perpendicular to the line $y = 3x + 4$. Graph both lines on the same plane.

■

Exercise 10.8 Determine whether the lines given by the equations $2x-3y+6=0$ and $3x+2y-12=0$ intersect, and if so, calculate the intersection point. Graph the lines and verify the intersection point graphically.

By solving these exercises, you deepen your understanding of how slopes and intersections determine the position and orientation of lines in the plane.

> R The analysis of slopes and intersections is fundamental not only in analytic geometry but also in calculus, physics, and engineering, where lines and their properties are used to model and solve complex problems.

Understanding how slopes and intersections interact allows us to predict behaviors, find solutions to systems of equations, and analyze trends in graphical data.

10.3 Systems of Equations: Solution by Substitution and Elimination

10.3.1 Solution by Equalization and Elimination.

Solving systems of linear equations is a fundamental skill in mathematical reasoning. The methods of equalization and elimination are essential techniques that allow us to find exact solutions to these systems. Below, we will explore these methods in detail, providing definitions, theorems, and examples to facilitate understanding and application.

Definition 10.3.1 A *system of linear equations* is a set of two or more linear equations involving the same variables. The solution to the system is the set of values that simultaneously satisfy all the equations.

These systems frequently appear in fields such as physics, engineering, and economics, where it is necessary to model and solve problems with multiple interrelated variables.

Theorem 10.3.1 — Equalization Method. Consider a system of two linear equations with two variables:

$$\begin{cases} a_1x+b_1y=c_1, \\ a_2x+b_2y=c_2. \end{cases}$$

If the same variable is isolated in both equations and the resulting expressions are set equal, the system can be solved for the other variable.

Demostración. To demonstrate the equalization method, consider the system of two linear equations with two variables:

$$\begin{cases} a_1x+b_1y=c_1, \\ a_2x+b_2y=c_2. \end{cases}$$

1. **Isolate a variable in both equations:**
Choose one of the variables, for example, x, and isolate it in each equation. From the first equation, solve for x:

$$x=\frac{c_1-b_1y}{a_1}.$$

From the second equation, solve for x similarly:

$$x = \frac{c_2 - b_2 y}{a_2}.$$

2. **Equalize the expressions:**
Since both expressions represent x, set them equal:

$$\frac{c_1 - b_1 y}{a_1} = \frac{c_2 - b_2 y}{a_2}.$$

3. **Solve for y:**
Multiply both sides by $a_1 a_2$ to eliminate the denominators:

$$a_2(c_1 - b_1 y) = a_1(c_2 - b_2 y).$$

Expanding and grouping terms involving y, we obtain a linear equation in y that can be solved. Once y is found, substitute its value into either expression for x to find x. ■

This procedure shows how the equalization method allows solving the system for the other variable. This concludes the proof.
This method is efficient when it is easy to isolate a variable in both equations. Let's see an example to illustrate its application.

■ **Example 10.11** Solve the following system using the equalization method:

$$\begin{cases} 2x + 3y = 8, \\ x - y = 1. \end{cases}$$

Isolate x in the second equation:

$$x = y + 1.$$

Substitute into the first equation:

$$2(y+1) + 3y = 8 \implies 2y + 2 + 3y = 8 \implies 5y + 2 = 8 \implies 5y = 6 \implies y = \frac{6}{5}.$$

Now, find x:

$$x = \frac{6}{5} + 1 = \frac{6}{5} + \frac{5}{5} = \frac{11}{5}.$$

The solution is $(x, y) = \left(\frac{11}{5}, \frac{6}{5}\right)$. ■

R The equalization method is particularly useful when the equations are simplified or when one of the variables has a coefficient of 1 or -1, making isolation easier.

Now, let's consider the elimination method, also known as the reduction method.

Theorem 10.3.2 — **Elimination Method.** In a system of linear equations, if we multiply the equations by suitable constants so that the coefficients of one variable are opposites, adding or subtracting the equations eliminates that variable, allowing us to solve for the other variable.

Demostración. To demonstrate the elimination method, consider a system of two linear equations with two variables:

$$\begin{cases} a_1x + b_1y = c_1, \\ a_2x + b_2y = c_2. \end{cases}$$

1. **Multiplication by appropriate constants:**
To eliminate one of the variables, let's choose x and multiply each equation by a constant such that the coefficients of x in both equations become opposites. This means we seek constants k_1 and k_2 such that:

$$k_1a_1 = -k_2a_2.$$

Multiplying the first equation by k_1 and the second by k_2, we get:

$$\begin{cases} k_1a_1x + k_1b_1y = k_1c_1, \\ k_2a_2x + k_2b_2y = k_2c_2. \end{cases}$$

2. **Adding or subtracting the equations:**
Now, adding both equations cancels out the x terms, since their coefficients are opposites:

$$(k_1a_1 + k_2a_2)x + (k_1b_1 + k_2b_2)y = k_1c_1 + k_2c_2.$$

Since $k_1a_1 + k_2a_2 = 0$, we obtain an equation with only the variable y:

$$(k_1b_1 + k_2b_2)y = k_1c_1 + k_2c_2.$$

3. **Solving for y:**
We can solve this equation for y. Once we obtain the value of y, substitute it into one of the original equations to find the value of x.
This concludes the proof of the elimination method, which allows solving the system by eliminating one of the variables. ■

This method is very powerful, especially when the equations have coefficients that are multiples of each other.

■ **Example 10.12** Solve the following system using the elimination method:

$$\begin{cases} 3x + 2y = 16, \\ 5x - 2y = 4. \end{cases}$$

Notice that the coefficients of y are 2 and -2. Adding both equations eliminates y:

$$(3x + 2y) + (5x - 2y) = 16 + 4 \implies 8x = 20 \implies x = \frac{20}{8} = \frac{5}{2}.$$

Substitute x into the first equation:

$$3\left(\frac{5}{2}\right) + 2y = 16 \implies \frac{15}{2} + 2y = 16 \implies 2y = 16 - \frac{15}{2} = \frac{32}{2} - \frac{15}{2} = \frac{17}{2} \implies y = \frac{17}{4}.$$

The solution is $(x, y) = \left(\frac{5}{2}, \frac{17}{4}\right)$. ■

Corollary 10.3.3 The elimination method can be generalized to solve systems with three or more variables by eliminating one variable at a time using linear combinations of the equations.

It is important to know how to choose the most appropriate method depending on the system's characteristics.

In some cases, combining the substitution and elimination methods can simplify solving the system. Flexibility in approach is key for solving more complex systems.

Now, let's look at some exercises to apply these methods.

Exercise 10.9 Solve the following system using the method you consider most appropriate:

$$\begin{cases} 2x - 3y = 7, \\ 4x + y = 1. \end{cases}$$

■

Exercise 10.10 Use the elimination method to solve the system of three equations:

$$\begin{cases} x + 2y - z = 4, \\ 2x - y + 3z = -6, \\ -3x + 4y + 2z = 7. \end{cases}$$

■

These exercises help practice and consolidate the use of substitution and elimination methods in different contexts and complexity levels.

The ability to solve systems of linear equations is fundamental in many areas of mathematics and its applications, including linear algebra, analysis, and mathematical modeling.

Understanding and mastering these methods expand our tools for tackling more advanced problems and prepare us for deeper studies in mathematics and sciences.

10.3.2 Applications in Geometric Problems.

Linear equations are fundamental tools for solving geometric problems because they allow us to model and analyze relationships between points, lines, and figures in the Cartesian plane. Through algebraic reasoning, we can establish connections between geometric properties and algebraic equations.

Definition 10.3.2 The *midpoint* M of a segment defined by the points $A(x_1, y_1)$ and $B(x_2, y_2)$ is the point whose coordinates are:

$$M\left(\frac{x_1 + x_2}{2}, \frac{y_1 + y_2}{2}\right).$$

This definition allows us to find the point that divides a segment into two equal parts, which is useful in various geometric problems.

Theorem 10.3.4 The *distance* d between two points $A(x_1, y_1)$ and $B(x_2, y_2)$ in the Cartesian plane is:

$$d = \sqrt{(x_2 - x_1)^2 + (y_2 - y_1)^2}.$$

Demostración. To prove the formula for the distance between two points in the Cartesian plane, consider two points $A(x_1, y_1)$ and $B(x_2, y_2)$.
Applying the Pythagorean Theorem:
The distance d between points A and B can be interpreted as the length of the hypotenuse of a right triangle, where:
- The base is the difference in the x-coordinates, which is $|x_2 - x_1|$. - The height is the difference in the y-coordinates, which is $|y_2 - y_1|$.
Calculating the distance using the Pythagorean Theorem:
According to the Pythagorean Theorem, the hypotenuse d of this right triangle satisfies:

$$d = \sqrt{(x_2 - x_1)^2 + (y_2 - y_1)^2}.$$

This demonstrates that the distance between points $A(x_1, y_1)$ and $B(x_2, y_2)$ is:

$$d = \sqrt{(x_2 - x_1)^2 + (y_2 - y_1)^2}.$$

This concludes the proof. ■

Knowing the distance between points is essential for calculating perimeters, diagonals, and for analyzing the congruence and similarity of geometric figures.

■ **Example 10.13** Determine the equation of the perpendicular bisector of the segment joining the points $A(2,3)$ and $B(6,7)$.
Solution:
The perpendicular bisector of a segment is the line perpendicular to the segment that passes through its midpoint.
1. Calculate the midpoint M:

$$M\left(\frac{2+6}{2}, \frac{3+7}{2}\right) = M(4,5).$$

2. Calculate the slope m_{AB} of segment AB:

$$m_{AB} = \frac{y_2 - y_1}{x_2 - x_1} = \frac{7-3}{6-2} = \frac{4}{4} = 1.$$

3. The slope m of the perpendicular bisector is the negative reciprocal of m_{AB}:

$$m = -\frac{1}{m_{AB}} = -1.$$

4. Use the point-slope form of the equation of a line:

$$y - y_0 = m(x - x_0),$$

where $(x_0, y_0) = M(4,5)$ and $m = -1$.
5. Therefore, the equation of the perpendicular bisector is:

$$y - 5 = -1(x-4) \implies y - 5 = -x + 4 \implies y = -x + 9.$$

■

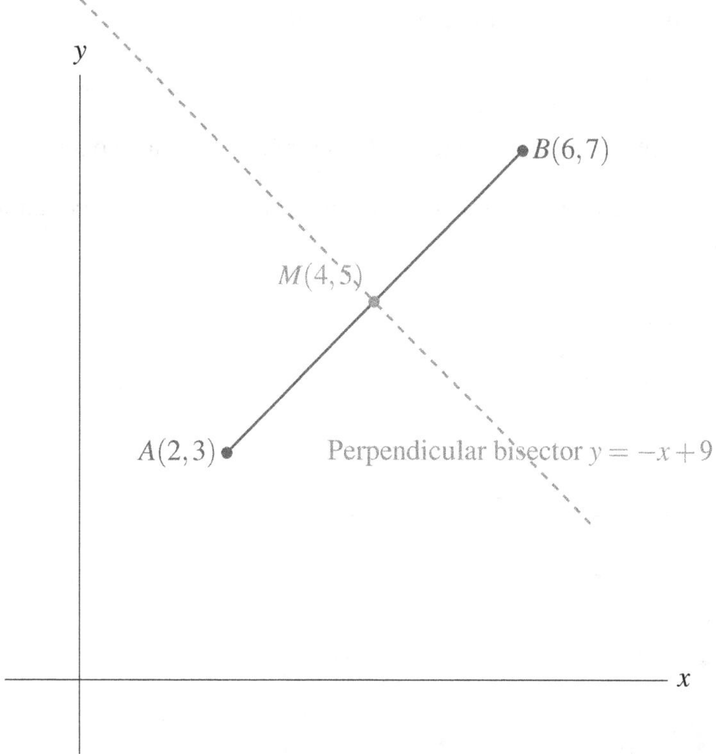

Figura 10.3.1: *Graph of segment AB and its perpendicular bisector.*

In this example, we used the relationship between perpendicular slopes and the midpoint to find the equation of the perpendicular bisector, a direct geometric application of linear equations.

Lema 10.3.1 The slope of a line perpendicular to another line with slope m is $m' = -\frac{1}{m}$, provided $m \neq 0$.

Demostración. If two lines are perpendicular, the product of their slopes is -1: $m \cdot m' = -1$. Solving for m' yields $m' = -\frac{1}{m}$. ■

This lemma is fundamental for solving problems involving perpendicular lines, as seen in the previous example.

Theorem 10.3.5 The general equation of a circle with center (h,k) and radius r is:

$$(x-h)^2 + (y-k)^2 = r^2.$$

Demostración. Consider a circle with center at the point (h,k) and radius r. By definition, any point (x,y) on the circle is at a distance r from the center (h,k).
Using the formula for the distance between two points, the distance d between (x,y) and (h,k) is:

$$d = \sqrt{(x-h)^2 + (y-k)^2}.$$

For the point (x,y) to lie on the circle, this distance must be equal to r. Therefore,

$$\sqrt{(x-h)^2 + (y-k)^2} = r.$$

Squaring both sides to eliminate the square root, we get:

$$(x-h)^2+(y-k)^2=r^2.$$

This is the equation of the circle with center (h,k) and radius r, which concludes the proof. ■

■ **Example 10.14** Find the points of intersection between the circle with equation $(x-2)^2+(y-3)^2=25$ and the line $y=x+1$.

Solution:

1. Substitute y into the circle's equation:

$$(x-2)^2+((x+1)-3)^2=25.$$

2. Simplify:

$$(x-2)^2+(x-2)^2=25 \implies 2(x-2)^2=25.$$

3. Solve for x:

$$(x-2)^2=\frac{25}{2} \implies x-2=\pm\sqrt{\frac{25}{2}}=\pm\frac{5}{\sqrt{2}}.$$

4. Thus, $x=2\pm\frac{5}{\sqrt{2}}$.

5. Calculate y for each x value:

$$y=x+1.$$

For $x=2+\frac{5}{\sqrt{2}}$:

$$y=2+\frac{5}{\sqrt{2}}+1=3+\frac{5}{\sqrt{2}}.$$

For $x=2-\frac{5}{\sqrt{2}}$:

$$y=2-\frac{5}{\sqrt{2}}+1=3-\frac{5}{\sqrt{2}}.$$

6. The points of intersection are:

$$\left(2+\frac{5}{\sqrt{2}},3+\frac{5}{\sqrt{2}}\right), \quad \left(2-\frac{5}{\sqrt{2}},3-\frac{5}{\sqrt{2}}\right).$$

■

This example shows how linear equations can be used alongside quadratic equations to solve more complex geometric problems.

Exercise 10.11 Determine the equation of the line that is tangent to the circle with equation $(x-1)^2+(y+2)^2=10$ at the point $P(4,0)$. ■

Exercise 10.12 Find the coordinates of the point C that divides the segment AB, where $A(-2,1)$ and $B(4,5)$, in a ratio of 2:1, i.e., $\frac{AC}{CB}=\frac{2}{1}$. ■

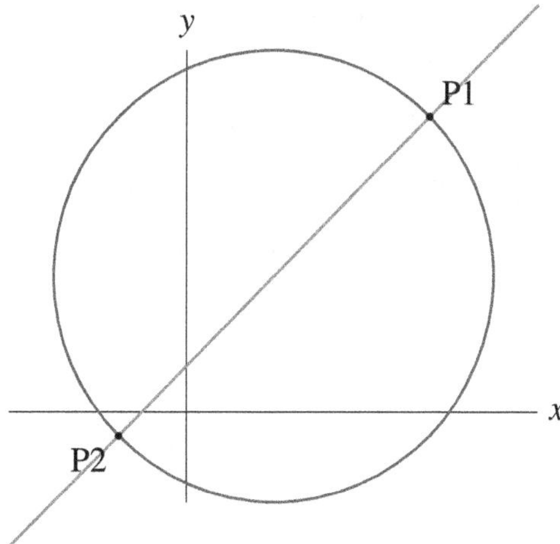

Figura 10.3.2: *Intersection between the circle and the line.*

■

R Geometric problems often require combining algebra and geometry concepts. Using linear equations to model lines, perpendicular bisectors, tangents, and other geometric figures allows us to solve problems analytically and precisely.

The ability to translate geometric properties into algebraic equations is a key skill in advanced mathematics, as it facilitates the analysis and resolution of complex problems.

Through these examples and exercises, we have explored how linear equations can be applied in geometric problems, demonstrating the power of algebraic reasoning in understanding and solving geometric questions.

10.4 Solved Exercises

Exercise 10.13 A train travels at a constant speed of 90 km/h. How long will it take to cover a distance of 270 km? ■

Demostración. We know that the speed $v = \frac{d}{t}$, where d is the distance and t is the time. Solving for time:

$$t = \frac{d}{v} = \frac{270 \text{ km}}{90 \text{ km/h}} = 3 \text{ hours}.$$

Therefore, the train will take 3 hours to cover the distance of 270 km. ■

Exercise 10.14 Solve the system of linear equations:

$$\begin{cases} 2x + 3y = 13, \\ 4x - y = 5. \end{cases}$$

■

Demostración. To solve this system, we use the substitution method. Solve for y in the first equation:

$$3y = 13 - 2x \Rightarrow y = \frac{13 - 2x}{3}.$$

Substitute y into the second equation:

$$4x - \frac{13-2x}{3} = 5.$$

Multiply by 3 to eliminate the denominator:

$$12x - (13-2x) = 15 \Rightarrow 12x - 13 + 2x = 15 \Rightarrow 14x = 28 \Rightarrow x = 2.$$

Substitute $x = 2$ into the equation for y:

$$y = \frac{13-2(2)}{3} = \frac{13-4}{3} = 3.$$

The solution is $(x,y) = (2,3)$. ■

Exercise 10.15 Find the area of a triangle with vertices at the points $(1,2)$, $(4,6)$, and $(7,2)$. ■

Demostración. We use the formula for the area of a triangle with vertices (x_1,y_1), (x_2,y_2), and (x_3,y_3):

$$A = \frac{1}{2}\left|x_1(y_2-y_3)+x_2(y_3-y_1)+x_3(y_1-y_2)\right|.$$

Substituting the values:

$$A = \frac{1}{2}\left|1(6-2)+4(2-2)+7(2-6)\right| = \frac{1}{2}\left|1\cdot 4+4\cdot 0+7\cdot(-4)\right|.$$

$$A = \frac{1}{2}|4-28| = \frac{1}{2}|-24| = \frac{24}{2} = 12.$$

The area of the triangle is 12 square units. ■

Exercise 10.16 Calculate the value of x if $3^x = 81$. ■

Demostración. We know that $81 = 3^4$. Therefore, we can write the equation as:

$$3^x = 3^4.$$

Equating the exponents, we get:

$$x = 4.$$

The value of x is 4. ■

Exercise 10.17 Determine the slope of the line passing through the points $A(2,-3)$ and $B(5,4)$. ■

Demostración. The slope m of a line passing through two points (x_1,y_1) and (x_2,y_2) is calculated using the formula:

$$m = \frac{y_2-y_1}{x_2-x_1}.$$

Substituting the values:

$$m = \frac{4-(-3)}{5-2} = \frac{4+3}{3} = \frac{7}{3}.$$

The slope of the line is $\frac{7}{3}$. ■

10.5 Ejercicios Propuestos

10.5.1 Ecuaciones lineales: planteamiento de problemas y modelado

Exercise 10.18 Un ciclista recorre 120 km a una velocidad constante. Si le toma 4 horas, ¿cuál es su velocidad? ■

Exercise 10.19 Un automóvil viaja a 80 km/h. ¿Cuánto tiempo tardará en recorrer una distancia de 200 km? ■

Exercise 10.20 Dos trenes salen de la misma estación en direcciones opuestas. Uno viaja a 90 km/h y el otro a 110 km/h. ¿Qué distancia habrá entre ellos después de 3 horas? ■

Exercise 10.21 Una empresa fabrica productos a un costo fijo de $500 y un costo variable de $10 por unidad. Si vende cada producto a $25, ¿cuántas unidades debe vender para alcanzar el punto de equilibrio? ■

Exercise 10.22 Un corredor que va a una velocidad de 12 km/h intenta alcanzar a otro que va a 9 km/h y que partió 30 minutos antes. ¿Cuánto tiempo tardará en alcanzarlo? ■

10.5.2 Resolución gráfica y algebraica: representación en el plano cartesiano

Exercise 10.23 Dibuja la gráfica de la ecuación $y = 2x + 3$ e identifica su pendiente e intersección con el eje y. ■

Exercise 10.24 Encuentra la ecuación de la recta que pasa por los puntos $(1,2)$ y $(4,6)$. ■

Exercise 10.25 Determina si las rectas $y = 3x + 2$ y $y = -\frac{1}{3}x + 5$ son perpendiculares. ■

Exercise 10.26 Calcula la distancia entre los puntos $(2,3)$ y $(5,7)$ en el plano cartesiano. ■

Exercise 10.27 Encuentra el punto de intersección entre las rectas $y = x + 2$ y $y = -x + 4$. ■

10.6 Proposed Exercises

10.6.1 Linear Equations: Problem Formulation and Modeling

Exercise 10.28 A cyclist covers 120 km at a constant speed. If it takes 4 hours, what is his speed? ■

Exercise 10.29 A car travels at 80 km/h. How long will it take to cover a distance of 200 km? ■

Exercise 10.30 Two trains leave the same station in opposite directions. One travels at 90 km/h, and the other at 110 km/h. What distance will be between them after 3 hours? ■

Exercise 10.31 A company manufactures products with a fixed cost of \$500 and a variable cost of \$10 per unit. If each product is sold for \$25, how many units must be sold to break even? ■

Exercise 10.32 A runner traveling at a speed of 12 km/h tries to catch up with another runner moving at 9 km/h who left 30 minutes earlier. How long will it take to catch up? ■

10.6.2 Graphical and Algebraic Solutions: Representation in the Cartesian Plane

Exercise 10.33 Draw the graph of the equation $y = 2x + 3$ and identify its slope and y-intercept. ■

Exercise 10.34 Find the equation of the line passing through the points $(1,2)$ and $(4,6)$. ■

Exercise 10.35 Determine if the lines $y = 3x + 2$ and $y = -\frac{1}{3}x + 5$ are perpendicular. ■

Exercise 10.36 Calculate the distance between the points $(2,3)$ and $(5,7)$ in the Cartesian plane. ■

Exercise 10.37 Find the point of intersection of the lines $y = x + 2$ and $y = -x + 4$. ■

10.6.3 Systems of Equations: Solution by Substitution and Elimination

Exercise 10.38 Solve the following system of equations:

$$\begin{cases} x + 2y = 5, \\ 3x - y = 4. \end{cases}$$

■

Exercise 10.39 Find the values of x and y that satisfy the system:

$$\begin{cases} 2x + 3y = 12, \\ x - y = 2. \end{cases}$$

■

Exercise 10.40 Solve the system using the elimination method:

$$\begin{cases} 4x + 5y = 20, \\ 2x - 3y = -4. \end{cases}$$

■

Exercise 10.41 Use the substitution method to solve the system:

$$\begin{cases} 3x + 4y = 10, \\ x - 2y = -1. \end{cases}$$

■

Exercise 10.42 Determine if the system has a solution and, if so, solve it:

$$\begin{cases} x + y = 7, \\ x - y = 3. \end{cases}$$

■

11. Modeling with Quadratic Equations

11.1 Quadratic Equations: Factoring and the Quadratic Formula

11.1.1 Completing the Square Method

The method of completing the square is a fundamental algebraic technique that allows transforming a quadratic expression into a form that facilitates its analysis and solution. This method is particularly useful for solving quadratic equations, analyzing quadratic functions, and proving related theorems.

Definition 11.1.1 A *quadratic expression* is a second-degree polynomial function of the form:

$$f(x) = ax^2 + bx + c,$$

where a, b, and c are real coefficients and $a \neq 0$.

The goal of completing the square is to rewrite the quadratic expression in the canonical or standard form:

$$f(x) = a(x-h)^2 + k,$$

where (h,k) is the vertex of the parabola represented by $f(x)$. This form is particularly useful for identifying geometric properties of the function.

Theorem 11.1.1 Any quadratic expression $f(x) = ax^2 + bx + c$ can be expressed in the form $f(x) = a(x-h)^2 + k$, where:

$$h = -\frac{b}{2a}, \quad k = f(h) = c - \frac{b^2}{4a}.$$

Demostración. Consider the quadratic expression $f(x) = ax^2 + bx + c$. Our goal is to express this function in the form $f(x) = a(x-h)^2 + k$, where $h = -\frac{b}{2a}$ and $k = f(h) = c - \frac{b^2}{4a}$. To achieve this, we complete the square.

Factor a out of the first two terms:

$$f(x) = a\left(x^2 + \frac{b}{a}x\right) + c.$$

Complete the square inside the parentheses: take the linear term $\frac{b}{a}x$ and find $\left(\frac{b}{2a}\right)^2 = \frac{b^2}{4a^2}$. Add and subtract $\frac{b^2}{4a^2}$ inside the parentheses:

$$f(x) = a\left(x^2 + \frac{b}{a}x + \frac{b^2}{4a^2} - \frac{b^2}{4a^2}\right) + c.$$

Rewrite the expression inside the parentheses as a perfect square and simplify:

$$f(x) = a\left(\left(x + \frac{b}{2a}\right)^2 - \frac{b^2}{4a^2}\right) + c.$$

Expanding and simplifying, we obtain:

$$f(x) = a\left(x + \frac{b}{2a}\right)^2 - \frac{b^2}{4a} + c.$$

Now the expression is in the desired form:

$$f(x) = a(x-h)^2 + k,$$

where $h = -\frac{b}{2a}$ and $k = c - \frac{b^2}{4a}$. This concludes the proof. ■

This result is crucial for understanding the structure of quadratic functions and facilitates the analysis of their graphical properties.

■ **Example 11.1** Complete the square for the quadratic expression $f(x) = 2x^2 + 8x + 5$ and express $f(x)$ in canonical form.

Solution:

1. Factor the coefficient a:

$$f(x) = 2\left(x^2 + 4x\right) + 5.$$

2. Add and subtract $\left(\frac{4}{2}\right)^2 = 4$ inside the parentheses:

$$f(x) = 2\left(x^2 + 4x + 4 - 4\right) + 5.$$

3. Rewrite inside the parentheses:

$$f(x) = 2\left((x+2)^2 - 4\right) + 5.$$

4. Distribute 2:

$$f(x) = 2(x+2)^2 - 8 + 5 = 2(x+2)^2 - 3.$$

Thus, the canonical form is:

$$f(x) = 2(x+2)^2 - 3.$$

The vertex of the parabola is $(-2, -3)$.

R Expressing $f(x)$ in canonical form allows us to easily identify the vertex and the direction of the parabola's opening, which is useful for graphing and solving optimization problems.

■

The method of completing the square is also a powerful tool for solving quadratic equations.

Lema 11.1.1 The quadratic equation $ax^2+bx+c=0$ can be solved by completing the square, yielding:

$$x = -\frac{b}{2a} \pm \sqrt{\frac{b^2-4ac}{4a^2}}.$$

Demostración. Following the process of completing the square as in the previous proof, we get:

$$a\left(x+\frac{b}{2a}\right)^2 = \frac{b^2-4ac}{4a}.$$

Divide both sides by a:

$$\left(x+\frac{b}{2a}\right)^2 = \frac{b^2-4ac}{4a^2}.$$

Take the square root of both sides:

$$x+\frac{b}{2a} = \pm\frac{\sqrt{b^2-4ac}}{2a}.$$

Solving for x:

$$x = -\frac{b}{2a} \pm \frac{\sqrt{b^2-4ac}}{2a}.$$

■

We observe that this result is equivalent to the quadratic formula, demonstrating the connection between both methods.

Corollary 11.1.2 The method of completing the square allows deriving the quadratic formula and understanding the nature of the solutions in terms of the discriminant $D=b^2-4ac$.

In addition, completing the square is useful in the study of conic sections and in solving integrals in calculus.

■ **Example 11.2** Solve the equation $x^2+6x+5=0$ by completing the square.
Solution:
1. The equation is in the form $x^2+6x+5=0$.
2. Move the constant term to the other side:

$$x^2+6x=-5.$$

3. Complete the square by adding $\left(\frac{6}{2}\right)^2=9$ to both sides:

$$x^2+6x+9=-5+9 \implies (x+3)^2=4.$$

4. Take the square root of both sides:

$$x+3=\pm 2.$$

5. Solve for x:

$$x = -3 \pm 2.$$

6. The solutions are:

$$x = -3 + 2 = -1, \quad x = -3 - 2 = -5.$$

■

(R) The method of completing the square is especially useful when the coefficient of the quadratic term is 1, but it can be applied more generally, as shown in the previous examples.

Exercise 11.1 Complete the square for the expression $f(x) = -x^2 + 4x - 1$ and express $f(x)$ in vertex form. Determine the vertex of the parabola and its direction of opening.

■

Exercise 11.2 Using the method of completing the square, solve the equation $3x^2 - 12x + 9 = 0$. Classify the roots and analyze their multiplicity.

■

The method of completing the square is a versatile technique that not only facilitates solving quadratic equations but is also essential in function analysis and various areas of advanced mathematics.

(R) The ability to complete the square is fundamental in the study of integral calculus, especially when solving integrals involving quadratic expressions in the denominator or under a square root.

Understanding and mastering this method expands our mathematical tools and deepens our algebraic reasoning.

11.1.2 Application of the Quadratic Formula

The quadratic equation is a fundamental tool in mathematics, especially in algebra and analysis. The *quadratic formula* allows us to find the roots of any second-degree equation and is essential for solving problems involving quadratic expressions.

Definition 11.1.2 A *quadratic equation* is a second-degree polynomial equation of the form:

$$ax^2 + bx + c = 0,$$

where a, b, and c are real coefficients with $a \neq 0$.

The general solution of a quadratic equation is obtained using the quadratic formula, which is derived from the process of completing the square.

Theorem 11.1.3 — Quadratic Formula. The roots of the quadratic equation $ax^2 + bx + c = 0$ are given by:

$$x = \frac{-b \pm \sqrt{b^2 - 4ac}}{2a}.$$

Demostración. We start with the general equation:

$$ax^2 + bx + c = 0.$$

Divide both sides by a to make the leading coefficient 1:

$$x^2 + \frac{b}{a}x + \frac{c}{a} = 0.$$

Proceed to complete the square:

$$x^2 + \frac{b}{a}x = -\frac{c}{a}.$$

Add and subtract $\left(\frac{b}{2a}\right)^2$ on the left-hand side:

$$x^2 + \frac{b}{a}x + \left(\frac{b}{2a}\right)^2 - \left(\frac{b}{2a}\right)^2 = -\frac{c}{a}.$$

Rewrite as:

$$\left(x + \frac{b}{2a}\right)^2 = \frac{b^2}{4a^2} - \frac{c}{a}.$$

Simplify the right-hand side:

$$\left(x + \frac{b}{2a}\right)^2 = \frac{b^2 - 4ac}{4a^2}.$$

Take the square root of both sides:

$$x + \frac{b}{2a} = \pm\frac{\sqrt{b^2 - 4ac}}{2a}.$$

Finally, solve for x:

$$x = \frac{-b \pm \sqrt{b^2 - 4ac}}{2a}.$$

■

The expression $D = b^2 - 4ac$ is called the *discriminant* and determines the nature of the roots of the quadratic equation.

Definition 11.1.3 The *discriminant* of the quadratic equation $ax^2 + bx + c = 0$ is:

$$D = b^2 - 4ac.$$

Lema 11.1.2 Let D be the discriminant of a quadratic equation:

- If $D > 0$, the equation has two distinct real roots.
- If $D = 0$, the equation has a double real root.
- If $D < 0$, the equation has two conjugate complex roots.

Demostración. The nature of the roots depends on the value under the square root in the quadratic formula:

- If $D > 0$, $\sqrt{D}$ is real and positive, resulting in two distinct real solutions.
- If $D = 0$, $\sqrt{D} = 0$, and both solutions coincide, yielding a double real root.
- If $D < 0$, $\sqrt{D}$ is a pure imaginary number, resulting in two conjugate complex roots.

■

Understanding the discriminant is crucial for predicting the type of solutions and analyzing the behavior of quadratic functions.

■ **Example 11.3** Solve the quadratic equation $x^2 - 6x + 8 = 0$ using the quadratic formula.
Solution:
Identify the coefficients:

$$a = 1, \quad b = -6, \quad c = 8.$$

Compute the discriminant:

$$D = (-6)^2 - 4(1)(8) = 36 - 32 = 4.$$

Since $D > 0$, there are two distinct real roots.
Apply the quadratic formula:

$$x = \frac{-(-6) \pm \sqrt{4}}{2(1)} = \frac{6 \pm 2}{2}.$$

The solutions are:

$$x_1 = \frac{6+2}{2} = \frac{8}{2} = 4, \quad x_2 = \frac{6-2}{2} = \frac{4}{2} = 2.$$

Therefore, the roots are $x = 4$ and $x = 2$.

(R) These roots can also be found through factorization, since the equation can be written as $(x-4)(x-2) = 0$.

■

This example shows the direct application of the quadratic formula and how it relates to other solution methods.

Corollary 11.1.4 If a quadratic equation has a negative discriminant, the roots are complex conjugates and can be expressed as:

$$x = \frac{-b}{2a} \pm \frac{\sqrt{|D|}}{2a} i,$$

where $i = \sqrt{-1}$.

Demostración. Consider a quadratic equation of the form $ax^2 + bx + c = 0$, with discriminant $D = b^2 - 4ac$.
If D is negative, then $D < 0$, which implies there is no real root for this equation. However, we can solve for complex roots using the quadratic formula:

$$x = \frac{-b \pm \sqrt{D}}{2a}.$$

When D is negative, we write $D = -|D|$ with $|D| > 0$, and substitute in the formula:

$$x = \frac{-b \pm \sqrt{-|D|}}{2a}.$$

Since $\sqrt{-|D|} = \sqrt{|D|} \cdot i$, where $i = \sqrt{-1}$, we get:

$$x = \frac{-b}{2a} \pm \frac{\sqrt{|D|}}{2a} i.$$

Thus, the roots are complex conjugates and can be expressed as

$$x = \frac{-b}{2a} \pm \frac{\sqrt{|D|}}{2a} i.$$

This completes the proof. ■

■ **Example 11.4** Solve the quadratic equation $2x^2 + 4x + 5 = 0$.
Solution:
Coefficients:

$$a = 2, \quad b = 4, \quad c = 5.$$

Discriminant:

$$D = 4^2 - 4(2)(5) = 16 - 40 = -24.$$

Since $D < 0$, the roots are complex conjugates.
Apply the quadratic formula:

$$x = \frac{-4 \pm \sqrt{-24}}{2 \times 2} = \frac{-4 \pm 2i\sqrt{6}}{4} = \frac{-4}{4} \pm \frac{2i\sqrt{6}}{4} = -1 \pm \frac{i\sqrt{6}}{2}.$$

The roots are $x = -1 + \frac{i\sqrt{6}}{2}$ and $x = -1 - \frac{i\sqrt{6}}{2}$.

■

This example illustrates how to handle complex solutions using the quadratic formula.

Exercise 11.3 Solve the quadratic equation $x^2 + 2x + 5 = 0$ and classify the roots. ■

Exercise 11.4 Determine the values of m for which the equation $x^2 + (m-3)x + m = 0$ has a double real root. ■

These exercises allow further practice in applying the quadratic formula and analyzing the discriminant in different contexts.

(R) The quadratic formula is not only useful for finding exact solutions but is also fundamental in the analysis of quadratic functions, such as determining vertices, axes of symmetry, and points of intersection with the coordinate axes.

Understanding and applying the quadratic formula is essential for advanced mathematical reasoning, as it lays the foundation for further studies in algebra, calculus, and other areas of mathematics.

11.2 Practical Applications: Problems Involving Areas and Trajectories.

11.2.1 Free-Fall Problems and Parabolic Trajectories.

In this section, we explore the mathematical foundations describing the motion of objects under the influence of gravity, specifically in free-fall situations and parabolic trajectories. These concepts are essential for understanding phenomena in physics and applied mathematics.

Definition 11.2.1 A *free-fall motion* is one in which an object moves solely under the influence of gravity, neglecting air resistance or other forces. Mathematically, it is modeled by the equation:

$$s(t) = s_0 + v_0 t - \frac{1}{2} g t^2,$$

where $s(t)$ is the position as a function of time, s_0 is the initial position, v_0 is the initial velocity, and g is the acceleration due to gravity.

Understanding this definition allows us to analyze everyday and scientific situations where gravity plays a crucial role.

■ **Example 11.5** If an object is dropped from a height of 50 meters ($s_0 = 50$) with zero initial velocity ($v_0 = 0$), its position as a function of time is:

$$s(t) = 50 - \frac{1}{2} g t^2.$$

By solving $s(t) = 0$, we can determine the time it takes to reach the ground. ■

Moving on to greater complexity, consider motion involving two dimensions.

Definition 11.2.2 A *parabolic trajectory* describes the motion of an object launched with an initial velocity forming an angle θ with the horizontal, under the influence of gravity. The parametric equations of motion are:

$$\begin{cases} x(t) = v_0 \cos\theta \, t, \\ y(t) = v_0 \sin\theta \, t - \frac{1}{2} g t^2. \end{cases}$$

These equations allow us to model the behavior of projectiles and other moving objects.

Theorem 11.2.1 The trajectory of a projectile launched in a uniform gravitational field is a parabola.

Demostración. Solving for t in the equation $x(t)$:

$$t = \frac{x}{v_0 \cos\theta},$$

and substituting into $y(t)$:

$$y = v_0 \sin\theta \left(\frac{x}{v_0 \cos\theta} \right) - \frac{1}{2} g \left(\frac{x}{v_0 \cos\theta} \right)^2.$$

Simplifying:

$$y = x \tan\theta - \frac{g x^2}{2 v_0^2 \cos^2\theta},$$

which is a quadratic equation in x, representing a parabola. ■

This result is fundamental for predicting and analyzing the behavior of objects in motion under gravity.

Lema 11.2.1 The *total flight time* of a projectile is:

$$T = \frac{2v_0 \sin\theta}{g}.$$

Demostración. The time to reach the maximum height is $t_{max} = \frac{v_0 \sin\theta}{g}$. Since the motion is symmetric, the total time is $T = 2t_{max}$. ■

Corollary 11.2.2 The *maximum horizontal range* is:

$$R = \frac{v_0^2 \sin 2\theta}{g}.$$

Demostración. The range is obtained by evaluating $x(T)$:

$$R = v_0 \cos\theta \, T = v_0 \cos\theta \left(\frac{2v_0 \sin\theta}{g} \right) = \frac{v_0^2 \sin 2\theta}{g}.$$

■

These results allow us to compute key parameters in projectile motion problems.

■ **Example 11.6** A basketball player throws a ball with an initial velocity of 8 m/s at an angle of 60° to the horizontal. Determine the maximum height and the horizontal range of the throw. ■

R Note that an angle of 45° maximizes the horizontal range for a given initial velocity, which is a common strategy in sports and military applications.

To consolidate these concepts, the following exercises are proposed.

Exercise 11.5 An object is launched vertically upwards with an initial velocity of 20 m/s. Calculate the time it takes to reach its maximum height and the maximum height attained. ■

Exercise 11.6 Show that for a projectile launched from the origin, the equation of its trajectory is $y = x\tan\theta - \frac{gx^2}{2v_0^2 \cos^2\theta}$. ■

These exercises facilitate the understanding and application of the theories presented in practical and complex situations.

11.2.2 Calculating Areas with Quadratic Equations.

In this section, we explore how to calculate areas under curves defined by quadratic equations. This topic is fundamental in mathematical analysis and has applications in various fields such as physics, engineering, and economics.

Definition 11.2.3 A *quadratic function* is a polynomial function of degree two of the form:

$$f(x) = ax^2 + bx + c,$$

where a, b, and c are constants and $a \neq 0$.

To calculate the area under the curve of a quadratic function between two points, we use definite integration techniques.

Theorem 11.2.3 The area under the curve of a quadratic function $f(x) = ax^2 + bx + c$ between $x = p$ and $x = q$ is:

$$A = \int_p^q f(x)\,dx = \left[\frac{a}{3}x^3 + \frac{b}{2}x^2 + cx\right]_p^q.$$

Demostración. To find the area under the curve of the quadratic function $f(x) = ax^2 + bx + c$ between $x = p$ and $x = q$, we compute the definite integral:

$$A = \int_p^q f(x)\,dx = \int_p^q (ax^2 + bx + c)\,dx.$$

We integrate term by term:

$$A = \int_p^q ax^2\,dx + \int_p^q bx\,dx + \int_p^q c\,dx.$$

Integrating each term, we obtain:

$$\int ax^2\,dx = \frac{a}{3}x^3, \quad \int bx\,dx = \frac{b}{2}x^2, \quad \int c\,dx = cx.$$

Thus,

$$A = \left[\frac{a}{3}x^3 + \frac{b}{2}x^2 + cx\right]_p^q.$$

Evaluating at the limits p and q:

$$A = \left(\frac{a}{3}q^3 + \frac{b}{2}q^2 + cq\right) - \left(\frac{a}{3}p^3 + \frac{b}{2}p^2 + cp\right).$$

This concludes the proof. ■

This result allows us to calculate exact areas under quadratic curves, which is essential in problems involving accumulation and change.

■ **Example 11.7** Calculate the area under the curve $f(x) = 2x^2 + 3x + 1$ between $x = 0$ and $x = 2$.
■

Solution. We use the previous theorem:

$$A = \left[\frac{2}{3}x^3 + \frac{3}{2}x^2 + x\right]_0^2 = \left(\frac{2}{3}(2)^3 + \frac{3}{2}(2)^2 + 2\right) - (0).$$

Calculate:

$$A = \left(\frac{16}{3} + 6 + 2\right) = \frac{16}{3} + 8.$$

Sum:

$$A = \frac{16}{3} + \frac{24}{3} = \frac{40}{3}.$$

Therefore, the area is $\frac{40}{3}$ square units. ■

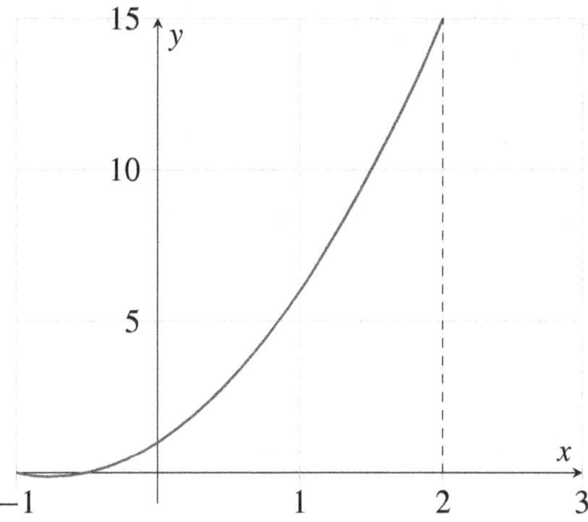

Figura 11.2.1: *Area under the curve* $f(x) = 2x^2 + 3x + 1$ *between* $x = 0$ *and* $x = 2$.

It is helpful to visualize the region whose area we have calculated. Below is a graph of the function and the area under the curve between $x = 0$ and $x = 2$.
Another interesting application is calculating the area between two quadratic functions.

Definition 11.2.4 The *area between two curves* $f(x)$ and $g(x)$ on the interval $[a,b]$ is:

$$A = \int_a^b |f(x) - g(x)|\, dx.$$

This concept allows us to calculate more complex areas, as shown in the following theorem.

Theorem 11.2.4 Let $f(x) = ax^2 + bx + c$ and $g(x) = dx^2 + ex + f$ be two continuous quadratic functions on $[p,q]$. The area between $f(x)$ and $g(x)$ is:

$$A = \int_p^q |(a-d)x^2 + (b-e)x + (c-f)|\, dx.$$

Demostración. To find the area between the two quadratic functions $f(x) = ax^2 + bx + c$ and $g(x) = dx^2 + ex + f$ on the interval $[p,q]$, we compute the integral of the absolute value of their difference, as the area between two curves is the integral of the vertical distance between them.
The difference between $f(x)$ and $g(x)$ is:

$$f(x) - g(x) = (a-d)x^2 + (b-e)x + (c-f).$$

Thus, the area A between $f(x)$ and $g(x)$ is:

$$A = \int_p^q |(a-d)x^2 + (b-e)x + (c-f)|\, dx.$$

This integral calculates the magnitude of the difference between $f(x)$ and $g(x)$ over the interval $[p,q]$, giving the total area between the two curves.
This concludes the proof. ■

To consolidate this concept, let's consider a practical example.

■ **Example 11.8** Calculate the area between the curves $f(x) = x^2$ and $g(x) = x + 2$ from $x = -1$ to $x = 2$. ■

Solution. First, we find the points where $f(x) = g(x)$ to determine if the functions intersect within the given interval.
Set the functions equal:

$$x^2 = x+2 \implies x^2 - x - 2 = 0.$$

Solve the quadratic equation:

$$x = \frac{1 \pm \sqrt{1+8}}{2} = \frac{1 \pm 3}{2}.$$

We get $x = 2$ and $x = -1$.
In the interval $[-1,2]$, the functions intersect at the endpoints, so the area between the curves is:

$$A = \int_{-1}^{2} |x^2 - x - 2| \, dx.$$

We compute the absolute value of $x^2 - x - 2$ in the interval:
Notice that $x^2 - x - 2$ is negative on $[-1,1]$ and positive on $[1,2]$.
Split the integral:

$$A = \int_{-1}^{1} -(x^2 - x - 2) \, dx + \int_{1}^{2} (x^2 - x - 2) \, dx.$$

Compute each integral separately.
First integral:

$$A_1 = -\int_{-1}^{1} (x^2 - x - 2) \, dx = -\left[\frac{x^3}{3} - \frac{x^2}{2} - 2x\right]_{-1}^{1}.$$

Evaluate:

$$A_1 = -\left(\left(\frac{1}{3} - \frac{1}{2} - 2\right) - \left(\frac{-1}{3} - \frac{1}{2} + 2\right)\right).$$

Simplify:

$$A_1 = -\left(-\frac{7}{6} - \frac{7}{6}\right) = -\left(-\frac{14}{6}\right) = \frac{14}{6} = \frac{7}{3}.$$

Second integral:

$$A_2 = \int_{1}^{2} (x^2 - x - 2) \, dx = \left[\frac{x^3}{3} - \frac{x^2}{2} - 2x\right]_{1}^{2}.$$

Evaluate:

$$A_2 = \left(\frac{8}{3} - 2 - 4\right) - \left(\frac{1}{3} - \frac{1}{2} - 2\right) = -\frac{10}{3} + \frac{7}{6}.$$

Compute:

$$A_2 = -\frac{20}{6} + \frac{7}{6} = -\frac{13}{6}.$$

Take the absolute value:

$$A_2 = \frac{13}{6}.$$

Add both areas:

$$A = A_1 + A_2 = \frac{7}{3} + \frac{13}{6} = \frac{14}{6} + \frac{13}{6} = \frac{27}{6} = \frac{9}{2}.$$

Therefore, the area between the curves is $\frac{9}{2}$ square units. ■

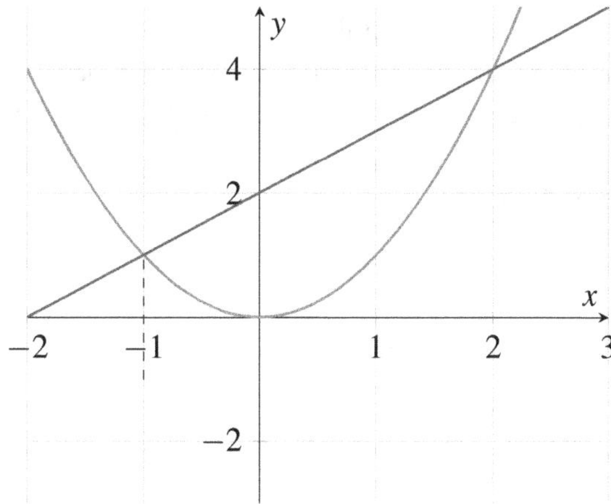

Figura 11.2.2: *Area between the curves $f(x) = x^2$ and $g(x) = x+2$ from $x = -1$ to $x = 2$.*

The graphical representation of these functions and the area between them is shown below.

> R It is important to note that when calculating the area between two curves, we must consider the intersection points and check if the functions cross within the given interval. This ensures the area calculation is correct and the absolute value is applied appropriately.

To reinforce the acquired knowledge, the following exercises are proposed.

Exercise 11.7 Calculate the area under the curve $f(x) = -x^2 + 4$ between $x = 0$ and $x = 2$. ■

Exercise 11.8 Determine the area between the curves $f(x) = x^2 - 4$ and $g(x) = -x$ in the interval $x \in [-2, 2]$. ■

These exercises will help apply the learned techniques and deepen the understanding of area calculations involving quadratic equations.

11.3 Graphing Quadratic Functions: Vertex, Axis of Symmetry, and Roots

11.3.1 Calculating the Vertex Using the General Formula

The quadratic function is fundamental in the study of mathematics, particularly in analysis and analytic geometry. To better understand its properties, it is essential to know how to determine the vertex of its graph from its general form.

Definition 11.3.1 A **quadratic function** is a function of the form $f(x) = ax^2 + bx + c$, where a, b, and c are real numbers and $a \neq 0$.

The vertex of the parabola represented by a quadratic function is a critical point that indicates the maximum or minimum of the function.

Theorem 11.3.1 The **vertex** of the parabola given by the quadratic function $f(x) = ax^2 + bx + c$ is located at the point

$$V\left(-\frac{b}{2a}, f\left(-\frac{b}{2a}\right)\right).$$

Demostración. To find the vertex of the parabola given by the quadratic function $f(x) = ax^2 + bx + c$, consider that the vertex is the point where the parabola reaches its maximum or minimum value. This occurs at the x-value where the derivative of $f(x)$ equals zero, as the tangent's slope is horizontal at the vertex.
We calculate the derivative of $f(x)$:

$$f'(x) = 2ax + b.$$

Setting the derivative equal to zero to find the critical point:

$$2ax + b = 0 \Rightarrow x = -\frac{b}{2a}.$$

This is the x-coordinate of the vertex. To find the y-coordinate of the vertex, we evaluate f at $x = -\frac{b}{2a}$:

$$f\left(-\frac{b}{2a}\right) = a\left(-\frac{b}{2a}\right)^2 + b\left(-\frac{b}{2a}\right) + c.$$

Simplifying, we obtain the y-coordinate of the vertex. Thus, the vertex of the parabola is the point

$$V\left(-\frac{b}{2a}, f\left(-\frac{b}{2a}\right)\right).$$

This concludes the proof. ■

R This result is essential for studying quadratic functions and is widely used in optimization problems and mathematical modeling.

■ **Example 11.9** Calculate the vertex of the quadratic function $f(x) = 2x^2 - 4x + 1$ and represent it graphically.
Identify the coefficients:

$$a = 2, \quad b = -4, \quad c = 1.$$

Calculate the x-coordinate of the vertex:

$$x_v = -\frac{b}{2a} = -\frac{-4}{2 \cdot 2} = 1.$$

Calculate the y-coordinate by evaluating x_v in $f(x)$:

$$y_v = f(1) = 2(1)^2 - 4(1) + 1 = -1.$$

Therefore, the vertex is $V(1, -1)$.

■

Figure 11.3.1 illustrates the parabola and highlights the vertex, facilitating the analysis of the function's behavior.

Exercise 11.9 Determine the vertex of the quadratic function $f(x) = -3x^2 + 6x - 2$ and draw its graph. ■

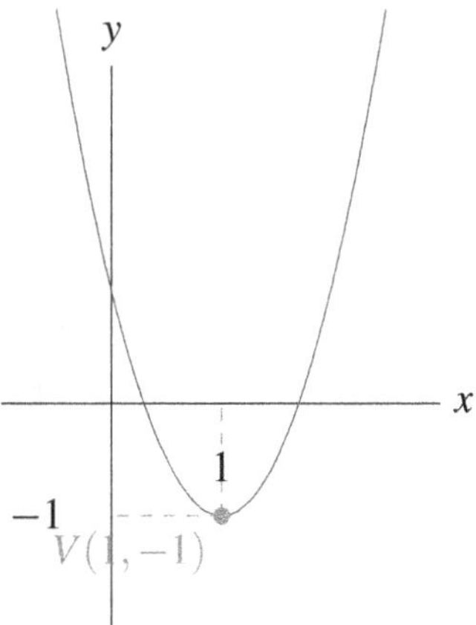

Figura 11.3.1: *Graph of the function $f(x) = 2x^2 - 4x + 1$ showing the vertex.*

Exercise 11.10 Find the vertex of $f(x) = x^2 - 4x + 7$ and analyze whether the function has a maximum or a minimum. Justify your answer. ■

Ⓡ Observe that the sign of the coefficient a in $f(x) = ax^2 + bx + c$ determines the concavity of the parabola:

- If $a > 0$, the parabola opens upwards, and the vertex is a **minimum**.
- If $a < 0$, the parabola opens downwards, and the vertex is a **maximum**.

This property is crucial in problems involving the optimization of quadratic functions.

In conclusion, knowing the method to calculate the vertex from the general formula is a powerful tool in the analysis of quadratic functions and their applications in various fields of mathematics and physics.
subsectionGraphs in Relation to Physical Problems.
In this section, we explore how mathematical graphs are integrated into the interpretation and resolution of physical problems. Graphs are essential tools that allow us to visualize relationships between physical variables and understand the dynamic behavior of systems.

Definition 11.3.2 A *graph of a function* is the geometric representation of the set of ordered pairs $(x, f(x))$, where x belongs to the domain of the function f. In physics, these graphs illustrate how one physical quantity depends on another.

Understanding graphs is fundamental to interpreting phenomena such as motion, energy, and forces. Let us consider an example relating a graph to a physical problem.

■ **Example 11.10** Consider a simple harmonic oscillator with the equation of motion $x(t) = A\cos(\omega t + \phi)$, where:

- A is the amplitude,
- ω is the angular frequency,
- ϕ is the initial phase.

The graph of $x(t)$ as a function of time t shows the periodic motion of the system.
Below is the graph for $A = 1$ m, $\omega = \pi$ rad/s, and $\phi = 0$ rad.

■

This example illustrates how the graph allows us to visualize oscillatory behavior and predict future positions of the system.

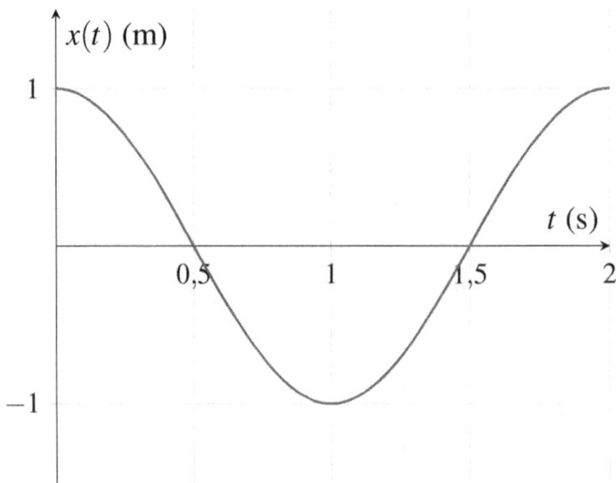

Figura 11.3.2: *Motion of a simple harmonic oscillator with $A = 1$ m, $\omega = \pi$ rad/s, and $\phi = 0$ rad.*

Theorem 11.3.2 If a function $f : \mathbb{R} \to \mathbb{R}$ is continuous and differentiable on an interval I, then its graph on I represents a physical behavior without discontinuities or abrupt changes, which is essential for modeling real-world physical systems.

Demostración. The continuity of f ensures there are no jumps in the graph, meaning that for all $x \in I$, the values of $f(x)$ vary without interruptions. Differentiability ensures that the rate of change of $f(x)$ with respect to x is defined at every point in I, implying a smooth change in the physical quantity represented. ■

This theorem is crucial in physics, where many quantities change continuously and differentiably with respect to time or space.

Corollary 11.3.3 If f is twice differentiable on I, then its second derivative $f''(x)$ exists on I, enabling the analysis of accelerations in physical systems where $f(x)$ represents position or velocity.

Demostración. Since f is differentiable on I and its derivative f' is also differentiable, the second derivative f'' exists and is continuous on I. This is essential for calculating accelerations in classical mechanics. ■

The ability to graph f, f', and f'' provides a comprehensive view of the system's dynamic behavior.

R In physical problems, the slope of the graph of position versus time represents velocity, and the slope of the graph of velocity versus time represents acceleration. This geometric interpretation is essential for motion analysis.

Let us consider another example demonstrating the application of graphs in a physical context.

■ **Example 11.11** Let us analyze the motion of a projectile launched from the origin with an initial velocity $v_0 = 20$ m/s at an angle $\theta = 45°$ to the horizontal. The parametric equations of motion are:

$$\begin{cases} x(t) = v_0 \cos\theta \, t, \\ y(t) = v_0 \sin\theta \, t - \dfrac{1}{2} g t^2, \end{cases}$$

where $g = 9{,}8$ m/s^2 is the acceleration due to gravity.
The trajectory of the projectile is a parabola, and its graph is shown below.

■

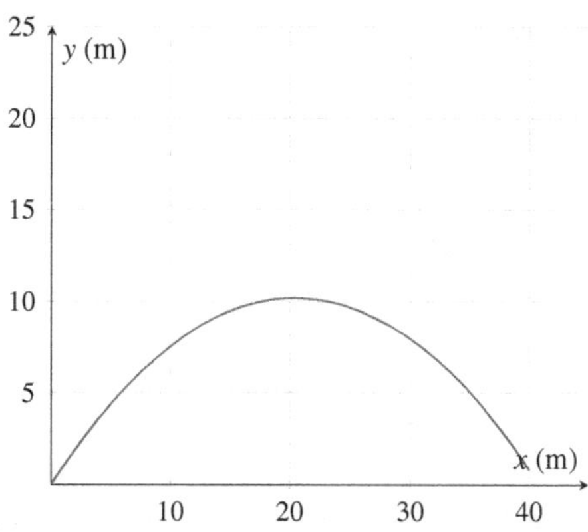

Figura 11.3.3: *Trajectory of a projectile with* $v_0 = 20$ *m/s and* $\theta = 45°$.

The graph allows us to visualize the trajectory and determine characteristics such as the maximum range and the maximum height reached by the projectile.

Lema 11.3.1 The slope of the position graph with respect to time, $x(t)$, represents the instantaneous velocity $v(t) = \frac{dx}{dt}$.

Demostración. By the definition of a derivative, the instantaneous rate of change of position with respect to time is the velocity:

$$v(t) = \lim_{\Delta t \to 0} \frac{x(t+\Delta t) - x(t)}{\Delta t} = \frac{dx}{dt}.$$

Geometrically, this rate of change corresponds to the slope of the tangent line to the curve $x(t)$ at the point t. ■

(R) The interpretation of derivatives and slopes in graphs is fundamental to understanding concepts such as velocity and acceleration in physics. This facilitates the prediction and analysis of the motion of particles and objects.

Let us now see how these ideas apply to practical exercises.

Exercise 11.11 Given the position function $x(t) = 5t^3 - 2t^2 + t$, perform the following:

1. Graph the position $x(t)$, velocity $v(t)$, and acceleration $a(t)$ in the interval $t \in [0,2]$.
2. Physically interpret the behavior of the object in this interval.

Exercise 11.12 An object moves in a straight line with velocity $v(t) = 4\cos(2\pi t)$ m/s.

1. Graph the velocity as a function of time for $t \in [0,1]$ s.
2. Determine the instances when the object changes direction.

These exercises allow the application of theoretical concepts to concrete situations, reinforcing the understanding of the relationship between graphs and physical problems.

(R) In Figure 11.3.9, the points where $v(t) = 0$ indicate changes in the direction of the object's motion. This occurs when $\cos(2\pi t) = 0$, i.e., at $t = \frac{1}{4}$ s and $t = \frac{3}{4}$ s.

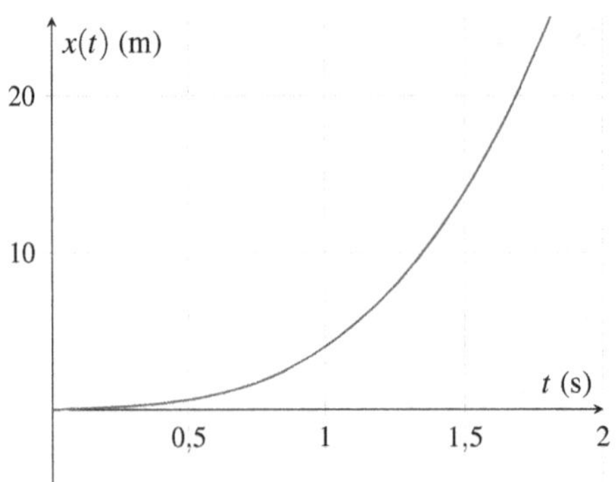

Figura 11.3.4: *Graph of position $x(t) = 5t^3 - 2t^2 + t$ in the interval $t \in [0,2]$.*

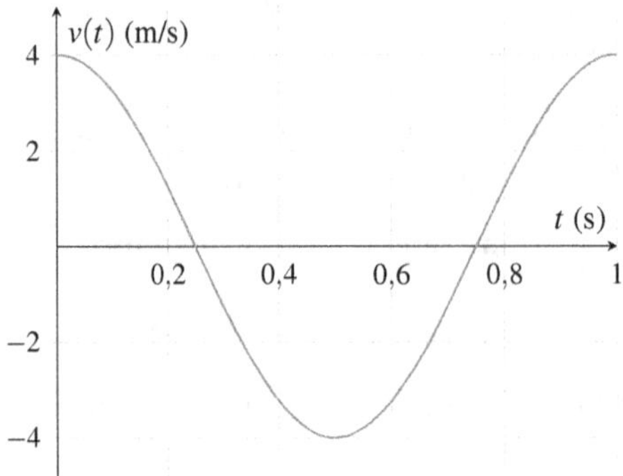

Figura 11.3.5: *Graph of velocity $v(t) = 4\cos(2\pi t)$ m/s for $t \in [0,1]$ s.*

11.3.2 Graphs in Relation to Physical Problems.

In this section, we explore how mathematical graphs are integrated into the interpretation and resolution of physical problems. Graphs are essential tools that allow us to visualize relationships between physical variables and understand the dynamic behavior of systems.

Definition 11.3.3 A *graph of a function* is the geometric representation of the set of ordered pairs $(x, f(x))$, where x belongs to the domain of the function f. In physics, these graphs illustrate how one physical quantity depends on another.

Understanding graphs is fundamental to interpreting phenomena such as motion, energy, and forces. Let us consider an example relating a graph to a physical problem.

■ **Example 11.12** Consider a simple harmonic oscillator with the equation of motion $x(t) = A\cos(\omega t + \phi)$, where:

- A is the amplitude,
- ω is the angular frequency,
- ϕ is the initial phase.

The graph of $x(t)$ as a function of time t shows the periodic motion of the system.
Below is the graph for $A = 1$ m, $\omega = \pi$ rad/s, and $\phi = 0$ rad.

■

This example illustrates how the graph allows us to visualize oscillatory behavior and predict future positions of the system.

Theorem 11.3.4 If a function $f : \mathbb{R} \to \mathbb{R}$ is continuous and differentiable on an interval I, then its graph on I represents a physical behavior without discontinuities or abrupt changes, which is

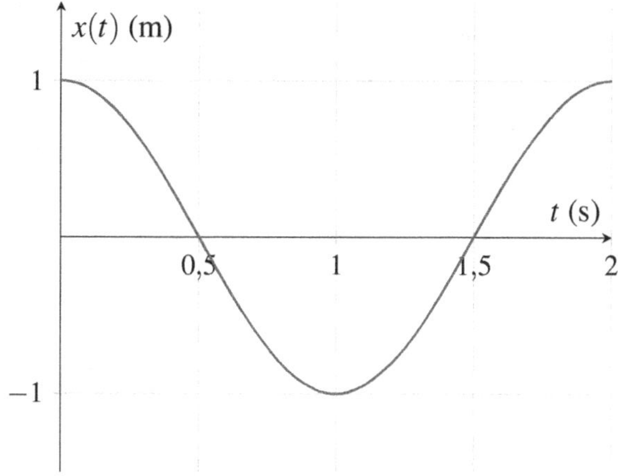

Figura 11.3.6: *Motion of a simple harmonic oscillator with* $A = 1$ *m,* $\omega = \pi$ *rad/s, and* $\phi = 0$ *rad.*

essential for modeling real-world physical systems.

Demostración. The continuity of f ensures there are no jumps in the graph, meaning that for all $x \in I$, the values of $f(x)$ vary without interruptions. Differentiability ensures that the rate of change of $f(x)$ with respect to x is defined at every point in I, implying a smooth change in the physical quantity represented. ■

This theorem is crucial in physics, where many quantities change continuously and differentiably with respect to time or space.

Corollary 11.3.5 If f is twice differentiable on I, then its second derivative $f''(x)$ exists on I, enabling the analysis of accelerations in physical systems where $f(x)$ represents position or velocity.

Demostración. Since f is differentiable on I and its derivative f' is also differentiable, the second derivative f'' exists and is continuous on I. This is essential for calculating accelerations in classical mechanics. ■

The ability to graph f, f', and f'' provides a comprehensive view of the system's dynamic behavior.

R In physical problems, the slope of the graph of position versus time represents velocity, and the slope of the graph of velocity versus time represents acceleration. This geometric interpretation is essential for motion analysis.

Let us consider another example demonstrating the application of graphs in a physical context.

■ **Example 11.13** Let us analyze the motion of a projectile launched from the origin with an initial velocity $v_0 = 20$ m/s at an angle $\theta = 45°$ to the horizontal. The parametric equations of motion are:

$$\begin{cases} x(t) = v_0 \cos\theta \, t, \\ y(t) = v_0 \sin\theta \, t - \dfrac{1}{2} g t^2, \end{cases}$$

where $g = 9{,}8$ m/s^2 is the acceleration due to gravity.
The trajectory of the projectile is a parabola, and its graph is shown below.

■

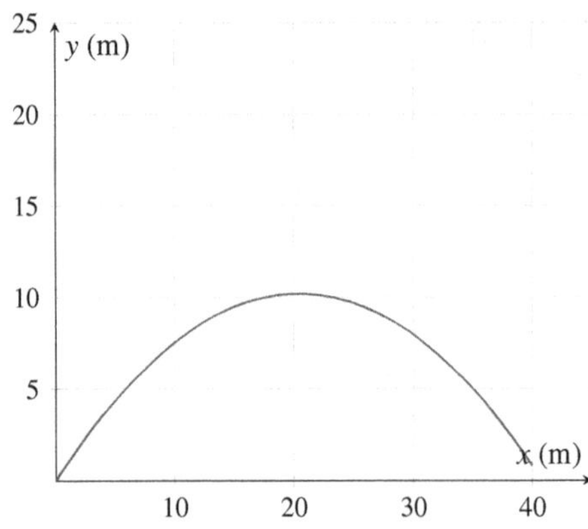

Figura 11.3.7: *Trajectory of a projectile with* $v_0 = 20$ *m/s and* $\theta = 45°$.

The graph allows us to visualize the trajectory and determine characteristics such as the maximum range and the maximum height reached by the projectile.

Lema 11.3.2 The slope of the position graph with respect to time, $x(t)$, represents the instantaneous velocity $v(t) = \frac{dx}{dt}$.

Demostración. By the definition of a derivative, the instantaneous rate of change of position with respect to time is the velocity:

$$v(t) = \lim_{\Delta t \to 0} \frac{x(t+\Delta t) - x(t)}{\Delta t} = \frac{dx}{dt}.$$

Geometrically, this rate of change corresponds to the slope of the tangent line to the curve $x(t)$ at the point t. ■

(R) The interpretation of derivatives and slopes in graphs is fundamental to understanding concepts such as velocity and acceleration in physics. This facilitates the prediction and analysis of the motion of particles and objects.

Let us now see how these ideas apply to practical exercises.

Exercise 11.13 Given the position function $x(t) = 5t^3 - 2t^2 + t$, perform the following:

1. Graph the position $x(t)$, velocity $v(t)$, and acceleration $a(t)$ in the interval $t \in [0,2]$.
2. Physically interpret the behavior of the object in this interval.

■

Exercise 11.14 An object moves in a straight line with velocity $v(t) = 4\cos(2\pi t)$ m/s.

1. Graph the velocity as a function of time for $t \in [0,1]$ s.
2. Determine the instances when the object changes direction.

■

These exercises allow the application of theoretical concepts to concrete situations, reinforcing the understanding of the relationship between graphs and physical problems.

(R) In Figure 11.3.9, the points where $v(t) = 0$ indicate changes in the direction of the object's motion. This occurs when $\cos(2\pi t) = 0$, i.e., at $t = \frac{1}{4}$ s and $t = \frac{3}{4}$ s.

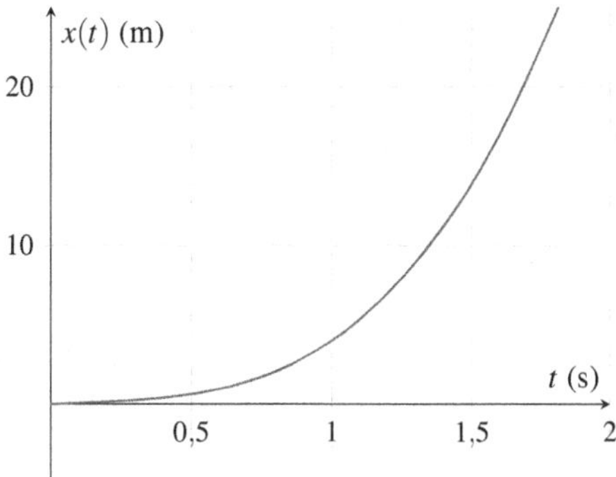

Figura 11.3.8: *Graph of position $x(t) = 5t^3 - 2t^2 + t$ in the interval $t \in [0,2]$.*

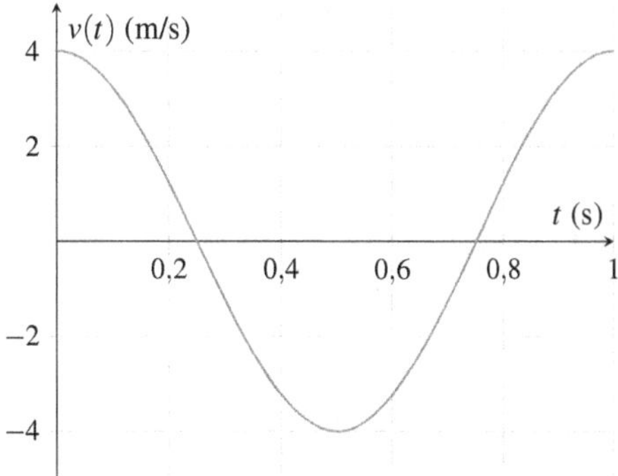

Figura 11.3.9: *Graph of velocity $v(t) = 4\cos(2\pi t)$ m/s for $t \in [0,1]$ s.*

11.4 Solved Exercises

Exercise 11.15 Solve the quadratic equation $x^2 - 6x + 8 = 0$ using factorization.

Demostración. To solve $x^2 - 6x + 8 = 0$ by factorization, we look for two numbers that multiply to 8 and add to -6. These numbers are -4 and -2. Thus, we can factorize as:

$$x^2 - 6x + 8 = (x-4)(x-2) = 0.$$

Solving each factor set to zero, we get:

$$x - 4 = 0 \Rightarrow x = 4,$$

$$x - 2 = 0 \Rightarrow x = 2.$$

Therefore, the solutions are $x = 4$ and $x = 2$. ■

Exercise 11.16 Solve the quadratic equation $3x^2 + 12x + 12 = 0$ using the quadratic formula.

Demostración. To solve $3x^2 + 12x + 12 = 0$, we use the quadratic formula:

$$x = \frac{-b \pm \sqrt{b^2 - 4ac}}{2a}.$$

Here, $a = 3$, $b = 12$, and $c = 12$. Calculate the discriminant:

$$b^2 - 4ac = 12^2 - 4 \cdot 3 \cdot 12 = 144 - 144 = 0.$$

Since the discriminant is 0, there is one real solution:

$$x = \frac{-12}{2 \cdot 3} = \frac{-12}{6} = -2.$$

Thus, the solution is $x = -2$. ■

Exercise 11.17 Complete the square to express the quadratic function $f(x) = x^2 + 6x + 5$ in its canonical form. ■

Demostración. To complete the square for $f(x) = x^2 + 6x + 5$, we take the coefficient of x, which is 6, divide it by 2, and square it:

$$\left(\frac{6}{2}\right)^2 = 9.$$

We rewrite the function as:

$$f(x) = (x^2 + 6x + 9) - 9 + 5 = (x+3)^2 - 4.$$

The canonical form is:

$$f(x) = (x+3)^2 - 4.$$

Therefore, the vertex of the parabola is $(-3, -4)$. ■

Exercise 11.18 Calculate the area under the curve $f(x) = 2x^2 + 3x$ in the interval $[0, 2]$. ■

Demostración. To calculate the area under the curve $f(x) = 2x^2 + 3x$ in the interval $[0, 2]$, we integrate the function over this interval:

$$A = \int_0^2 (2x^2 + 3x)\, dx.$$

Integrating term by term, we get:

$$A = \left[\frac{2}{3}x^3 + \frac{3}{2}x^2\right]_0^2.$$

Evaluate at the limits:

$$A = \left(\frac{2}{3}(2)^3 + \frac{3}{2}(2)^2\right) - \left(\frac{2}{3}(0)^3 + \frac{3}{2}(0)^2\right).$$

Simplify:

$$A = \frac{2}{3} \cdot 8 + \frac{3}{2} \cdot 4 = \frac{16}{3} + 6 = \frac{34}{3}.$$

Thus, the area is $\frac{34}{3}$ square units. ■

Exercise 11.19 Determine the vertex and the roots of the quadratic function $f(x) = -x^2+4x-3$. ■

Demostración. To find the vertex of $f(x) = -x^2+4x-3$, we use the vertex formula:

$$x_v = -\frac{b}{2a}.$$

Here, $a = -1$ and $b = 4$, so:

$$x_v = -\frac{4}{2\cdot(-1)} = 2.$$

Evaluate $f(2)$ to find the y-coordinate of the vertex:

$$f(2) = -(2)^2+4\cdot 2-3 = -4+8-3 = 1.$$

Thus, the vertex is $(2,1)$.
To find the roots, solve $-x^2+4x-3 = 0$ using the quadratic formula:

$$x = \frac{-b\pm\sqrt{b^2-4ac}}{2a}.$$

With $a = -1$, $b = 4$, and $c = -3$:

$$x = \frac{-4\pm\sqrt{16+12}}{-2} = \frac{-4\pm\sqrt{28}}{-2} = \frac{-4\pm 2\sqrt{7}}{-2}.$$

Divide each term by -2:

$$x = 2\pm\sqrt{7}.$$

Thus, the roots are $x = 2+\sqrt{7}$ and $x = 2-\sqrt{7}$. ■

11.5 Proposed Exercises

11.5.1 Quadratic Equations: Factorization and General Formula

Exercise 11.20 Solve the quadratic equation $x^2-5x+6 = 0$ using the factorization method. ■

Exercise 11.21 Find the roots of the equation $2x^2+7x+3 = 0$ by applying the quadratic formula. ■

Exercise 11.22 Solve the equation $x^2+4x+4 = 0$ using the method of completing the square. ■

Exercise 11.23 Determine the values of m for which the equation $x^2+mx+16 = 0$ has real roots. ■

Exercise 11.24 Solve the quadratic equation $3x^2-x-4 = 0$ and classify the roots by their nature. ■

11.5.2 Practical Applications: Problems Involving Areas and Trajectories

Exercise 11.25 An object is launched vertically upwards from a height of 20 meters with an initial velocity of 15 m/s. Write the equation that models its position as a function of time and determine how long it takes to reach the ground. ■

Exercise 11.26 Calculate the area under the curve of the function $f(x) = x^2 + 2x$ between $x = 0$ and $x = 3$. ■

Exercise 11.27 A projectile is launched with a velocity of 30 m/s at an angle of 45°. Determine the equation of its trajectory and calculate the maximum range. ■

Exercise 11.28 A ball is thrown from a height of 1.5 meters with an initial velocity of 10 m/s. How long does it take to reach its maximum height, and what is that height? ■

Exercise 11.29 Find the area between the functions $f(x) = x^2$ and $g(x) = 4x - x^2$ in the interval $[0,2]$. ■

11.5.3 Graphing Quadratic Functions: Vertex, Axis of Symmetry, and Roots

Exercise 11.30 Determine the vertex, roots, and axis of symmetry of the quadratic function $f(x) = -2x^2 + 8x - 3$. ■

Exercise 11.31 Find the vertex and graph the quadratic function $f(x) = x^2 - 4x + 5$. ■

Exercise 11.32 Plot the graph of the quadratic function $f(x) = 3x^2 - 6x + 2$ and calculate its intersections with the axes. ■

Exercise 11.33 For the function $f(x) = -x^2 + 6x - 8$, determine the vertex and the interval where the function is decreasing. ■

Exercise 11.34 Analyze the concavity and determine the vertex of the function $f(x) = \frac{1}{2}x^2 - 3x + 7$. Does it have a maximum or a minimum? ■

12. Mathematical Modeling with Inequalities

12.1 Linear and Quadratic Inequalities: Resolution and Representation on the Number Line.

12.1.1 Solving Inequalities with One Variable.

In this section, we will address the resolution of inequalities with one real variable. Inequalities are mathematical expressions that establish an order relationship between two algebraic expressions. The ability to solve inequalities is fundamental in various areas of mathematics and its applications.

Definition 12.1.1 An *inequality* is an expression of the form $f(x) < g(x)$, $f(x) \leq g(x)$, $f(x) > g(x)$ or $f(x) \geq g(x)$, where $f(x)$ and $g(x)$ are functions or algebraic expressions. Solving an inequality involves finding all values of x that satisfy the established relationship.

It is essential to understand the fundamental properties of inequalities to manipulate them correctly and obtain precise solutions.

Theorem 12.1.1 — **Properties of Inequalities.** Let a, b, and c be real numbers.

1. If $a < b$, then $a+c < b+c$.
2. If $a < b$ and $c > 0$, then $ac < bc$.
3. If $a < b$ and $c < 0$, then $ac > bc$.
4. If $a < b$ and $b < c$, then $a < c$ (transitivity).

Demostración. We will prove each property separately:
1. If $a < b$, then $a+c < b+c$:
Adding c to both sides of the inequality $a < b$, we get $a+c < b+c$, since adding the same quantity to both sides of an inequality does not alter the order.
2. If $a < b$ and $c > 0$, then $ac < bc$:
Multiplying both sides of $a < b$ by $c > 0$, the direction of the inequality remains unchanged, resulting in $ac < bc$.
3. If $a < b$ and $c < 0$, then $ac > bc$:
Multiplying both sides of $a < b$ by $c < 0$, the inequality reverses its direction, leading to $ac > bc$.

4. If $a < b$ and $b < c$, then $a < c$ (transitivity):
Given $a < b$ and $b < c$, the transitive property of inequalities allows us to conclude that $a < c$.
This concludes the proof of the properties of inequalities. ■

These properties are indispensable tools for solving inequalities, especially when multiplying or dividing both sides by a quantity that may be negative.

■ **Example 12.1** Solve the linear inequality $3x - 5 < 7$. ■

Solution. We proceed to isolate the variable x:

$$
\begin{aligned}
3x - 5 &< 7 \\
3x &< 7 + 5 \\
3x &< 12 \\
x &< \frac{12}{3} \\
x &< 4.
\end{aligned}
$$

The solution is the set of all real numbers x such that $x < 4$. ■

The solution can be graphically represented on the number line for better visualization of the solution set.

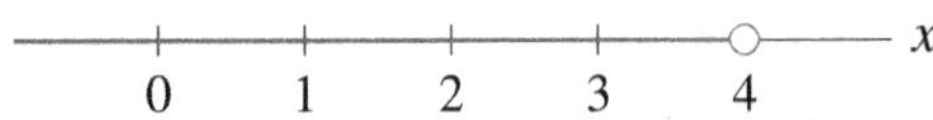

Figura 12.1.1: *Graphical representation of the solution $x < 4$.*

Moving on to more complex inequalities, let us consider those involving quadratic expressions.

Definition 12.1.2 A *quadratic inequality* is an inequality of the form $ax^2 + bx + c \square 0$, where a, b, c are real numbers, $a \neq 0$, and $\square$ is one of the symbols $<, \leq, >$ or $\geq$.

To solve quadratic inequalities, it is useful to analyze the sign of the corresponding quadratic function and its roots.

Theorem 12.1.2 Let $f(x) = ax^2 + bx + c$ be a quadratic function, and let x_1 and x_2 be the real roots of the equation $f(x) = 0$ (assuming they exist and are distinct). Then:

1. If $a > 0$, then $f(x) > 0$ for $x < x_1$ and $x > x_2$, and $f(x) < 0$ for $x_1 < x < x_2$.
2. If $a < 0$, then $f(x) < 0$ for $x < x_1$ and $x > x_2$, and $f(x) > 0$ for $x_1 < x < x_2$.

Demostración. Consider the quadratic function $f(x) = ax^2 + bx + c$, and let $f(x) = 0$ be the associated quadratic equation, whose distinct real roots are x_1 and x_2. The factorization of $f(x)$ can be written as:

$$f(x) = a(x - x_1)(x - x_2).$$

Let us analyze the sign of $f(x)$ in the intervals determined by x_1 and x_2.

1. Case $a > 0$:
   - For $x < x_1$: in this interval, both terms $(x - x_1)$ and $(x - x_2)$ are negative, so their product $(x - x_1)(x - x_2)$ is positive. Since $a > 0$, this implies that $f(x) > 0$. - For $x_1 < x < x_2$: in this interval, $(x - x_1)$ is positive and $(x - x_2)$ is negative, making their product negative. Since

$a > 0$, we have $f(x) < 0$. - For $x > x_2$: in this interval, both terms $(x - x_1)$ and $(x - x_2)$ are positive, so their product is positive. Since $a > 0$, we have $f(x) > 0$.

2. Case $a < 0$:
   - For $x < x_1$: in this interval, both terms $(x - x_1)$ and $(x - x_2)$ are negative, so their product is positive. Since $a < 0$, this implies that $f(x) < 0$. - For $x_1 < x < x_2$: in this interval, $(x - x_1)$ is positive and $(x - x_2)$ is negative, making their product negative. Since $a < 0$, we have $f(x) > 0$. - For $x > x_2$: in this interval, both terms $(x - x_1)$ and $(x - x_2)$ are positive, so their product is positive. Since $a < 0$, we have $f(x) < 0$.

This concludes the proof. ■

This property allows us to solve quadratic inequalities by analyzing the sign of the function in different intervals defined by its roots.

■ **Example 12.2** Solve the inequality $x^2 - 5x + 6 > 0$. ■

Solution. First, find the roots of the associated quadratic equation $x^2 - 5x + 6 = 0$:

$$x^2 - 5x + 6 = 0 \implies (x-2)(x-3) = 0 \implies x = 2 \text{ and } x = 3.$$

These roots divide the real line into three intervals: $(-\infty, 2)$, $(2, 3)$, and $(3, \infty)$. Let us evaluate the sign of $f(x) = x^2 - 5x + 6$ in each interval:

- For $x < 2$, take $x = 1$:

$$f(1) = 1^2 - 5 \cdot 1 + 6 = 1 - 5 + 6 = 2 > 0.$$

- For $2 < x < 3$, take $x = 2{,}5$:

$$f(2{,}5) = (2{,}5)^2 - 5 \cdot 2{,}5 + 6 = 6{,}25 - 12{,}5 + 6 = -0{,}25 < 0.$$

- For $x > 3$, take $x = 4$:

$$f(4) = 4^2 - 5 \cdot 4 + 6 = 16 - 20 + 6 = 2 > 0.$$

Thus, $f(x) > 0$ in $(-\infty, 2)$ and $(3, \infty)$, and $f(x) < 0$ in $(2, 3)$. Since we are looking for $f(x) > 0$, the solution is:

$$x \in (-\infty, 2) \cup (3, \infty).$$

■

The graphical representation of the solution is as follows:

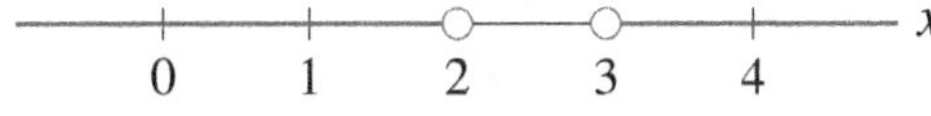

Figura 12.1.2: *Graphical representation of the solution $x \in (-\infty, 2) \cup (3, \infty)$.*

(R) It is important to consider whether the roots are included in the solution. In this case, the inequality is strict ($>$), so $x = 2$ and $x = 3$ are not included in the solution set.

For more complex inequalities, we can use the sign table method.

Lema 12.1.1 — **Sign Table Method.** To solve the inequality $f(x)\square 0$, where $f(x)$ is a factored expression, follow these steps:

1. Completely factorize $f(x)$.
2. Identify the zeros of each factor (critical points).
3. Divide the real line into intervals based on these critical points.
4. Determine the sign of $f(x)$ in each interval.
5. Select the intervals that satisfy the inequality.

▪ **Example 12.3** Solve the inequality $(x-1)(x+2)(x-3) \le 0$. ▪

Solution. Identify the critical points by setting each factor equal to zero:

$$x-1=0 \implies x=1,$$
$$x+2=0 \implies x=-2,$$
$$x-3=0 \implies x=3.$$

Arrange the critical points: $x=-2$, $x=1$, $x=3$. Divide the real line into intervals:

1. $(-\infty,-2)$
2. $(-2,1)$
3. $(1,3)$
4. $(3,\infty)$

Analyze the sign of $f(x)$ in each interval:

- **Interval** $(-\infty,-2)$: Take $x=-3$:

$$f(-3)=(-3-1)(-3+2)(-3-3)=(-4)(-1)(-6)=-24<0.$$

- **Interval** $(-2,1)$: Take $x=0$:

$$f(0)=(-1)(2)(-3)=6>0.$$

- **Interval** $(1,3)$: Take $x=2$:

$$f(2)=(2-1)(2+2)(2-3)=(1)(4)(-1)=-4<0.$$

- **Interval** $(3,\infty)$: Take $x=4$:

$$f(4)=(4-1)(4+2)(4-3)=(3)(6)(1)=18>0.$$

The sign of $f(x)$ is negative in $(-\infty,-2)$ and $(1,3)$, and positive in $(-2,1)$ and $(3,\infty)$. Since we are looking for $f(x)\le 0$, we select the intervals where $f(x)$ is negative or zero.
Let us verify if the critical points satisfy the inequality:

$$f(-2)=(-2-1)(-2+2)(-2-3)=(-3)(0)(-5)=0,$$
$$f(1)=(1-1)(1+2)(1-3)=(0)(3)(-2)=0,$$
$$f(3)=(3-1)(3+2)(3-3)=(2)(5)(0)=0.$$

Since $f(x)=0$ at $x=-2$, 1, 3, and the inequality includes equality ($\le$), these points are included in the solution.
Therefore, the solution is:

$$x\in(-\infty,-2]\cup[1,3].$$

■

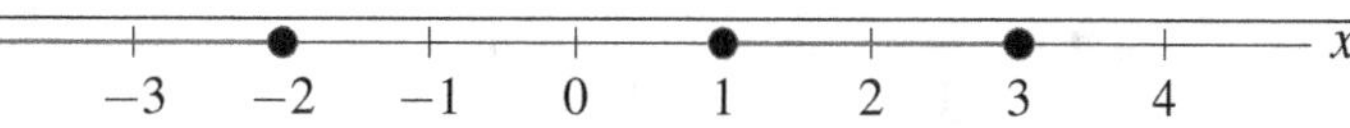

Figura 12.1.3: *Graphical representation of the solution* $x \in (-\infty, -2] \cup [1, 3]$.

The graphical representation of the solution is:

> The sign table method is especially useful for inequalities involving products or quotients of linear factors, allowing a systematic analysis of the signs in each interval.

Finally, let us address inequalities involving absolute values, which require special consideration due to the nature of absolute value.

Theorem 12.1.3 To solve an inequality of the form $|f(x)| \leq k$, with $k \geq 0$, it can be rewritten as $-k \leq f(x) \leq k$.

Demostración. Consider the inequality $|f(x)| \leq k$ with $k \geq 0$. By the definition of absolute value, $|f(x)| \leq k$ means that $f(x)$ is at most a distance of k from the origin. This translates to the condition:

$$-k \leq f(x) \leq k.$$

To justify this, note that $|f(x)| \leq k$ implies $-k \leq f(x) \leq k$, because if $f(x) > k$ or $f(x) < -k$, then $|f(x)|$ would exceed k, contradicting the given inequality. Thus, we can rewrite $|f(x)| \leq k$ as

$$-k \leq f(x) \leq k.$$

This concludes the proof. ■

■ **Example 12.4** Solve the inequality $|2x - 3| \leq 5$. ■

Solution. We apply the previous theorem:

$$-5 \leq 2x - 3 \leq 5.$$

Solving the inequalities simultaneously:

$$\begin{aligned} -5+3 \leq 2x &\leq 5+3 \\ -2 \leq 2x &\leq 8 \\ \frac{-2}{2} \leq x &\leq \frac{8}{2} \\ -1 \leq x &\leq 4. \end{aligned}$$

Thus, the solution is:

$$x \in [-1, 4].$$

■

The graphical representation is:

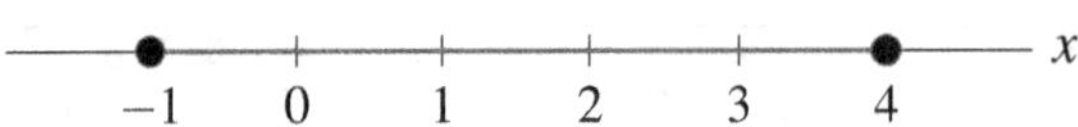

Figura 12.1.4: *Graphical representation of the solution $x \in [-1,4]$.*

Exercise 12.1 Solve the inequality $\dfrac{x-4}{x+1} > 0$ and represent the solution on the number line. ■

Exercise 12.2 Determine all values of x that satisfy the inequality $|x^2 - 9| \geq 0$. ■

The techniques and concepts presented in this section are fundamental for solving inequalities with one variable. A deep understanding of these methods will allow addressing more complex problems and applying these skills in various areas of mathematics and their applications.

12.1.2 Graphical Representation of Quadratic Inequalities.

In this section, we will explore how to graphically represent the solutions of quadratic inequalities. This topic is fundamental to understanding the relationship between algebraic expressions and their geometric interpretation on the number line and Cartesian plane.

Definition 12.1.3 A *quadratic inequality* is an inequality of the form:

$$ax^2 + bx + c \square 0,$$

where a, b, $c \in \mathbb{R}$, $a \neq 0$, and $\square$ is one of the symbols $<$, $\leq$, $>$, or $\geq$.

Solving quadratic inequalities involves determining the values of x that satisfy the given inequality. For this purpose, it is useful to analyze the associated quadratic function and its graph.

Theorem 12.1.4 Let $f(x) = ax^2 + bx + c$ be a quadratic function, and let x_1 and x_2 be the real roots of the equation $f(x) = 0$ (assuming they exist and are distinct). Then, the graph of $f(x)$ is a parabola that intersects the x-axis at x_1 and x_2. The parabola opens upwards if $a > 0$ and downwards if $a < 0$.

Demostración. Consider the quadratic function $f(x) = ax^2 + bx + c$ and the equation $f(x) = 0$, whose distinct real roots are x_1 and x_2. This implies that we can factorize $f(x)$ as

$$f(x) = a(x - x_1)(x - x_2).$$

The graph of $f(x)$ is a parabola. The roots x_1 and x_2 represent the points where $f(x) = 0$, i.e., the points where the parabola intersects the x-axis.
To determine the direction in which the parabola opens, we observe the coefficient a:
1. If $a > 0$, the term ax^2 dominates for large values of $|x|$, and since it is positive, the parabola opens upwards.
2. If $a < 0$, the term ax^2 dominates for large values of $|x|$, and since it is negative, the parabola opens downwards.
Thus, the parabola intersects the x-axis at x_1 and x_2, opens upwards if $a > 0$, and downwards if $a < 0$.
This concludes the proof. ■

This theorem allows us to graphically interpret the solutions of a quadratic inequality based on the position of the parabola with respect to the x-axis.

■ **Example 12.5** Solve the inequality $x^2 - 4x + 3 \geq 0$ and graphically represent the solution. ■

Solution. First, we find the roots of the associated equation $x^2 - 4x + 3 = 0$:

$$x^2 - 4x + 3 = 0 \implies (x-1)(x-3) = 0 \implies x = 1 \text{ and } x = 3.$$

The roots divide the real line into three intervals:
1. $(-\infty, 1]$ 2. $[1, 3]$ 3. $[3, \infty)$
Since $a = 1 > 0$, the parabola opens upwards. We evaluate the sign of $f(x)$ in each interval:
- For $x < 1$, take $x = 0$:

$$f(0) = (0)^2 - 4(0) + 3 = 3 > 0$$

- For $1 < x < 3$, take $x = 2$:

$$f(2) = (2)^2 - 4(2) + 3 = 4 - 8 + 3 = -1 < 0$$

- For $x > 3$, take $x = 4$:

$$f(4) = (4)^2 - 4(4) + 3 = 16 - 16 + 3 = 3 > 0$$

We look for the values of x where $f(x) \geq 0$, i.e., where the parabola is above or touches the x-axis. Therefore, the solution is:

$$x \in (-\infty, 1] \cup [3, \infty)$$

The graphical representation is:

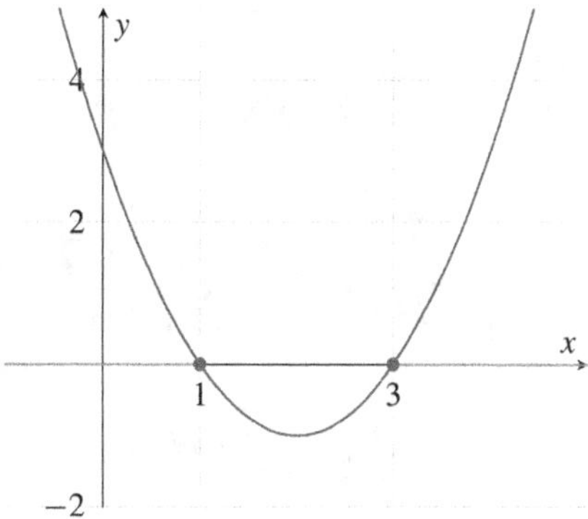

Figura 12.1.5: *Graph of $y = x^2 - 4x + 3$ and solution of $x^2 - 4x + 3 \geq 0$.*

In Figure 12.1.5, the regions where the parabola is above or touches the x-axis correspond to the solution of the inequality. ■

(R) The graphical representation facilitates understanding the solutions of quadratic inequalities, allowing visualization of the intervals where the function satisfies the inequality.

To generalize this approach, consider the following lemma.

Lema 12.1.2 Let $f(x) = a(x - x_1)(x - x_2)$, with $a \neq 0$, and $x_1 \leq x_2$. Then, for the inequality $f(x)\square 0$:
- If $a > 0$ and $\square$ is $\geq$ or $>$, the solution is: - $x \in (-\infty, x_1) \cup (x_2, \infty)$ for $f(x) > 0$ - $x \in (-\infty, x_1] \cup [x_2, \infty)$ for $f(x) \geq 0$ - If $a > 0$ and $\square$ is $\leq$ or $<$, the solution is: - $x \in (x_1, x_2)$ for $f(x) < 0$ - $x \in [x_1, x_2]$ for $f(x) \leq 0$
Similarly, if $a < 0$, the intervals are inverted.

Demostración. The function $f(x)$ is a parabola whose concavity is determined by the sign of a. The signs of $f(x)$ in the intervals defined by the roots x_1 and x_2 follow from the parabola's behavior, and the properties of quadratic inequalities are applied. ■

This lemma provides a systematic method for solving and graphically representing quadratic inequalities.

■ **Example 12.6** Solve the inequality $-2x^2 + 8x - 6 > 0$ and represent the solution graphically. ■

Solution. First, we factorize the associated quadratic equation $-2x^2 + 8x - 6 = 0$:
Divide by -2 (remembering that the inequality sign reverses when multiplying or dividing by a negative number, but as this is an auxiliary equation, it does not affect the solution):

$$-2x^2 + 8x - 6 = 0 \implies x^2 - 4x + 3 = 0 \implies (x-1)(x-3) = 0$$

Thus, the roots are $x = 1$ and $x = 3$.
Since $a = -2 < 0$, the parabola opens downward. Based on the lemma and the fact that we are looking for $f(x) > 0$, the solution is:
- $x \in (1, 3)$
The graphical representation is:

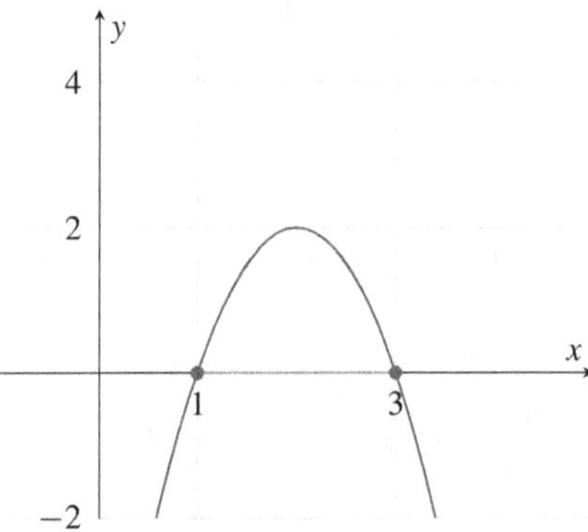

Figura 12.1.6: *Graph of $y = -2x^2 + 8x - 6$ and solution of $-2x^2 + 8x - 6 > 0$.*

In Figure 12.1.6, the parabola is above the x-axis in the interval $(1, 3)$, which is the solution to the inequality. ■

Corollary 12.1.5 The solution to a quadratic inequality can be an open, closed, or semi-open interval, depending on whether the inequality is strict ($<$, $>$) or non-strict ($\leq$, $\geq$), and whether the roots are included.

(R) It is crucial to consider the direction of the inequality and the sign of the quadratic coefficient a when determining the solution of a quadratic inequality and its graphical representation.

Exercise 12.3 Solve the inequality $x^2 + 2x - 8 \leq 0$ and represent the solution graphically. ∎

Exercise 12.4 Determine the values of x that satisfy the inequality $x^2 \geq 9$ and represent the solution on the number line. ∎

(R) In the case of quadratic inequalities without real roots (the discriminant $D = b^2 - 4ac < 0$), the quadratic function does not intersect the x-axis. Depending on the sign of a, the function will always be positive or always negative. Thus, the solution to the inequality will be either all $\mathbb{R}$ or the empty set.

∎ **Example 12.7** Solve the inequality $x^2 + x + 1 > 0$. ∎

Solution. Calculate the discriminant:

$$D = b^2 - 4ac = (1)^2 - 4(1)(1) = 1 - 4 = -3 < 0$$

Since $D < 0$, the equation $x^2 + x + 1 = 0$ has no real roots. Additionally, $a = 1 > 0$, so the parabola is always above the x-axis.
Therefore, $f(x) > 0$ for all $x \in \mathbb{R}$. The solution to the inequality is:

$$x \in \mathbb{R}$$

It is not necessary to graph in this case, but if we did, we would see that the parabola never touches or crosses the x-axis and remains entirely in the positive half-plane. ∎

(R) When the quadratic function is always positive or always negative, the solution to the inequality is immediate and depends only on the direction of the inequality.

In conclusion, the graphical representation of quadratic inequalities is a powerful tool for visualizing and understanding solutions. By analyzing the graph of the associated quadratic function, we can determine the intervals where the inequality holds and represent these solutions on the number line or Cartesian plane.

12.2 Intervals and Notation: Definition of Intervals and Inequalities.

12.2.1 Open and Closed Intervals.

In this section, we study intervals on the real number line, fundamental in mathematical analysis and set theory. We will understand the differences between open and closed intervals, exploring their properties and graphical representations.

Definition 12.2.1 Let $a, b \in \mathbb{R}$ with $a < b$. An *interval* is a subset of $\mathbb{R}$ that contains all real numbers between a and b. The most common types of intervals are:

- **Closed Interval**: $[a, b] = \{x \in \mathbb{R} \mid a \leq x \leq b\}$.
- **Open Interval**: $(a, b) = \{x \in \mathbb{R} \mid a < x < b\}$.
- **Half-Open or Half-Closed Interval**:
  - $[a, b) = \{x \in \mathbb{R} \mid a \leq x < b\}$.
  - $(a, b] = \{x \in \mathbb{R} \mid a < x \leq b\}$.

These intervals are graphically represented on the number line, facilitating their visualization and understanding.

Example 12.8 Graphically represent the following intervals:

1. $[1,4]$
2. $(1,4)$
3. $[1,4)$
4. $(1,4]$

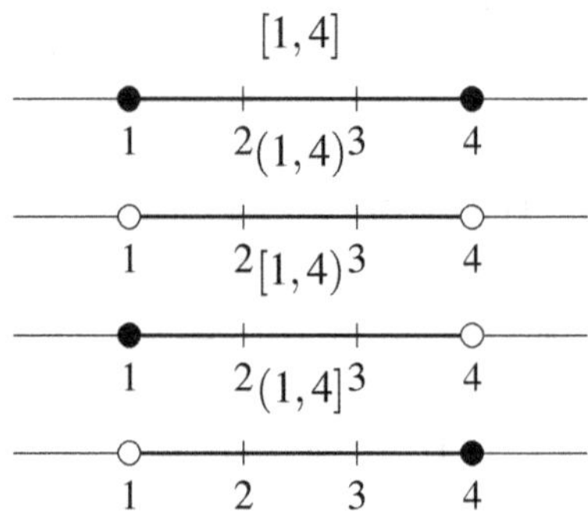

Figura 12.2.1: *Graphical representation of different types of intervals.*

■

As shown in Figure 12.2.1, intervals are distinguished by the inclusion or exclusion of the endpoints a and b, represented by filled (included) or hollow (excluded) points.

Theorem 12.2.1 In the set of real numbers $\mathbb{R}$, open and closed intervals have the following properties:

1. Closed intervals $[a,b]$ are *compact.*
2. Open intervals (a,b) are *open* in the topological sense.

Demostración. We will prove each property separately.

1. Closed intervals $[a,b]$ are compact:
   In the set of real numbers $\mathbb{R}$, a set is compact if it is closed and bounded. The interval $[a,b]$ is bounded because all its points x satisfy $a \leq x \leq b$. Moreover, it is closed because it includes its endpoints a and b. Therefore, the interval $[a,b]$ is compact in $\mathbb{R}$.
2. Open intervals (a,b) are open in the topological sense:
   In topology, a set is open if for every point in the set, there exists a neighborhood entirely contained within the set. For every point $x \in (a,b)$, we can find a neighborhood $(x-\varepsilon, x+\varepsilon) \subset (a,b)$ such that x is the center of the interval and ε is sufficiently small to ensure the neighborhood is entirely contained in (a,b). This implies that (a,b) is an open set in the topological sense.

This concludes the proof. ■

These properties are fundamental in analysis, especially in topics such as continuity and convergence.

Lema 12.2.1 The intersection of a finite number of closed intervals is a closed interval or empty.

Demostración. Let $\{[a_i,b_i]\}_{i=1}^{n}$ be a finite family of closed intervals. Then,

$$\bigcap_{i=1}^{n} [a_i,b_i] = \left[\max_{1\leq i\leq n} a_i, \min_{1\leq i\leq n} b_i \right],$$

provided that $\text{máx}\, a_i \leq \text{mín}\, b_i$. Otherwise, the intersection is the empty set. ■

Corollary 12.2.2 The union of a finite number of closed intervals is a closed interval if the intervals are adjacent or overlap.

■ **Example 12.9** Consider the intervals $[1,3]$ and $[2,5]$. The intersection and union of these intervals are:

- Intersection: $[1,3] \cap [2,5] = [2,3]$
- Union: $[1,3] \cup [2,5] = [1,5]$

Let us graphically represent these results.

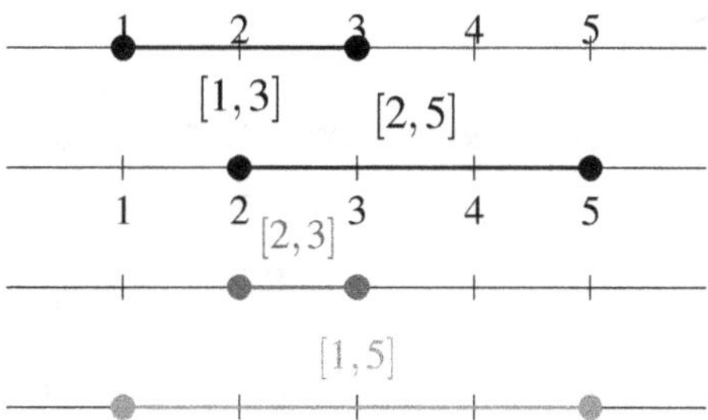

Figura 12.2.2: *Graphical representation of the intersection and union of* $[1,3]$ *and* $[2,5]$.

■

(R) The operations of union and intersection of intervals are fundamental in set theory and real analysis. Understanding how intervals interact allows us to manipulate subsets of real numbers effectively.

Theorem 12.2.3 — Nested intervals theorem. Let $\{[a_n,b_n]\}_{n=1}^{\infty}$ be a sequence of closed intervals such that $[a_{n+1},b_{n+1}] \subseteq [a_n,b_n]$ for all $n \in \mathbb{N}$. Then, the intersection of all intervals is non-empty, that is,

$$\bigcap_{n=1}^{\infty} [a_n,b_n] \neq \emptyset.$$

Demostración. Consider the sequence of closed intervals $\{[a_n,b_n]\}_{n=1}^{\infty}$ such that $[a_{n+1},b_{n+1}] \subseteq [a_n,b_n]$ for all $n \in \mathbb{N}$. This implies that the intervals are nested, i.e., each interval contains the next. Since each $[a_n,b_n]$ is closed and bounded, we can examine the sequences $\{a_n\}$ and $\{b_n\}$:
1. The sequence $\{a_n\}$ is increasing because each interval is contained within the previous one, so $a_n \leq a_{n+1}$ for all n. 2. The sequence $\{b_n\}$ is decreasing, as $b_{n+1} \leq b_n$ for all n.
Given that $\{a_n\}$ is increasing and bounded above by b_1, and $\{b_n\}$ is decreasing and bounded below by a_1, both sequences converge (by the monotone convergence theorem). Let

$$a = \lim_{n\to\infty} a_n \quad \text{and} \quad b = \lim_{n\to\infty} b_n.$$

Since $a_n \leq b_n$ for all n, it follows that $a \leq b$.
Finally, any point $x \in [a,b]$ belongs to all intervals $[a_n,b_n]$, because $a_n \leq a \leq x \leq b \leq b_n$ for all n. This ensures that

$$\bigcap_{n=1}^{\infty}[a_n,b_n] \neq \emptyset.$$

This concludes the proof. ■

This theorem has significant implications in analysis, particularly in the construction of real numbers and in convergence theory.

Exercise 12.5 Determine whether the following statement is true or false: The union of an infinite number of open intervals may not be an open set. Justify your answer. ■

Exercise 12.6 Let $I_n = \left(-\frac{1}{n}, \frac{1}{n}\right)$ for $n \in \mathbb{N}$. Compute the intersection of all I_n and represent it graphically. ■

(R) The understanding of open and closed intervals is essential in studying the topology of $\mathbb{R}$, as it lays the foundation for more advanced concepts such as continuity, differentiability, and integrability of functions.

In summary, open and closed intervals are fundamental tools in mathematics, allowing us to describe and analyze subsets of real numbers with precision. Graphical representations and associated topological properties facilitate their understanding and application in various fields.

12.2.2 Solving Inequalities with Absolute Value

The absolute value is a fundamental tool in mathematics that measures the distance of a real number from the origin on the number line. Before addressing the solution of inequalities involving absolute values, it is essential to understand its definition and basic properties.

Definition 12.2.2 The **absolute value** of a real number x, denoted by $|x|$, is defined as:

$$|x| = \begin{cases} x, & \text{if } x \geq 0, \\ -x, & \text{if } x < 0. \end{cases}$$

A key property of absolute value is its geometric interpretation as the distance between the point x and the origin on the real number line. This leads us to consider inequalities involving absolute values.

Theorem 12.2.4 Let a be a positive real number. The inequality $|x| < a$ is equivalent to $-a < x < a$. Similarly, the inequality $|x| > a$ is equivalent to $x < -a$ or $x > a$.

Demostración. Let us prove the first equivalence:
If $|x| < a$, then $-a < x < a$.
By the definition of absolute value, $|x| < a$ implies that x is at a distance less than a from the origin, i.e., x lies between $-a$ and a.
Conversely, if $-a < x < a$, then $|x| < a$ because x is within the open interval $(-a,a)$. ■

This theorem is fundamental for solving inequalities with absolute value since it allows us to transform an absolute value inequality into a simpler inequality without absolute value.

Example 12.10 Solve the inequality $|2x-3| \leq 5$. ■

Applying the theorem above, the inequality $|2x-3| \leq 5$ is equivalent to:

$$-5 \leq 2x-3 \leq 5.$$

Add 3 to all parts:

$$-5+3 \leq 2x \leq 5+3 \implies -2 \leq 2x \leq 8.$$

Divide by 2:

$$-\frac{2}{2} \leq x \leq \frac{8}{2} \implies -1 \leq x \leq 4.$$

Thus, the solution is the interval $[-1,4]$.
To visualize this solution, consider the graph of the function $y = |2x-3|$ and the line $y = 5$.

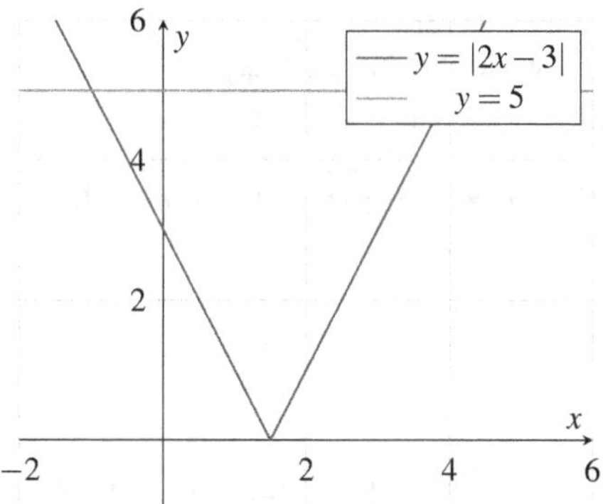

Figura 12.2.3: *Graph of* $y = |2x-3|$ *and* $y = 5$.

In the graph, we observe that the intersections of $y = |2x-3|$ and $y = 5$ occur at $x = -1$ and $x = 4$. The region where $|2x-3| \leq 5$ corresponds to the values of x between -1 and 4.

(R) The solution of inequalities of the form $|ax+b| \leq c$ follows the same procedure, provided that $c \geq 0$. If $c < 0$, the inequality has no solution because the absolute value is always non-negative.

Extending these concepts, we can address more complex inequalities.

Exercise 12.7 Solve the inequality $|x^2-4| > 5$. ■

First, establish the equivalences:
$|x^2-4| > 5$ is equivalent to $x^2-4 < -5$ or $x^2-4 > 5$.
Solve each case:
1. $x^2-4 < -5 \implies x^2 < -1$.
This has no real solution since $x^2 \geq 0$ always.
2. $x^2-4 > 5 \implies x^2 > 9 \implies x < -3$ or $x > 3$.
Thus, the solution is $(-\infty,-3) \cup (3,\infty)$.
The graph confirms that the regions where $|x^2-4| > 5$ correspond to $x < -3$ and $x > 3$.

Exercise 12.8 Determine all real values of x that satisfy $|3x+2| \geq 7$. ■

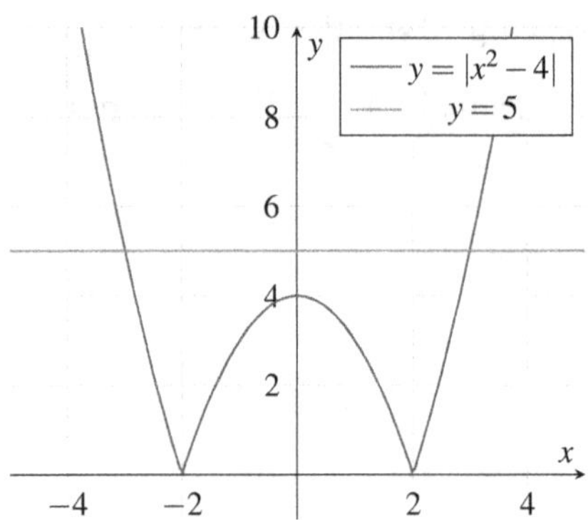

Figura 12.2.4: *Graph of $y = |x^2 - 4|$ and $y = 5$.*

The inequality $|3x+2| \geq 7$ can be broken down into two cases:
1. $3x+2 \geq 7 \implies 3x \geq 5 \implies x \geq \frac{5}{3}$.
2. $3x+2 \leq -7 \implies 3x \leq -9 \implies x \leq -3$.
Therefore, the solution is $x \leq -3$ or $x \geq \frac{5}{3}$.
These techniques allow solving a wide variety of inequalities involving absolute values, which are fundamental in mathematical analysis and its applications.

Corollary 12.2.5 If $a > 0$, then the solution of the inequality $|x-c| < a$ is the interval $(c-a, c+a)$.

Demostración. Applying the initial theorem, $|x-c| < a$ implies $-a < x-c < a$, which results in $c-a < x < c+a$. ■

This corollary is especially useful for solving inequalities centered around any point c.

■ **Example 12.11** Find all x that satisfy $|x-1| < 3$. ■

Applying the corollary:
The solution is $1-3 < x < 1+3 \implies -2 < x < 4$.
This interval represents all real numbers that are within a distance of less than 3 from the number 1 on the real line.

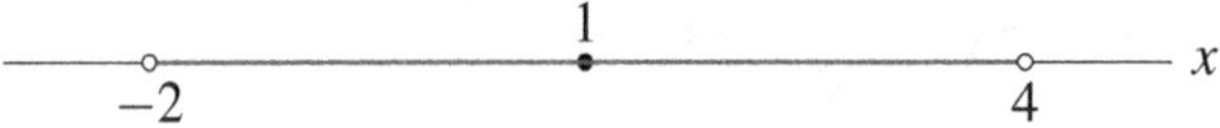

Figura 12.2.5: *Representation on the real line of $-2 < x < 4$.*

The graph illustrates the solution interval on the number line.

Lema 12.2.2 For any real number x, it holds that $|x| \geq 0$, and $|x| = 0$ if and only if $x = 0$.

Demostración. By definition, $|x|$ is always greater than or equal to zero, since it is x if $x \geq 0$ and $-x$ if $x < 0$, both of which are non-negative. Moreover, $|x| = 0$ implies that $x = 0$. ■

This property is essential in many mathematical arguments, particularly in analysis and algebra.

Theorem 12.2.6 For any real numbers x and y, the triangle inequality holds:

$$|x+y| \leq |x| + |y|.$$

Demostración. According to the definition of absolute value and the properties of real numbers, we have:

$$|x+y| \leq |x|+|y|.$$

This inequality is a direct consequence of the triangle inequality in the context of distance on the real line. ■

The triangle inequality is fundamental in many areas of mathematics, including analysis and the theory of metric spaces.

(R) Techniques for solving inequalities with absolute values are also applicable to functions of several variables and vector spaces, expanding their utility in advanced mathematics.

In conclusion, understanding and applying the properties of absolute value and techniques for solving related inequalities are essential tools in advanced mathematical reasoning.

12.3 Applications: Optimization Problems with Constraints

12.3.1 Application in Resource Maximization

Resource maximization is a central problem in applied mathematics, particularly in fields such as economics, engineering, and management. It involves determining the best way to allocate limited resources to achieve maximum benefit or performance. These types of problems are modeled and solved using optimization techniques, with *linear programming* being a fundamental tool.

Definition 12.3.1 A **linear programming problem** is one in which the goal is to maximize or minimize a linear objective function subject to a set of linear constraints. Formally, it is expressed as:

$$\text{Optimize} \quad Z = c_1x_1 + c_2x_2 + \cdots + c_nx_n,$$
$$\text{subject to} \quad \begin{cases} a_{11}x_1 + a_{12}x_2 + \cdots + a_{1n}x_n \leq b_1, \\ a_{21}x_1 + a_{22}x_2 + \cdots + a_{2n}x_n \leq b_2, \\ \vdots \\ a_{m1}x_1 + a_{m2}x_2 + \cdots + a_{mn}x_n \leq b_m, \\ x_i \geq 0, \quad i = 1, \ldots, n. \end{cases}$$

In this context, the *objective function* Z represents the efficiency or benefit measure to be optimized, and the *constraints* reflect resource limitations.
Understanding how these mathematical formulations help solve real-world resource maximization problems is essential.

Theorem 12.3.1 — Fundamental Theorem of Linear Programming. If the feasible set of a linear programming problem is non-empty and bounded, then there exists at least one optimal solution that occurs at a vertex of the feasible polyhedron.

Demostración. Consider a linear programming problem where the objective is to maximize or minimize a linear objective function over a feasible set, which is the set of all solutions that satisfy the problem's linear constraints.

Suppose the feasible set is non-empty and bounded. In linear programming, the feasible set of solutions can be represented as a polyhedron in the variable space. This polyhedron is bounded by the points where the constraints intersect, forming the vertices of the polyhedron.
Since the objective function is linear, it takes constant values in any direction within the space. This means that by moving in any direction from a point inside the polyhedron, the objective function can increase or decrease until it reaches an extreme value at one of the vertices (if the polyhedron is bounded).
Because the feasible set is bounded and non-empty, the objective function will reach its optimal value at some point within the polyhedron. By the properties of polyhedra and linear functions, this optimal value must occur at at least one of the vertices of the feasible set.
This concludes the proof of the Fundamental Theorem of Linear Programming. ■

This theorem greatly simplifies the search for optimal solutions by reducing the problem to a finite analysis.

■ **Example 12.12** A factory produces two types of products, P_1 and P_2. Each unit of P_1 requires 1 hour of machine time and 3 hours of labor. Each unit of P_2 requires 2 hours of machine time and 1 hour of labor. The factory has 80 hours of machine time and 60 hours of labor available per week. If the profits per unit are \$50 for P_1 and \$40 for P_2, how many units of each product should be produced to maximize profit? ■

Demostración. **Definition of Variables**
Let:

$$x = \text{number of units of } P_1 \text{ to produce}, \quad y = \text{number of units of } P_2 \text{ to produce}.$$

Objective Function
We want to maximize the total profit:

$$Z = 50x + 40y.$$

Constraints
1. Machine hours:

$$x + 2y \leq 80.$$

2. Labor hours:

$$3x + y \leq 60.$$

3. Non-negativity:

$$x \geq 0, \quad y \geq 0.$$

Solution of the Problem
We proceed by graphing the constraints to find the feasible region.
Step 1: Find Intercepts of the Constraints
- Machine constraint ($x + 2y \leq 80$):
- When $x = 0$:

$$0 + 2y = 80 \implies y = 40.$$

- When $y = 0$:

$$x + 0 = 80 \implies x = 80.$$

- Labor constraint ($3x+y \leq 60$):
- When $x=0$:

$$3 \times 0 + y = 60 \implies y = 60.$$

- When $y=0$:

$$3x + 0 = 60 \implies x = 20.$$

Step 2: Finding the Points of Intersection
- Intersection of the Two Constraints:
Solve the system:

$$\begin{cases} x+2y=80, \\ 3x+y=60. \end{cases}$$

Multiply the first equation by 1 and the second by 2 to match the coefficients of y:

$$\begin{cases} x+2y=80, \\ 6x+2y=120. \end{cases}$$

Subtract the first equation from the second:

$$(6x+2y)-(x+2y) = 120-80 \implies 5x = 40 \implies x = 8.$$

Substitute $x=8$ into $x+2y=80$:

$$8+2y = 80 \implies 2y = 72 \implies y = 36.$$

Intersection Point: $(8,36)$.
Step 3: Graphing the Feasible Region

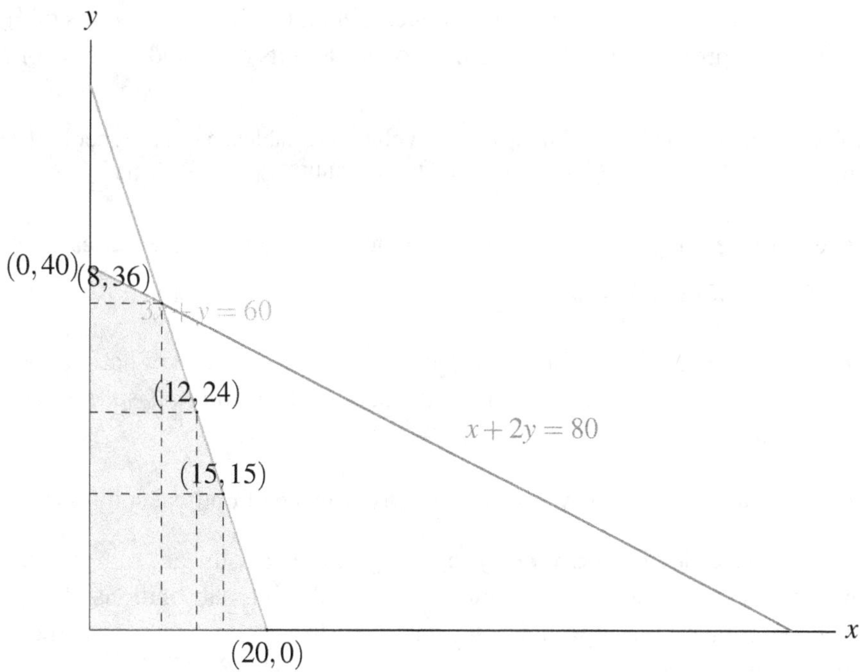

Figura 12.3.1: *Feasible region for the profit maximization problem.*

Step 4: Determining the Vertices of the Feasible Region
The vertices of the feasible region are:
1. $(0,40)$: Intersection of $x=0$ and $x+2y=80$.
2. $(8,36)$: Intersection of $x+2y=80$ and $3x+y=60$.
3. $(20,0)$: Intersection of $y=0$ and $3x+y=60$.
Additional Points Within the Feasible Region:
- $(12,24)$: Point within the feasible region that satisfies both constraints.
- $(15,15)$: Another point within the feasible region.

Step 5: Calculating the Objective Function at Each Vertex
1. At $(0,40)$:

$$Z = 50(0)+40(40) = \$1600.$$

2. At $(8,36)$:

$$Z = 50(8)+40(36) = 400+1440 = \$1840.$$

3. At $(12,24)$:

$$Z = 50(12)+40(24) = 600+960 = \$1560.$$

4. At $(15,15)$:

$$Z = 50(15)+40(15) = 750+600 = \$1350.$$

5. At $(20,0)$:

$$Z = 50(20)+40(0) = \$1000.$$

Conclusion
The maximum profit of \$1840 is obtained by producing 8 units of P_1 and 36 units of P_2.
Answer: To maximize profits, the factory should produce 8 units of P_1 and 36 units of P_2.

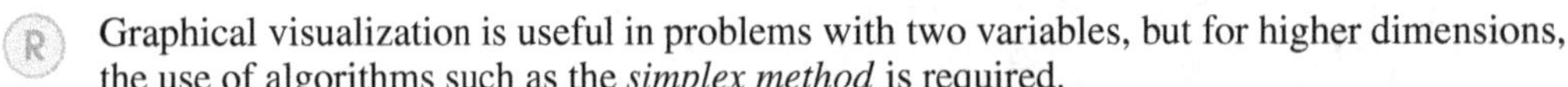
Graphical visualization is useful in problems with two variables, but for higher dimensions, the use of algorithms such as the *simplex method* is required.

To delve deeper into these techniques, it is important to understand the theoretical foundations.

Lema 12.3.1 The set of feasible solutions of a linear programming problem is a convex set.

Demostración. Linear constraints define half-spaces, and the intersection of half-spaces is a convex set. Therefore, the feasible set, which is the intersection of all the half-spaces defined by the constraints, is convex. ■

The convexity of the feasible set is key to ensuring the existence of optimal solutions at the vertices.

Theorem 12.3.2 — **Duality in Linear Programming.** For every linear programming problem (primal), there exists a corresponding problem (dual), and the optimal solutions of both problems are related. If both have finite optimal solutions, then the optimal values of their objective functions are equal.

This duality theorem is fundamental for analyzing and solving complex optimization problems.

■ **Example 12.13** Consider the following primal problem:

$$\text{Maximize} \quad Z = 3x_1 + 2x_2,$$
$$\text{subject to} \quad \begin{cases} x_1 + x_2 \leq 4, \\ 2x_1 + x_2 \leq 5, \\ x_1, x_2 \geq 0. \end{cases}$$

Formulate the dual problem and find the optimal solutions for both problems. ■

Demostración. The dual problem is:

$$\text{Minimize} \quad W = 4y_1 + 5y_2,$$
$$\text{subject to} \quad \begin{cases} y_1 + 2y_2 \geq 3, \\ y_1 + y_2 \geq 2, \\ y_1, y_2 \geq 0. \end{cases}$$

We solve the primal problem using the graphical method.

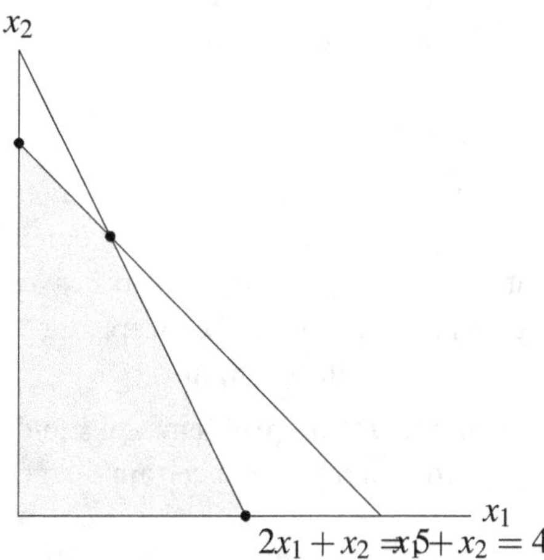

Figura 12.3.2: *Feasible region of the primal problem.*

We compute Z at the vertices:

$$Z(0,4) = 3(0) + 2(4) = 8,$$
$$Z(1,3) = 3(1) + 2(3) = 9,$$
$$Z(2{,}5,0) = 3(2{,}5) + 2(0) = 7{,}5.$$

The optimal solution is $Z = 9$ at $(1,3)$.
For the dual problem, we can verify that the optimal solution is $W = 9$.

■

This example demonstrates the relationship between primal and dual problems and how their optimal solutions coincide.

Exercise 12.9 A company wants to minimize the production cost of two products, A and B. The cost per unit of A is \$30 and of B is \$20. Each product requires machine hours and labor hours according to the following table:

	Machine (hours)	Labor (hours)
A	2	1
B	1	2

The company has 100 machine hours and 80 labor hours available. Additionally, it needs to produce at least 20 units in total. Formulate the problem and determine the optimal quantities of each product to minimize the cost. ■

Exercise 12.10 A farmer has 50 hectares of land and wants to decide how much to plant of wheat and corn. Each hectare of wheat requires \$100 of investment and yields a profit of \$2000. Each hectare of corn requires \$200 of investment and yields a profit of \$3000. The farmer has \$8000 to invest. How should he distribute his land to maximize his profits? ■

The techniques and theorems presented are powerful tools for solving resource maximization problems in various fields. A deep understanding of these concepts allows tackling complex problems and making optimal decisions based on rigorous mathematical analysis.

12.3.2 Minimization Problems in Geometry.

Minimization problems in geometry are fundamental in mathematics, as they involve finding the optimal configuration of geometric figures that satisfy certain conditions. These problems are not only theoretically significant but also find applications in physics, engineering, and other sciences. To address these problems, it is essential to understand concepts such as distances, areas, perimeters, and how these can be optimized under certain constraints.

Definition 12.3.2 A **geometric minimization problem** seeks to determine the figure or geometric configuration that minimizes (or maximizes) a certain quantity (such as distance, area, or volume) under given conditions.

A classic example of such problems is finding the minimum distance from a point to a line or between sets of points.

Theorem 12.3.3 — Fermat's Problem. Given a triangle with vertices A, B, and C, there exists a point P in the plane that minimizes the sum of the distances from P to each of the vertices. This point is known as the **Fermat point** of the triangle.

Demostración. Consider a triangle with vertices A, B, and C. We aim to find a point P in the plane that minimizes the sum of the distances $PA + PB + PC$.
To solve this problem, there are two cases:
1. If the triangle has an angle of 120 degrees or more:
If one of the angles of the triangle is greater than or equal to 120 degrees, the Fermat point that minimizes the sum of the distances is the vertex opposite this obtuse angle. Placing P at this vertex ensures the minimal sum of distances.
2. If all angles are less than 120 degrees:
In this case, the Fermat point is inside the triangle and is the point such that the angles between the segments PA, PB, and PC are all 120 degrees. This point can be constructed as follows:

- Construct an equilateral triangle on each side of the given triangle. - The centers of these equilateral triangles form a new triangle. The lines connecting each vertex of the original triangle to the opposite vertex of the new triangle intersect at the Fermat point.
This point P, defined as the point where the segments form angles of 120 degrees with the vertices, minimizes the sum $PA+PB+PC$ and is unique.
This concludes the proof of Fermat's problem for a triangle. ■

■ **Example 12.14** Find the point P that minimizes the sum of the distances to the vertices of an equilateral triangle with side length a. ■

Demostración. In an equilateral triangle, all angles are 60°, so Fermat's theorem applies. The Fermat point coincides with the center of the triangle, which is also the center of its circumscribed circle.
The minimum sum of distances is:

$$S_{\min} = 3 \times \frac{a}{\sqrt{3}} = a\sqrt{3}.$$

■

To visualize this result, consider the following diagram:

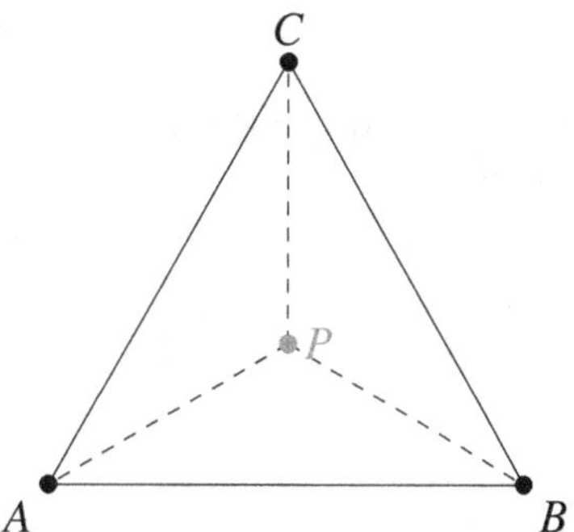

Figura 12.3.3: *The Fermat point P in an equilateral triangle.*

This example demonstrates how the Fermat point provides the optimal solution in terms of minimum total distance.

(R) Fermat's problem is a particular case of a more general problem known as the **Steiner problem**, which seeks to connect a set of points with the minimum total length of connection.

Another classic minimization problem in geometry is determining the figure with maximum area for a given perimeter, or vice versa.

Theorem 12.3.4 — **Isoperimetric Inequality in the Plane.** Among all planar figures with a given perimeter, the circle encloses the maximum area.

Demostración. Although a full proof of this theorem exceeds the scope of this text, it relies on techniques from variational calculus and analysis. The central idea is that the circle, due to its symmetry, distributes the perimeter uniformly around an area, maximizing it. ■

Corollary 12.3.5 For a given area, the circle has the minimum perimeter among all planar figures.

This result has practical implications in disciplines such as architecture and engineering, where efficiency in material usage is crucial.

■ **Example 12.15** Determine the shape of a rectangle with a fixed area A that has the minimum possible perimeter. ■

Demostración. Let a rectangle have sides of lengths x and y, such that $xy = A$. We aim to minimize the perimeter $P = 2x + 2y$.
Using the constraint $y = \frac{A}{x}$, the perimeter is expressed as:

$$P(x) = 2x + 2\left(\frac{A}{x}\right) = 2x + \frac{2A}{x}.$$

To find the minimum, we differentiate with respect to x and set it equal to zero:

$$P'(x) = 2 - \frac{2A}{x^2} = 0 \implies x^2 = A \implies x = \sqrt{A}.$$

Thus, $y = \sqrt{A}$, meaning the rectangle must be a square with side $\sqrt{A}$ to have the minimum perimeter. ■

This result aligns with the intuition that the square is the regular figure which, for a given area, has the minimum perimeter among rectangles.

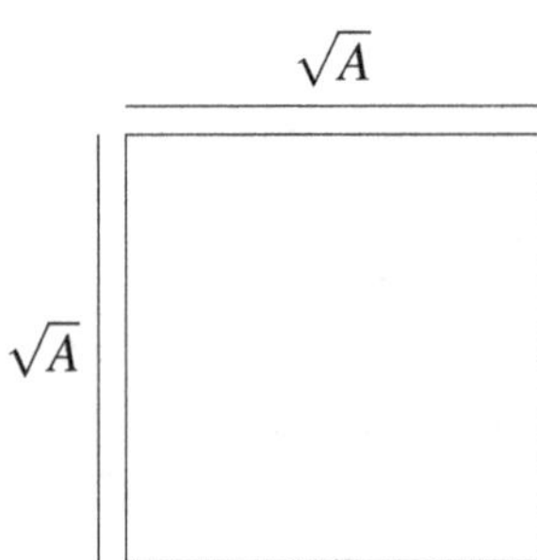

Figura 12.3.4: *The square with area A has the minimum perimeter among all rectangles of area A.*

Lema 12.3.2 Among all regular polygons with n sides inscribed in a given circle, the one with the maximum area is the regular n-sided polygon itself.

Demostración. The proof relies on the symmetry and properties of regular polygons. When a regular polygon is inscribed in a circle, each vertex touches the circle, and due to equal distances and angles, the area is maximized compared to any other n-sided polygon inscribed in the same circle. ■

This lemma reinforces the idea that regularity and symmetry in geometric figures lead to extremal properties.

Exercise 12.11 Prove that, among all cylinders that can be inscribed in a sphere of radius R, the cylinder with maximum volume has a height equal to the diameter of the sphere. ■

Exercise 12.12 Find the dimensions of the rectangle with the maximum area that can be inscribed in a circle of radius r. ■

To solve these exercises, differential calculus techniques and concepts from analytic geometry are applied.

Theorem 12.3.6 The triangle with the minimum area circumscribed around a given circle is an equilateral triangle.

Demostración. Consider all triangles circumscribed around a circle of radius r. The area A of a circumscribed triangle is given by:

$$A = \frac{abc}{4R},$$

where a, b, and c are the side lengths, and R is the radius of the circumscribed circle. To minimize A, under the constraint that R is constant, it is necessary that $a = b = c$, i.e., the triangle is equilateral. ■

This theorem again highlights how symmetry leads to optimal solutions in minimization problems.

R Minimization problems in geometry are often solved using derivatives and optimization techniques from differential calculus.

■ **Example 12.16** Find the point P on the line $y = 2x + 3$ that is closest to the origin. ■

Demostración. The distance from the origin to a point (x, y) is $D = \sqrt{x^2 + y^2}$. We aim to minimize D subject to $y = 2x + 3$.
Substituting y:

$$D(x) = \sqrt{x^2 + (2x+3)^2}.$$

To minimize D, it suffices to minimize D^2:

$$D^2(x) = x^2 + (2x+3)^2 = x^2 + 4x^2 + 12x + 9 = 5x^2 + 12x + 9.$$

Differentiating with respect to x:

$$\frac{d}{dx} D^2(x) = 10x + 12.$$

Setting the derivative to zero:

$$10x + 12 = 0 \implies x = -\frac{6}{5}.$$

Then,

$$y = 2\left(-\frac{6}{5}\right) + 3 = -\frac{12}{5} + 3 = -\frac{12}{5} + \frac{15}{5} = \frac{3}{5}.$$

The closest point is $\left(-\frac{6}{5}, \frac{3}{5}\right)$.

■

This example demonstrates how to apply calculus techniques to solve minimization problems in analytic geometry.

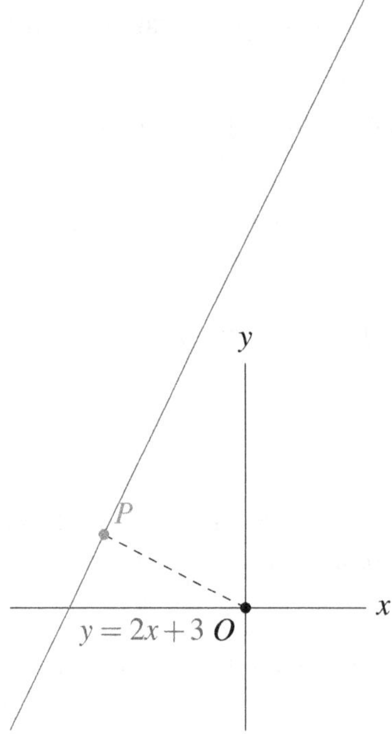

Figura 12.3.5: *The point P on the line $y = 2x + 3$ closest to the origin O.*

Corollary 12.3.7 The minimum distance from a point P_0 to a line L in the plane is perpendicular to L.

Demostración. In the example above, we observed that the line joining the origin to the point of minimum distance is perpendicular to the given line. This is a consequence of the orthogonal projection minimizing the distance between a point and a line. ■

Lema 12.3.3 In a right triangle, the leg opposite the 45° angle equals the hypotenuse divided by $\sqrt{2}$, minimizing the lengths of the legs for a given hypotenuse.

Demostración. In an isosceles right triangle, both legs are equal. If the hypotenuse is h, then:

$$c = \frac{h}{\sqrt{2}}.$$

This minimizes the sum of the leg lengths for a given hypotenuse. ■

Exercise 12.13 Determine the point on the circle $x^2 + y^2 = r^2$ that minimizes the function $f(x,y) = x + y$. ■

Exercise 12.14 Find the dimensions of the right cone with maximum volume that can be inscribed in a sphere of radius R. ■

These exercises invite the reader to apply concepts of calculus and geometry to solve more complex minimization problems.
In summary, minimization problems in geometry combine concepts from various mathematical fields, including geometry, calculus, and optimization. A deep understanding of these concepts enables solving both practical and theoretical problems, showcasing the elegance and utility of advanced mathematics.

12.4 Solved Exercises

Exercise 12.15 Solve the inequality $3x+5 > 2x+8$.

Demostración. First, isolate x:

$$3x+5 > 2x+8$$

Subtract $2x$ from both sides:

$$x+5 > 8$$

Subtract 5 from both sides:

$$x > 3$$

The solution is $x > 3$. ■

Exercise 12.16 Find the values of x that satisfy the quadratic inequality $x^2 - 5x + 6 \leq 0$.

Demostración. First, factorize the trinomial:

$$x^2 - 5x + 6 = (x-2)(x-3)$$

Now, set up the inequality $(x-2)(x-3) \leq 0$ and find the critical points $x = 2$ and $x = 3$. Divide the real line into intervals $(-\infty, 2]$, $(2,3)$, and $[3,\infty)$.
Test the sign in each interval: 1. For $x \in (-\infty, 2]$, $(x-2)(x-3) > 0$. 2. For $x \in (2,3)$, $(x-2)(x-3) < 0$. 3. For $x \in [3,\infty)$, $(x-2)(x-3) > 0$.
Since we seek $(x-2)(x-3) \leq 0$, the solution is $x \in [2,3]$. ■

Exercise 12.17 Solve the absolute value inequality $|2x-4| < 6$.

Demostración. Using the definition of absolute value, we have:

$$-6 < 2x-4 < 6$$

Add 4 to each side:

$$-2 < 2x < 10$$

Divide by 2:

$$-1 < x < 5$$

The solution is $x \in (-1,5)$. ■

Exercise 12.18 Find the point on the line $y = -\frac{1}{2}x + 3$ that is closest to the origin.

Demostración. The distance from the origin $(0,0)$ to a point (x,y) is $D=\sqrt{x^2+y^2}$. We aim to minimize D subject to $y=-\frac{1}{2}x+3$.
Substituting y, we have:

$$D(x)=\sqrt{x^2+\left(-\frac{1}{2}x+3\right)^2}$$

To minimize D, it suffices to minimize D^2:

$$D^2(x)=x^2+\left(-\frac{1}{2}x+3\right)^2=x^2+\frac{1}{4}x^2-3x+9$$

Combining terms:

$$D^2(x)=\frac{5}{4}x^2-3x+9$$

Differentiate with respect to x and set to zero:

$$\frac{d}{dx}D^2(x)=\frac{5}{2}x-3=0$$

$$x=\frac{6}{5}$$

Substituting $x=\frac{6}{5}$ into $y=-\frac{1}{2}x+3$, we get:

$$y=-\frac{1}{2}\cdot\frac{6}{5}+3=-\frac{3}{5}+3=\frac{12}{5}$$

The closest point is $\left(\frac{6}{5},\frac{12}{5}\right)$. ■

Exercise 12.19 Determine the interval of x such that $|x-4|\geq 7$.

Demostración. The inequality $|x-4|\geq 7$ splits into two cases: 1. $x-4\geq 7 \implies x\geq 11$. 2. $x-4\leq -7 \implies x\leq -3$.
Thus, the solution is $x\in(-\infty,-3]\cup(11,\infty)$. ■

12.5 Proposed Exercises

12.5.1 Linear and Quadratic Inequalities: Solving and Representation on the Number Line

Exercise 12.20 Solve the linear inequality $4x-7>9$ and represent the solution on the number line.

Exercise 12.21 Find the solution to the quadratic inequality $x^2-3x-10<0$ and represent it on the number line.

Exercise 12.22 Determine the solution set for the inequality $5-2x\leq 3x+10$ and show the graphical representation.

Exercise 12.23 Solve the inequality $(x-2)(x+3) \geq 0$ and represent the interval solution on the number line. ■

Exercise 12.24 Solve the quadratic inequality $2x^2 + 3x - 5 > 0$ and represent the solution set on the number line. ■

12.5.2 Intervals and Notation: Definition of Intervals and Inequalities

Exercise 12.25 Write the solution set of the inequality $-4 \leq x < 7$ in interval notation. ■

Exercise 12.26 Determine the intersection and union of the intervals $(-\infty, 2]$ and $[1, 5)$. ■

Exercise 12.27 Express the solution set of the inequality $3 < x \leq 9$ in interval notation and represent it on the number line. ■

Exercise 12.28 Convert the expression $x \in (-\infty, -3) \cup (2, \infty)$ into a compound inequality. ■

Exercise 12.29 Describe in words the interval $[a, b)$ and provide an example on the number line. ■

12.5.3 Applications: Optimization Problems with Constraints

Exercise 12.30 A manufacturer produces two products, A and B. Each unit of A requires 2 machine hours, and each unit of B requires 1 machine hour. The company has 100 machine hours available per week. How many units of each product can be produced if at least 10 units of each are required? ■

Exercise 12.31 A farm has 30 hectares available to plant wheat and corn. Each hectare of wheat generates a profit of \$2000, and each hectare of corn generates a profit of \$3000. If the investment cost is \$100 per hectare of wheat and \$150 per hectare of corn, and the farmer has a budget of \$4500, how should the land be distributed to maximize profits? ■

Exercise 12.32 A carpenter wants to make tables and chairs. Each table requires 3 hours of work, and each chair requires 2 hours. If they have 60 hours available per week and must make at least 5 tables and 8 chairs, how many can they produce of each? ■

Exercise 12.33 A store wants to maximize profits by selling two types of T-shirts, X and Y. Each T-shirt X generates a profit of \$5, and each T-shirt Y generates a profit of \$7. If the store has \$300 to spend on T-shirts, and each T-shirt X costs \$10 while each T-shirt Y costs \$15, how many units of each type should be purchased to maximize profits? ■

Exercise 12.34 A company wants to optimize the use of its resources to produce two products, C and D. Each product C generates a profit of \$8, and each product D generates a profit of \$10. Producing C requires 4 hours of labor, and producing D requires 5 hours. If the company has 100 hours of labor available and must produce at least 5 units of each product, what is the optimal combination to maximize profits? ■

IV Planar and Spatial GeometricReasoning

13. Proportionality and Similarity

13.1 Thales' Theorem: Applications in Similar Figures.

13.1.1 Application in Triangles.

The similarity of triangles is a fundamental tool in geometry that allows solving measurement problems and establishing proportionality relationships. Through the similarity criteria, we can analyze and compare different triangles to find unknown measures or demonstrate geometric properties.

Definition 13.1.1 Two triangles are **similar** if their corresponding angles are congruent and their corresponding sides are proportional.

This definition allows us to identify triangles that have the same shape but different sizes, which is essential in various geometric applications.

Theorem 13.1.1 — **AA Similarity Criterion.** If two triangles have two corresponding angles congruent, then the triangles are similar.

Demostración. Let $\triangle ABC$ and $\triangle DEF$ be two triangles such that two of their corresponding angles are congruent. Suppose:

$$\angle A = \angle D \quad \text{and} \quad \angle B = \angle E.$$

Since the sum of the interior angles of any triangle is 180 degrees, it follows that:

$$\angle C = 180^\circ - (\angle A + \angle B) = 180^\circ - (\angle D + \angle E) = \angle F.$$

Thus, the three corresponding angles of $\triangle ABC$ and $\triangle DEF$ are congruent: $\angle A = \angle D$, $\angle B = \angle E$, and $\angle C = \angle F$.
According to the similarity criterion for triangles, if the three corresponding angles of two triangles are congruent, then the triangles are similar. In this case, as we have demonstrated that the three angles are congruent, we conclude that $\triangle ABC \sim \triangle DEF$.
This concludes the proof of the AA similarity criterion. ∎

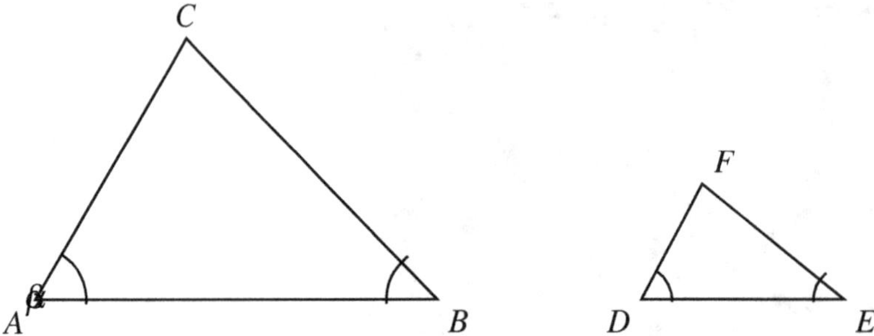

Figura 13.1.1: *AA Similarity Criterion*

This theorem is fundamental as it allows establishing the similarity of triangles with only two pairs of congruent angles.

■ **Example 13.1** In triangle $\triangle ABC$, a line parallel to side BC intersects side AB at D and side AC at E. If $AD = 4$ cm, $DB = 6$ cm, and $AE = 5$ cm, calculate the length of EC.

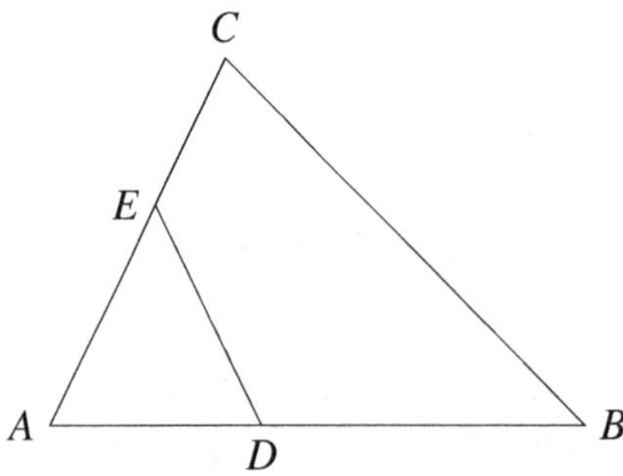

Figura 13.1.2: *Triangle with a line parallel to side BC intersecting at D and E.*

■

Since DE is parallel to BC, by **Thales' Theorem**, we have:

$$\frac{AD}{DB} = \frac{AE}{EC}.$$

Substituting the known values:

$$\frac{4}{6} = \frac{5}{EC} \implies EC = \frac{5 \times 6}{4} = 7{,}5 \text{ cm}.$$

In this example, we apply the proportionality established by Thales' Theorem to find an unknown length.

> Theorem 13.1.2 — **Thales' Theorem.** If a line parallel to one side of a triangle intersects the other two sides, then it divides those sides into proportional segments.

Demostración. Let $\triangle ABC$ be a triangle, and let DE be a line parallel to side BC that intersects sides AB and AC at points D and E, respectively.
We want to prove that:

$$\frac{AD}{DB} = \frac{AE}{EC}.$$

Since $DE \parallel BC$, the corresponding angles are equal. Therefore:

$$\angle ADE = \angle ABC \quad \text{and} \quad \angle AED = \angle ACB.$$

As the corresponding angles are equal, the triangles $\triangle ADE$ and $\triangle ABC$ are similar by the AA similarity criterion.
Since the triangles are similar, the proportion of their corresponding sides holds:

$$\frac{AD}{AB} = \frac{AE}{AC}.$$

Rewriting this proportion in terms of the segments DB and EC, we have:

$$\frac{AD}{DB} = \frac{AE}{EC}.$$

This proves that the line DE, being parallel to side BC, divides sides AB and AC into proportional segments.
This concludes the proof of Thales' Theorem. ■

This theorem is a key tool for solving problems involving proportional segments in triangles.

Lema 13.1.1 In a right triangle, the segment of the altitude drawn from the vertex of the right angle to the hypotenuse divides the triangle into two triangles similar to the original triangle and to each other.

Demostración. Let $\triangle ABC$ be a right triangle with $\angle C = 90°$, and let CD be the altitude to the hypotenuse AB. The triangles $\triangle ACD$ and $\triangle CBD$ are right triangles and share angles with the original triangle, hence they are similar to each other and to $\triangle ABC$.

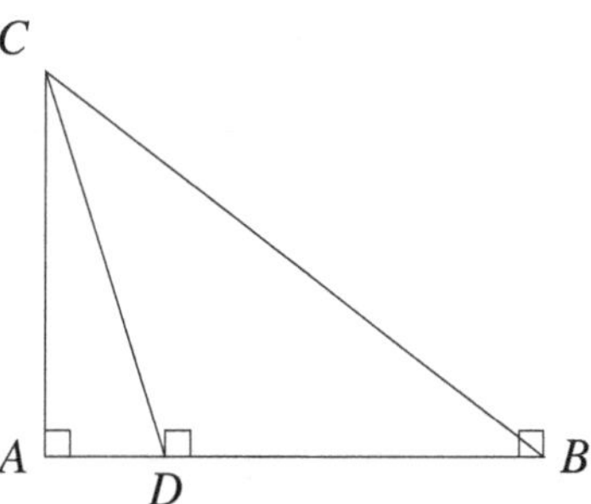

Figura 13.1.3: *Altitude CD in the right triangle $\triangle ABC$.*

■

This result establishes proportional relationships between the segments of a right triangle and its altitudes.

Theorem 13.1.3 In a right triangle, the square of the altitude relative to the hypotenuse equals the product of the projections of the legs on the hypotenuse:

$$h^2 = m \cdot n,$$

where h is the altitude relative to the hypotenuse, and m and n are the projections of the legs on the hypotenuse.

Demostración. Consider a right triangle $\triangle ABC$ with $\angle C = 90°$, and let h be the altitude from C to the hypotenuse AB, which divides AB into segments m and n, corresponding to the projections of the legs AC and BC on AB, respectively.
By the similarity theorem, the triangles $\triangle ABC$, $\triangle ACD$, and $\triangle CBD$ are similar to each other. Thus, the following proportions hold:

$$\frac{h}{m} = \frac{b}{c} \quad \text{and} \quad \frac{h}{n} = \frac{a}{c}.$$

Multiplying these two equations:

$$\frac{h^2}{m \cdot n} = \frac{b \cdot a}{c^2}.$$

Since $c = m + n$, the product of the projections equals the square of the altitude. Therefore,

$$h^2 = m \cdot n.$$

■

This theorem is useful for calculating altitudes and segments in right triangles when certain measurements are known.

■ **Example 13.2** Calculate the altitude relative to the hypotenuse in a right triangle with legs measuring 9 cm and 12 cm.

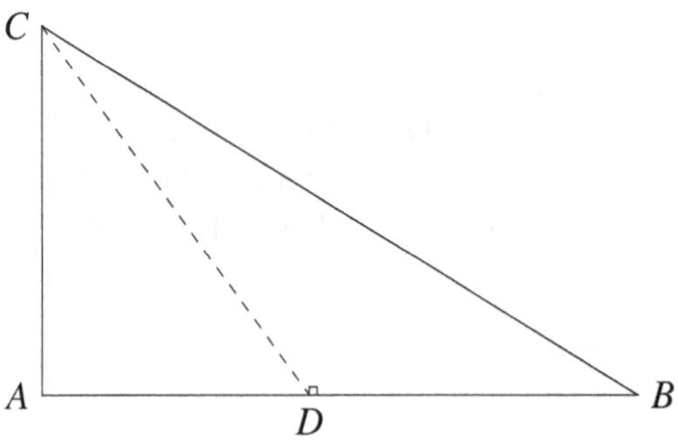

Figura 13.1.4: *Right triangle with altitude relative to the hypotenuse.*

■

First, calculate the hypotenuse using the Pythagorean Theorem:

$$c = \sqrt{9^2 + 12^2} = \sqrt{81 + 144} = \sqrt{225} = 15 \text{ cm}.$$

Next, find the projections m and n:

$$m = \frac{a^2}{c} = \frac{9^2}{15} = \frac{81}{15} = 5{,}4 \text{ cm},$$

$$n = \frac{b^2}{c} = \frac{12^2}{15} = \frac{144}{15} = 9{,}6 \text{ cm}.$$

Finally, apply the theorem:

$$h^2 = m \cdot n = 5{,}4 \times 9{,}6 = 51{,}84 \implies h = \sqrt{51{,}84} = 7{,}2 \text{ cm}.$$

This example demonstrates how to use similarity relationships and the theorem above to calculate altitudes in right triangles.

Exercise 13.1 In a right triangle, the hypotenuse measures 26 cm, and one leg measures 10 cm. Determine the altitude relative to the hypotenuse and the projections of the legs on the hypotenuse. ■

Exercise 13.2 Prove that in an isosceles triangle, the altitude relative to the base divides the triangle into two congruent triangles, which are thus similar to the original triangle. ■

These exercises apply the concepts learned and strengthen the understanding of triangle properties.

(R) The application of triangle similarity is essential not only in geometry but also in trigonometry and real-life problems, such as indirect measurements of distances and heights.

Corollary 13.1.4 In any right triangle, the geometric mean of the projections of the legs on the hypotenuse equals the altitude relative to the hypotenuse:

$$h = \sqrt{m \cdot n}.$$

Demostración. This result follows directly from the theorem stating that $h^2 = m \cdot n$. ■

This corollary highlights the relationship between the altitude and the projections in a right triangle, reinforcing the importance of proportions and similarities in geometry.
In summary, triangle similarity allows solving complex problems through proportional relationships and is an invaluable tool in advanced mathematical reasoning.

13.1.2 Applications in Shadows and Scales.

Shadows and scales are practical applications of triangle similarity and proportionality in geometry. These tools allow solving indirect measurement problems, where directly measuring a length or height is complicated or impossible.

Definition 13.1.2 A **scale** is the mathematical relationship between the dimensions represented on a plan or map and the actual dimensions. It is generally expressed as a ratio or proportion.

The use of scales is fundamental in cartography, architecture, and design, enabling the representation of large objects in manageable dimensions while maintaining proportionality.

Theorem 13.1.5 — **Principle of proportionality in shadows.** At the same moment and place, the ratio between the height of an object and the length of its shadow is constant for all vertical objects.

Demostración. Sunlight can be considered as coming from a distant source, implying parallel solar rays. This forms similar triangles between the objects and their shadows, making the ratio between the object's height and the shadow's length constant. ■

This principle allows calculating inaccessible heights by measuring shadows and applying proportions.

■ **Example 13.3** To determine the height of a building, a vertical pole of 1.5 meters projects a shadow of 2 meters, while the building projects a shadow of 30 meters. Calculate the building's height.

■

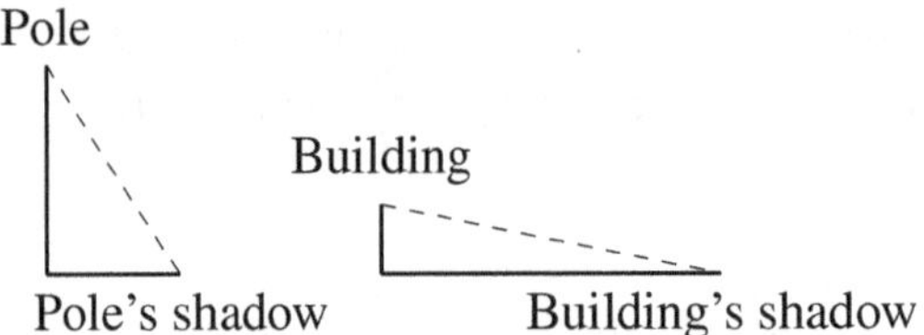

Figura 13.1.5: *Representation of the shadows of the pole and the building.*

Demostración. Applying the principle of proportionality:

$$\frac{\text{Height of the pole}}{\text{Shadow of the pole}} = \frac{\text{Height of the building}}{\text{Shadow of the building}} \implies \frac{1{,}5}{2} = \frac{h}{30}.$$

Solving for h:

$$h = \frac{1{,}5 \times 30}{2} = \frac{45}{2} = 22{,}5 \text{ meters.}$$

Thus, the building's height is 22.5 meters. ■

This example illustrates how shadows and proportionality can calculate inaccessible measurements indirectly.

Corollary 13.1.6 The height of an object can be calculated using the formula:

$$\text{Height of the object} = \left(\frac{\text{Height of the reference}}{\text{Length of the shadow of the reference}} \right) \times \text{Length of the shadow of the object.}$$

■ **Example 13.4** A tree projects a shadow of 12 meters at the same time a 4-meter pole projects a shadow of 3 meters. Determine the tree's height. ■

Demostración. Using the previous corollary:

$$\text{Height of the tree} = \left(\frac{4}{3}\right) \times 12 = \frac{4 \times 12}{3} = 16 \text{ meters.}$$

Thus, the tree is 16 meters tall. ■

Scales are also essential in representing objects in reduced-size drawings or models.

Definition 13.1.3 A **graphical scale** is a line or bar on a plan or map that indicates the relationship between represented distances and actual distances. It allows direct measurements on the map to be converted into real distances.

Lema 13.1.2 If two figures are similar, then their corresponding lengths are related by a scale factor k, and their corresponding areas are related by k^2.

Demostración. Let k be the scale factor between two similar figures. For any length L' in the scaled figure, $L' = kL$, where L is the corresponding length in the original figure. The area A' of the scaled figure is $A' = k^2 A$, where A is the original figure's area since the area is a two-dimensional quantity. ■

This lemma is essential for problems involving scales and areas.

Theorem 13.1.7 In similar figures, the relationship between their corresponding volumes is the cube of the scale factor k^3.

Demostración. Using reasoning similar to the previous lemma, since volume is a three-dimensional quantity, for two similar solids, the volume V' is given by $V' = k^3 V$, where V is the original solid's volume. ■

These relationships are fundamental in disciplines like physics and engineering, where scaled models are used.

Exercise 13.3 A building model is constructed at a scale of 1 : 50. If the model's height is 1.2 meters, what is the building's actual height? Additionally, calculate the ratio between the model's base area and the building's base area.

Demostración. The scale 1 : 50 indicates that each unit on the model corresponds to 50 units in reality.
Real height of the building:

$$\text{Real height} = 1{,}2\text{ m} \times 50 = 60\text{ m}.$$

Area ratio:
The scale factor is $k = \frac{1}{50}$.
Thus, the area ratio is:

$$\frac{A_{\text{model}}}{A_{\text{real}}} = k^2 = \left(\frac{1}{50}\right)^2 = \frac{1}{2500}.$$

Therefore, the base area of the model is $\frac{1}{2500}$ of the base area of the real building. ■

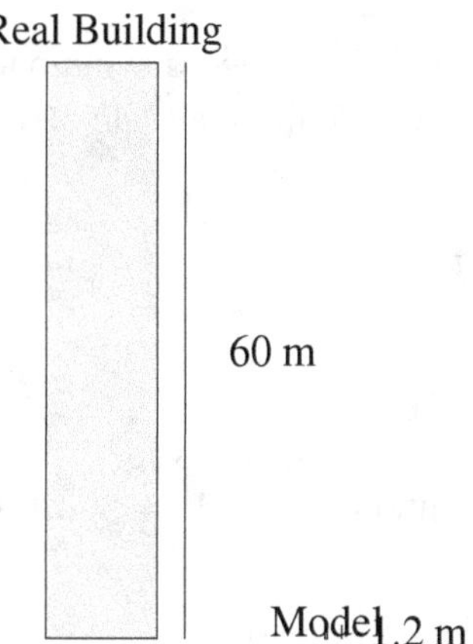

Figura 13.1.6: *Comparison between the real building and the model.*

This exercise demonstrates how to apply the scale factor to linear dimensions and areas.

(R) It is important to note that when working with scales, linear dimensions, areas, and volumes do not scale in the same way. Lengths scale by k, areas by k^2, and volumes by k^3.

Exercise 13.4 A map has a scale of $1:100,000$. If two cities are separated by a distance of 8 cm on the map, what is the real distance between them in kilometers? Additionally, if a lake occupies an area of 5 cm^2 on the map, what is its real area in square kilometers? ■

Demostración. Real distance between the cities:

$$\text{Real distance} = 8 \text{ cm} \times 100,000 = 800,000 \text{ cm} = 8 \text{ km}.$$

Real area of the lake:
The scale factor is $k = \frac{1}{100,000}$.
Area ratio:

$$\text{Real area} = \text{Map area} \times k^{-2} = 5 \text{ cm}^2 \times (100,000)^2 = 5 \times 10^{10} \text{ cm}^2.$$

Convert to square kilometers:

$$5 \times 10^{10} \text{ cm}^2 = 5 \times 10^{10} \times \left(\frac{1 \text{ m}}{100 \text{ cm}}\right)^2 \times \left(\frac{1 \text{ km}}{1000 \text{ m}}\right)^2 = 5 \text{ km}^2.$$

Thus, the real area of the lake is 5 square kilometers. ■

This exercise emphasizes the importance of correctly handling units when working with scales and areas.

Corollary 13.1.8 When scaling a figure on a plane by a factor of k, all lengths are multiplied by k, all areas by k^2, and all volumes by k^3. This holds true regardless of the shape of the figure.

Demostración. This result is a generalization of the concepts previously established in the lemma and theorem on similar figures. ■

■ **Example 13.5** An architect is designing a scaled-down model of a pool that in reality will have a capacity of 500,000 liters. If the model is built at a scale of $1:25$, what will be the volume of the model in liters? ■

Demostración. The scale factor is $k = \frac{1}{25}$.
Volume ratio:

$$V_{\text{model}} = V_{\text{real}} \times k^3 = 500,000 \text{ liters} \times \left(\frac{1}{25}\right)^3 = 500,000 \times \frac{1}{15,625} = 32 \text{ liters}.$$

Thus, the volume of the model will be 32 liters. ■

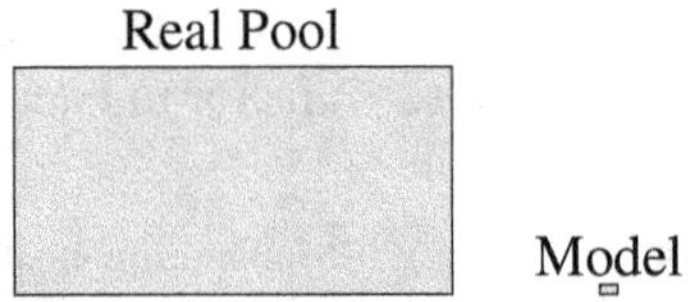

Figura 13.1.7: *Comparison between the real pool and the scaled model.*

This example shows the application of the scale factor to volumes and its relevance in designing and constructing models.
In conclusion, the application of scales and proportions is a practical manifestation of the principles of similarity and proportionality in geometry. These tools allow for solving indirect measurement problems and represent an essential component in fields such as cartography, architecture, and engineering.

13.2 Criteria for Triangle Similarity: AA, SAS, SSS

13.2.1 Applications in Maps and Designs.

Proportionality and similarity are fundamental concepts in map-making and blueprint design, allowing objects and areas to be represented at manageable scales while preserving essential geometric relationships. In this section, we will explore how these concepts are applied in cartography and design and how mathematics provides the necessary tools to work with scales and proportions rigorously.

Definition 13.2.1 A **scale** is the mathematical relationship between the dimensions of an object or area represented on a map or blueprint and its actual dimensions. It is generally expressed as a ratio or proportion, for example, $1:50,000$, which indicates that one unit on the map corresponds to $50,000$ units in reality.

The use of scales is essential in cartography and architectural design, enabling the precise representation of large distances or structures in a compact and practical format.

Theorem 13.2.1 If two figures are similar, then the corresponding side lengths are related by a constant scale factor k, and their areas are related by k^2.

Demostración. Let F be a figure and F' a figure similar to F with a scale factor k. By the definition of similarity, the corresponding lengths are related by $l' = kl$, where l and l' are corresponding lengths in F and F', respectively. The area of F' is then $A' = k^2 A$, as the area is a two-dimensional quantity and scales with the square of the scale factor. ■

This theorem is fundamental in working with scales, as it allows the calculation of areas and volumes in scaled models or representations.

■ **Example 13.6** An architect is designing a blueprint of a house at a scale of $1:100$. If a wall measures 12 meters in reality, what will its length be on the blueprint? Additionally, if the actual area of a room is 30 square meters, what will its area be on the blueprint? ■

Demostración. For the wall length:
The scale is $1:100$, meaning 1 unit on the blueprint represents 100 units in reality.
Thus, the length on the blueprint is:

$$l_{\text{blueprint}} = \frac{l_{\text{real}}}{k} = \frac{12\text{ m}}{100} = 0{,}12\text{ m} = 12\text{ cm}.$$

For the room area:
The area on the blueprint is calculated by scaling the actual area by the square of the scale factor:

$$A_{\text{blueprint}} = \frac{A_{\text{real}}}{k^2} = \frac{30\text{ m}^2}{100^2} = \frac{30}{10{,}000}\text{ m}^2 = 0{,}003\text{ m}^2 = 300\text{ cm}^2.$$

■

In this example, we applied the scale factor to both lengths and areas, demonstrating how dimensions are reduced in the blueprint according to the scale used.

It is important to note that when working with scales, units must be consistent. When converting between meters and centimeters, care must be taken to maintain precision in the calculations.

The representation of three-dimensional objects is also affected by scales, where volumes are related by the cube of the scale factor.

Theorem 13.2.2 If two solids are similar with a scale factor k, then their volumes are related by $V' = k^3 V$, where V and V' are the volumes of the original and scaled solids, respectively.

Demostración. Similar to the case of areas, volumes scale by the power of the scale factor corresponding to the three dimensions. If each linear dimension is multiplied by k, then the total volume is multiplied by k^3. ■

This result is crucial in fields such as engineering and architecture, where scaled models are built to study complex structures.

■ **Example 13.7** A scale model of a monument is built at a scale of $1:20$. If the actual volume of the monument is $8,000$ cubic meters, what is the volume of the model? ■

Demostración. Using the previous theorem, the volume of the model is:

$$V_{\text{model}} = \frac{V_{\text{real}}}{k^3} = \frac{8,000\text{ m}^3}{20^3} = \frac{8,000}{8,000}\text{ m}^3 = 1\text{ m}^3.$$

Thus, the volume of the model is 1 cubic meter. ■

This result demonstrates how scales significantly affect volumes, reducing them by much larger proportions than lengths or areas.

Lema 13.2.1 On a map, the distance between two points is proportional to the actual distance between those points, multiplied by the scale factor k. If the map uses a scale of $1:M$, then $k = \frac{1}{M}$.

Demostración. By the definition of scale in maps, the relationship between a distance on the map d_{map} and the actual distance d_{real} is:

$$d_{\text{map}} = k d_{\text{real}}, \quad \text{where} \quad k = \frac{1}{M}.$$

This shows that distances on the map are proportional to real distances by the factor k. ■

This concept is fundamental in cartography, allowing users to correctly interpret the distances represented.

■ **Example 13.8** On a map with a scale of $1:250,000$, the distance between two cities is 4 cm. Calculate the actual distance between the cities in kilometers. ■

Demostración. The scale factor is $k = \frac{1}{250{,}000}$.
The actual distance is:

$$d_{\text{real}} = \frac{d_{\text{map}}}{k} = 4\text{ cm} \times 250,000 = 1,000,000\text{ cm}.$$

Converting to kilometers:

$$1,000,000 \text{ cm} = 10,000 \text{ m} = 10 \text{ km}.$$

Therefore, the actual distance between the cities is 10 kilometers. ■

This example demonstrates the practical application of scales in cartography, enabling the conversion of measured distances on the map to real-world distances.

Exercise 13.5 An architectural plan is drawn at a scale of 1 : 75. If a room measures 6 cm by 4 cm on the plan, determine the actual dimensions of the room in meters. Additionally, calculate the actual area of the room.

Demostración. The actual dimensions are obtained by multiplying the plan dimensions by the scale factor:

$$\text{Actual length} = 6 \text{ cm} \times 75 = 450 \text{ cm} = 4{,}5 \text{ m},$$

$$\text{Actual width} = 4 \text{ cm} \times 75 = 300 \text{ cm} = 3 \text{ m}.$$

The actual area is:

$$A_{\text{actual}} = \text{Actual length} \times \text{Actual width} = 4{,}5 \text{ m} \times 3 \text{ m} = 13{,}5 \text{ m}^2.$$

■

This exercise reinforces the understanding of how to apply scales to derive actual dimensions and areas from a plan.

(R) In design and architecture, it is common to use different scales for varying levels of detail. It is essential to maintain consistency in units and scale factors to avoid errors in final dimensions.

Graphical representations are also fundamental in this context. For example, when designing maps or plans, it is useful to include graphical scales that allow users to measure distances directly on the document.

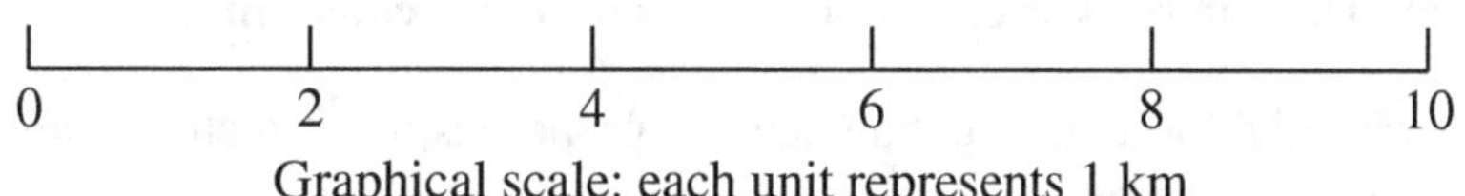

Figura 13.2.1: *Graphical scale for direct measurement on maps.*

In Figure 13.2.1, a graphical scale is presented to facilitate distance measurement on a map, visually representing the relationship between map distances and real-world distances.

Theorem 13.2.3 The precision of a scale on a map or plan determines the level of detail and accuracy with which geographic or structural features can be represented.

Demostración. A smaller scale (e.g., $1 : 1,000,000$) means that one unit on the map represents a large number of units in reality, limiting the level of detail that can be shown. Conversely, a larger scale (e.g., $1 : 1,000$) allows for finer and more precise details to be represented. The precision depends directly on the scale factor used. ■

This theorem is crucial to understanding how to choose the appropriate scale based on the purpose of the map or plan.

Exercise 13.6 A civil engineer is designing a road to be represented on a plan with a scale of $1:5,000$. If a curve on the road has a real radius of 250 meters, what will the radius of the curve be on the plan? Additionally, if the engineer needs to represent more precise details, should the scale factor be increased or decreased? ■

Demostración. The radius on the plan is:

$$r_{\text{plan}} = \frac{r_{\text{real}}}{k} = \frac{250 \text{ m}}{5,000} = 0{,}05 \text{ m} = 5 \text{ cm}.$$

To represent more precise details, the engineer should increase the scale factor (i.e., use a larger scale), for example, changing from $1:5,000$ to $1:1,000$, which would allow the curve's radius to be represented larger on the plan and show more details. ■

This exercise shows how the choice of scale affects the ability to represent details in a design or map.

Corollary 13.2.4 By reducing the scale factor (using larger scales), the ability to represent finer details is increased, which is essential in architectural plans and engineering designs.

In summary, the application of proportionality and similarity in maps and designs is crucial for the precise and practical representation of objects and areas. Mathematics provides the tools necessary to work with scales, ensuring that geometric relationships are maintained and real-world measurements can be accurately derived from reduced representations.

13.2.2 Proportions in Geometric Figures.

Understanding proportions in geometric figures is fundamental in advanced geometry studies and has applications in various areas of mathematics. Proportions establish relationships between the dimensions of similar figures and are essential for solving problems involving similarity, scaling, and indirect measurement.

Definition 13.2.2 Two quantities a and b are **proportional** if there exists a real number $k \neq 0$ such that $a = kb$. The number k is called the **proportionality constant**.

This definition is essential for understanding how the properties of geometric figures are preserved under scaling and transformation.

Theorem 13.2.5 In similar geometric figures, the corresponding lengths are proportional, the corresponding areas are proportional to the square of the similarity ratio, and the corresponding volumes are proportional to the cube of the similarity ratio.

Demostración. Let F and F' be two similar geometric figures with a similarity ratio k. Then, for any corresponding length l and l' in F and F', we have $l' = kl$.
For corresponding areas A and A', since area is a two-dimensional measure, we have $A' = k^2A$.
For corresponding volumes V and V', since volume is a three-dimensional measure, we have $V' = k^3V$. ■

This theorem is fundamental for studying how geometric properties change under scaling transformations.

■ **Example 13.9** Consider two squares, one with side $l = 4$ cm and the other with side $l' = 6$ cm. Determine the similarity ratio and calculate the relationships between their areas and perimeters. ■

Demostración. The similarity ratio is $k = \frac{l'}{l} = \frac{6}{4} = \frac{3}{2}$.
The perimeter of the first square is $P = 4l = 16$ cm, and the perimeter of the second square is $P' = 4l' = 24$ cm.
The relationship between the perimeters is:

$$\frac{P'}{P} = \frac{24}{16} = \frac{3}{2} = k.$$

The area of the first square is $A = l^2 = 16$ cm^2, and the area of the second square is $A' = (l')^2 = 36$ cm^2.
The relationship between the areas is:

$$\frac{A'}{A} = \frac{36}{16} = \frac{9}{4} = k^2.$$

Thus, the perimeters are proportional to k, and the areas are proportional to k^2. ■

This example illustrates how proportions affect measurements in similar geometric figures.

Lema 13.2.2 In two similar polygons, the measurements of the corresponding segments are proportional to the similarity ratio k.

Demostración. Since the polygons are similar, there is a one-to-one correspondence between their sides and angles, and each pair of corresponding sides has a proportionality ratio k. Thus, any segment in one of the polygons has a length $l' = kl$, where l is the length of the corresponding segment in the other polygon. ■

This lemma is useful for establishing relationships between different geometric elements in similar figures.

Theorem 13.2.6 In two circles with radii r and r', the ratio of their areas equals the square of the ratio of their radii, that is:

$$\frac{A'}{A} = \left(\frac{r'}{r}\right)^2.$$

Demostración. Let $A = \pi r^2$ be the area of the first circle with radius r, and $A' = \pi(r')^2$ be the area of the second circle with radius r'. Then, the ratio of their areas is

$$\frac{A'}{A} = \frac{\pi(r')^2}{\pi r^2} = \frac{(r')^2}{r^2} = \left(\frac{r'}{r}\right)^2.$$

■

This result shows how the areas of circles change with respect to their radii.

■ **Example 13.10** Two circles have radii $r = 5$ cm and $r' = 8$ cm. Calculate the ratio of their perimeters and the ratio of their areas. ■

Demostración. The ratio of the radii is:

$$k = \frac{r'}{r} = \frac{8}{5}.$$

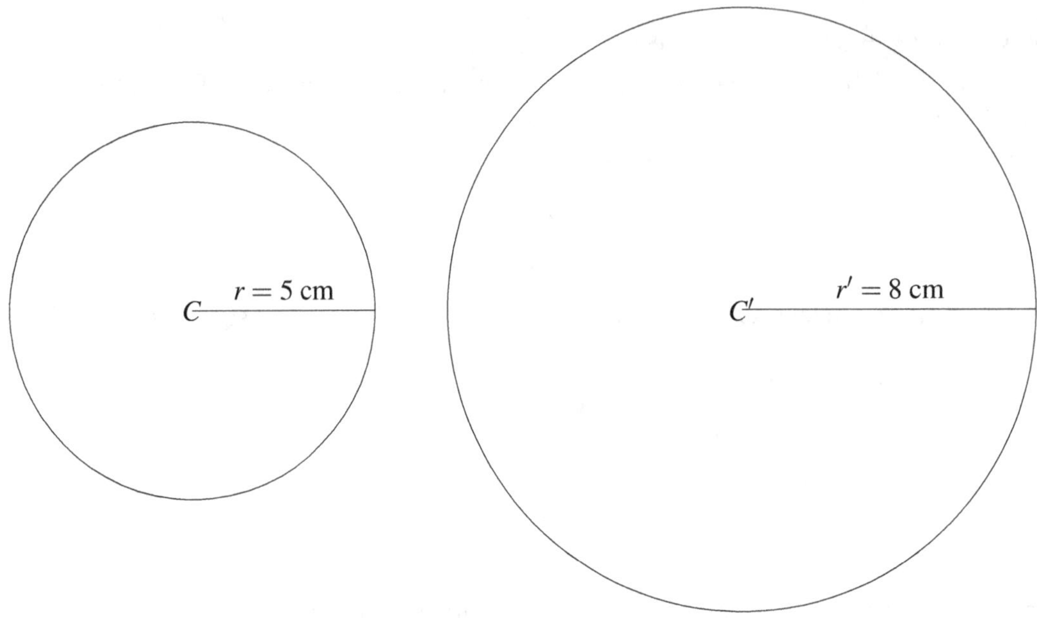

Figura 13.2.2: *Circles with radii $r = 5$ cm and $r' = 8$ cm.*

The perimeter of a circle is $P = 2\pi r$. Thus, the ratio of the perimeters is:

$$\frac{P'}{P} = \frac{2\pi r'}{2\pi r} = \frac{r'}{r} = k = \frac{8}{5}.$$

The area of a circle is $A = \pi r^2$. The ratio of the areas is:

$$\frac{A'}{A} = \left(\frac{r'}{r}\right)^2 = \left(\frac{8}{5}\right)^2 = \frac{64}{25}.$$

Therefore, the perimeters are proportional to the radii, and the areas are proportional to the square of the radii. ■

Figure **??** illustrates the two circles and their radii.

Corollary 13.2.7 In similar geometric figures, if the similarity ratio is k, then:

- The corresponding lengths are related by $l' = kl$.
- The corresponding areas are related by $A' = k^2 A$.
- The corresponding volumes are related by $V' = k^3 V$.

Demostración. This corollary is a direct consequence of the previous theorem on similar geometric figures and the definition of the similarity ratio. ■

(R) These relationships are essential in practical applications, such as the scaled design of models and map interpretation, where maintaining proportions is crucial to preserve geometric properties.

Now, let us consider how these proportions apply to geometric solids.

Theorem 13.2.8 For two spheres with radii r and r', the ratio of their volumes equals the cube of the ratio of their radii:

$$\frac{V'}{V} = \left(\frac{r'}{r}\right)^3.$$

Demostración. Let $V = \frac{4}{3}\pi r^3$ be the volume of a sphere with radius r, and $V' = \frac{4}{3}\pi(r')^3$ the volume of a sphere with radius r'. The ratio of their volumes is

$$\frac{V'}{V} = \frac{\frac{4}{3}\pi(r')^3}{\frac{4}{3}\pi r^3} = \frac{(r')^3}{r^3} = \left(\frac{r'}{r}\right)^3.$$

■

This theorem shows how the volumes of spheres change concerning their radii.

Exercise 13.7 A cone and a pyramid are similar, and the height of the cone is twice the height of the pyramid. If the volume of the pyramid is V, what is the volume of the cone in terms of V?

■

Demostración. Since the figures are similar and the height of the cone is twice the height of the pyramid, the similarity ratio is $k = 2$.
By the previous corollary, the volumes are related by:

$$V_{\text{cone}} = k^3 V_{\text{pyramid}} = 2^3 V = 8V.$$

Thus, the volume of the cone is $8V$. ■

This exercise illustrates how proportions affect the volumes of similar figures.

Exercise 13.8 Two similar rectangular prisms have heights of 5 cm and 15 cm. If the base area of the smaller prism is 20 cm^2, determine the base area of the larger prism and the ratio of their volumes.

■

Demostración. The similarity ratio is $k = \frac{15}{5} = 3$.
The base area of the larger prism is:

$$A' = k^2 A = 3^2 \times 20 \text{ cm}^2 = 9 \times 20 \text{ cm}^2 = 180 \text{ cm}^2.$$

The ratio of the volumes is:

$$\frac{V'}{V} = k^3 = 3^3 = 27.$$

Thus, the volume of the larger prism is 27 times the volume of the smaller prism. ■

This exercise demonstrates the application of proportions in geometric solids and how different measurements are related.

R It is important to remember that when working with similar geometric figures, it is crucial to correctly identify the similarity ratio and apply the appropriate powers (1 for lengths, 2 for areas, and 3 for volumes) when calculating proportions.

In conclusion, proportions in geometric figures are fundamental to understanding how the dimensions and properties of figures change under scaling. This knowledge is essential in fields such as architecture, engineering, and physical sciences, where scaled models and geometric transformations are frequently used.

13.3 Application Problems: Scales and Maps.

13.3.1 Reduction Scales in Blueprints.

Reduction scales are fundamental in disciplines such as engineering, architecture, and cartography, as they allow large objects to be represented on manageable-sized blueprints or maps without losing proportion or essential geometric relationships. To understand how these scales work, it is necessary to delve into the mathematical concepts underlying them.

Definition 13.3.1 A **reduction scale** is a proportionality relationship that indicates how many times an object or figure has been reduced in a blueprint or map compared to its actual dimensions. It is commonly expressed as a ratio $1:n$, where n is the reduction factor.

It is important to note that when applying a reduction scale, all linear dimensions of the object are reduced by the same proportion, thus maintaining geometric similarity.

Theorem 13.3.1 If a geometric figure is reduced using a scale factor $k = \frac{1}{n}$, then:

1. Lengths are reduced by the factor k.
2. Areas are reduced by the factor k^2.
3. Volumes are reduced by the factor k^3.

Demostración. Let an original figure have linear dimensions L, area A, and volume V. When applying a reduction scale with factor k, the new dimensions will be:

1. Lengths: $L' = kL$.
2. Areas: $A' = k^2 A$ (because area is two-dimensional).
3. Volumes: $V' = k^3 V$ (because volume is three-dimensional).

This is because linear dimensions are multiplied by k, and areas and volumes depend on linear dimensions raised to the second and third power, respectively. ■

This theorem is essential for accurately calculating measurements on scaled blueprints and models.

■ **Example 13.11** An architect designs a blueprint of a building at a scale of $1:100$. If a wall measures 30 meters in reality, what will its length be on the blueprint? Additionally, if the real area of a floor is 900 square meters, what will its area be on the blueprint?

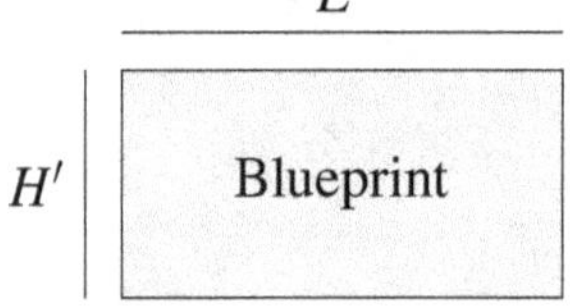

L'

H' Blueprint

Figura 13.3.1: *Representation of a wall on the blueprint at scale* $1:100$.

■

Demostración. The scale is $1:100$, so the reduction factor is $k = \frac{1}{100}$.
For the wall length on the blueprint:

$$L' = kL = \frac{1}{100} \times 30 \text{ m} = 0{,}3 \text{ m} = 30 \text{ cm}.$$

For the floor area on the blueprint:

$$A' = k^2 A = \left(\frac{1}{100}\right)^2 \times 900 \text{ m}^2 = \frac{1}{10,000} \times 900 \text{ m}^2 = 0{,}09 \text{ m}^2 = 900 \text{ cm}^2.$$

Therefore, on the blueprint, the wall measures 30 cm, and the floor area is 900 cm^2. ■

This example demonstrates how to apply the scale factor to obtain accurate measurements on a blueprint.

R It is crucial to maintain consistency in units when working with scales. If meters are used for real dimensions, it is advisable to convert the measurements to centimeters or millimeters on the blueprint, as needed.

In addition to practical applications, it is interesting to explore deeper mathematical properties related to reduction scales.

Lema 13.3.1 If two geometric figures are similar with a similarity ratio of k, then any proportion between their linear measurements, areas, or volumes is equal to k, k^2, or k^3, respectively.

Demostración. The similarity of figures implies that all linear dimensions are related by the factor k, so:

1. Lengths: $\frac{L'}{L} = k$.
2. Areas: $\frac{A'}{A} = k^2$.
3. Volumes: $\frac{V'}{V} = k^3$.

■

This lemma reinforces the understanding of how a figure's dimensions change when applying a reduction scale.

■ **Example 13.12** A map is created at a scale of $1 : 50,000$. If the distance between two cities on the map is 5 cm, calculate the real distance between them. Additionally, if a lake has an area of 10 cm^2 on the map, determine its real area.

■

Demostración. For the real distance between the cities:

$$\text{Real Distance} = \frac{1}{k} \times \text{Distance on the map} = 50,000 \times 5 \text{ cm} = 250,000 \text{ cm} = 2{,}5 \text{ km}.$$

For the real area of the lake:

$$\text{Real Area} = \frac{1}{k^2} \times \text{Area on the map} = 50,000^2 \times 10 \text{ cm}^2 = 2{,}5 \times 10^{11} \text{ cm}^2 = 25 \text{ km}^2.$$

Therefore, the real distance is 2,5 km, and the real area of the lake is 25 km^2. ■

This example highlights the importance of considering the square of the scale factor when calculating areas.

Theorem 13.3.2 In a scaled blueprint, the ratio between real linear measurements and those represented on the blueprint is the inverse of the reduction factor k.

Demostración. If L is the real measurement and L' is the measurement on the blueprint, then:

$$L' = kL \implies \frac{L}{L'} = \frac{1}{k}.$$

Thus, the ratio $\frac{L}{L'}$ is the inverse of the reduction factor k. ■

This theorem is useful for converting measurements between the blueprint and real dimensions, and vice versa.

Exercise 13.9 A civil engineer needs to represent a rectangular plot measuring 200 m in length and 150 m in width on a blueprint at a scale of $1 : 2,000$. What will be the dimensions of the plot on the blueprint? Additionally, what will be the area of the plot on the blueprint in square centimeters? ■

Demostración. Reduction factor: $k = \frac{1}{2,000}$.
Dimensions on the blueprint:

$$\text{Length on the blueprint} = k \times 200 \text{ m} = \frac{1}{2,000} \times 200 \text{ m} = 0{,}1 \text{ m} = 10 \text{ cm}.$$

$$\text{Width on the blueprint} = k \times 150 \text{ m} = \frac{1}{2,000} \times 150 \text{ m} = 0{,}075 \text{ m} = 7{,}5 \text{ cm}.$$

Area on the blueprint:

$$A' = \text{Length on the blueprint} \times \text{Width on the blueprint} = 10 \text{ cm} \times 7{,}5 \text{ cm} = 75 \text{ cm}^2.$$

Therefore, the dimensions on the blueprint are 10 cm in length and 7,5 cm in width, and the area is 75 cm^2. ■

Exercise 13.10 A designer creates a $1 : 500$ scale model of a park that, in reality, has a volume of $75,000 \text{ m}^3$. What is the volume of the model? Provide the answer in cubic meters and cubic centimeters. ■

Demostración. Reduction factor: $k = \frac{1}{500}$.
Volume of the model:

$$V' = k^3 V = \left(\frac{1}{500}\right)^3 \times 75,000 \text{ m}^3 = \frac{1}{125,000,000} \times 75,000 \text{ m}^3 = 0{,}0006 \text{ m}^3.$$

Convert to cubic centimeters ($1 \text{ m}^3 = 1,000,000 \text{ cm}^3$):

$$0{,}0006 \text{ m}^3 \times 1,000,000 \frac{\text{cm}^3}{\text{m}^3} = 600 \text{ cm}^3.$$

Therefore, the volume of the model is $0{,}0006 \text{ m}^3$ or 600 cm^3. ■

To deepen the study of reduction scales, it is interesting to analyze their effects on three-dimensional figures.

Corollary 13.3.3 In a reduction scale applied to solids, the weight of the model is proportional to the cube of the reduction factor if the same material is used.

Demostración. Since weight is directly proportional to volume and the material's density remains constant, we have:

$$\text{Weight of the model} = k^3 \times \text{Real weight}.$$

■

This corollary is relevant in engineering and design, especially when constructing physical models.

(R) When constructing scaled models, it is common to use different materials to avoid excessive weight, which introduces variations in density and requires adjustments in calculations.

■ **Example 13.13** A company manufactures a $1:10$ scale replica of a statue whose original weight is $1,000$ kg. If the replica is made of the same material, what will its weight be? ■

Demostración. Applying the previous corollary:

$$\text{Weight of the replica} = k^3 \times \text{Original weight} = \left(\frac{1}{10}\right)^3 \times 1,000 \text{ kg} = \frac{1}{1,000} \times 1,000 \text{ kg} = 1 \text{ kg}.$$

Thus, the replica will weigh 1 kg. ■

This example illustrates how the weight drastically decreases when reducing the scale of a three-dimensional object.
To conclude this section, let us consider a theorem related to scale in inclined planes.

Theorem 13.3.4 In an inclined plane represented at a reduced scale, the angle of inclination remains invariant with respect to the real plane.

Demostración. The angle of inclination θ of a plane depends solely on the ratio between the height and the base:

$$\tan\theta = \frac{\text{Height}}{\text{Base}}.$$

When applying a reduction scale, both measurements are reduced by the same factor k, so:

$$\tan\theta' = \frac{k \times \text{Height}}{k \times \text{Base}} = \frac{\text{Height}}{\text{Base}} = \tan\theta.$$

Thus, $\theta' = \theta$. ■

This result is significant when designing models or blueprints where the representation of the angle is crucial.

Exercise 13.11 A ramp has a length of 12 m and a height of 3 m. If it is represented on a blueprint at a scale of 1 : 100, determine the length and height on the blueprint, and verify that the angle of inclination is the same on the real ramp and the blueprint.

Demostración. Reduction factor: $k = \frac{1}{100}$.
Length on the blueprint:

$$L' = kL = \frac{1}{100} \times 12 \text{ m} = 0{,}12 \text{ m} = 12 \text{ cm}.$$

Height on the blueprint:

$$H' = kH = \frac{1}{100} \times 3 \text{ m} = 0{,}03 \text{ m} = 3 \text{ cm}.$$

Angle of inclination on the real ramp:

$$\tan\theta = \frac{3}{12} = \frac{1}{4}.$$

Angle of inclination on the blueprint:

$$\tan\theta' = \frac{3 \text{ cm}}{12 \text{ cm}} = \frac{1}{4}.$$

Thus, $\theta' = \theta$, and the angle of inclination is the same in both cases. ■

This exercise confirms the theorem and demonstrates its practical application.
In conclusion, reduction scales in blueprints are essential tools that, grounded in solid mathematical principles, enable the precise and manageable representation of large objects and spaces. Understanding proportional relationships and their effects on different measurements is fundamental for their correct application in various professional fields.

13.3.2 Proportions in Cartography.

Cartography is the science and art of representing the Earth's surface on maps. To achieve accurate and useful representations, it is essential to understand and correctly apply proportions and scales in cartography. These proportions allow the relationship of distances and areas on the map with their real-world counterparts.

Definition 13.3.2 A **cartographic scale** is the mathematical relationship expressing how much the Earth's surface has been reduced to be represented on a map. It is defined as the ratio between a distance on the map and the corresponding real-world distance, usually expressed as $1 : n$, where n is the *denominator of the scale*.

Scales can be numerical, graphical, or verbal, and their correct interpretation is crucial for measuring distances and areas on maps.

Theorem 13.3.5 In a cartographic projection without linear distortions, the real distance D_{real} between two points is proportional to the distance D_{map} between the corresponding points on the map, multiplied by the scale denominator n:

$$D_{\text{real}} = n \times D_{\text{map}}.$$

Demostración. By the definition of cartographic scale:

$$\frac{D_{\text{map}}}{D_{\text{real}}} = \frac{1}{n} \implies D_{\text{real}} = n \times D_{\text{map}}.$$

This establishes a direct proportionality between map and real-world distances. ■

It is important to note that this relationship holds true for maps with minimal or negligible distortions, such as large-scale maps covering small areas.

■ **Example 13.14** On a map with a scale of $1 : 100,000$, the distance between two cities is 7 cm. Calculate the real-world distance between the cities in kilometers. ■

Demostración. Using the previous theorem:

$$D_{\text{real}} = n \times D_{\text{map}} = 100,000 \times 7 \text{ cm} = 700,000 \text{ cm}.$$

Converting to kilometers:

$$700,000 \text{ cm} = 7,000 \text{ m} = 7 \text{ km}.$$

Therefore, the real-world distance between the cities is 7 km. ■

R When working with cartographic scales, it is essential to maintain unit consistency and convert them properly to obtain correct results.

In addition to linear distances, proportions in cartography also affect areas represented on maps.

Theorem 13.3.6 The real area A_{real} of a region is proportional to the square of the scale denominator n^2 multiplied by the map area A_{map} of the region:

$$A_{\text{real}} = n^2 \times A_{\text{map}}.$$

Demostración. The linear scale affects each linear dimension by a factor of n. Since the area is two-dimensional, the scale affects the area by the square of the scale factor:

$$A_{\text{real}} = (n \times L_{\text{map}}) \times (n \times W_{\text{map}}) = n^2 \times (L_{\text{map}} \times W_{\text{map}}) = n^2 \times A_{\text{map}}.$$

■

■ **Example 13.15** On a map with a scale of $1 : 50000$, a lake is represented with an area of $12\,\text{cm}^2$. Calculate the real area of the lake in square kilometers.

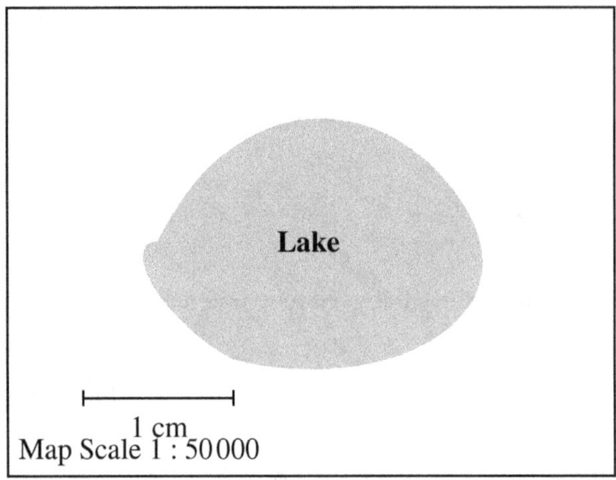

Figura 13.3.2: *Schematic representation of the lake on the map.*

■

Demostración. To calculate the real area of the lake, follow these steps:
Step 1: Understand the map scale
The scale 1 : 50 000 indicates that 1 cm on the map represents 50,000 cm in reality.
Step 2: Calculate the area scale factor
The linear scale factor is 50 000. Since the area is two-dimensional, square this factor:

$$\text{Area scale factor} = (50\,000)^2 = 2\,500\,000\,000.$$

Step 3: Calculate the real area in square centimeters
Multiply the lake's area on the map by the area scale factor:

$$\text{Real area} = 12\,\text{cm}^2 \times 2\,500\,000\,000 = 30\,000\,000\,000\,\text{cm}^2.$$

Step 4: Convert the area to square kilometers
Since:

$$1\,\text{km} = 100\,000\,\text{cm} \implies 1\,\text{km}^2 = (100\,000\,\text{cm})^2 = 10\,000\,000\,000\,\text{cm}^2,$$

the area in square kilometers is:

$$\text{Area in km}^2 = \frac{30\,000\,000\,000\,\text{cm}^2}{10\,000\,000\,000\,\text{cm}^2/\text{km}^2} = 3\,\text{km}^2.$$

Answer: The real area of the lake is $3\,\text{km}^2$.

■

This example demonstrates how areas on a map relate to real-world areas through the square of the scale denominator.

Lema 13.3.2 The **graphical scale** is a visual representation of the numerical scale that allows direct measurement on the map to obtain real distances without additional calculations.

Demostración. The graphical scale consists of a graduated bar on the map that indicates the correspondence between distances on the map and real distances. By measuring a distance on the map with a ruler and comparing it to the graphical scale, the real distance is directly obtained, as the graphical scale already incorporates the scale factor. ■

The graphical scale is particularly useful when the map is enlarged or reduced, as the graphical scale adjusts proportionally, maintaining its validity.

Theorem 13.3.7 In cartography, when representing large areas of the Earth's surface, map projections introduce distortions that affect the proportions of distances, areas, and angles. These distortions are unavoidable and must be considered when performing measurements on maps.

Demostración. The Earth is a spherical (or ellipsoidal) surface, and when projecting it onto a flat surface to create a map, it is impossible to preserve all metric properties (distances, areas, angles) simultaneously. According to Gauss's Theorem on the impossibility of an isometric representation of a spherical surface on a plane, distortions are inevitable. Different map projections aim to minimize or control these distortions depending on the map's purpose. ■

R It is important to select the appropriate map projection based on the type of analysis to be performed, as some projections preserve areas (equivalent), others preserve local shapes (conformal), and others distances from specific points (equidistant).

■ **Example 13.16** A world map uses the Mercator projection, which is conformal but distorts areas in high latitudes. For instance, in such maps, Greenland may appear the same size as Africa, whereas Africa is approximately 14 times larger.

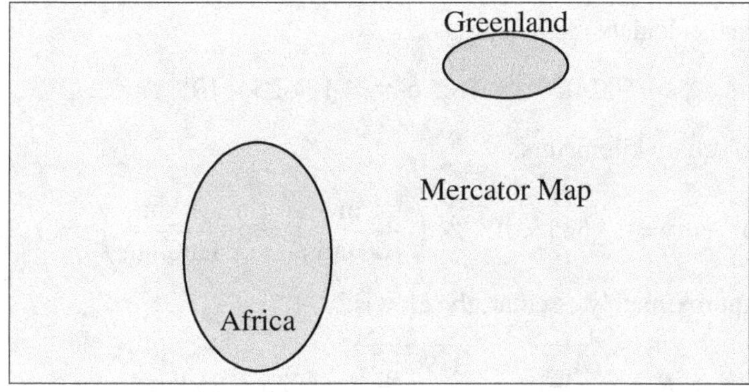

Figura 13.3.3: *Area distortion in the Mercator projection.*

■

Demostración. The Mercator projection expands areas as they approach the poles, preserving local shapes but distorting relative dimensions. Thus, Greenland, located in high latitudes, appears disproportionately large compared to Africa, which is near the equator. In reality, Africa's area is much larger than Greenland's, but the projection causes them to appear similar in size on the map. This distortion results from the mathematical properties of the projection and underscores the importance of choosing the appropriate projection to faithfully represent areas on a map. ■

This example illustrates the importance of considering distortions introduced by map projections when interpreting proportions on maps.

Corollary 13.3.8 When using small-scale maps (large denominators), projection distortions become more significant, so measurements of distances and areas must be made with caution.

Exercise 13.12 On a map with a $1:250,000$ scale, the straight-line distance between two cities is 12 cm. However, due to topography, the actual route between the cities is 15 cm on the map. Calculate the real distance of the route and compare it to the straight-line distance. ■

Demostración. Straight-line distance:

$$D_{\text{straight}} = n \times D_{\text{map}} = 250,000 \times 12 \text{ cm} = 3,000,000 \text{ cm} = 30 \text{ km}.$$

Route distance:

$$D_{\text{route}} = n \times D_{\text{map}} = 250,000 \times 15 \text{ cm} = 3,750,000 \text{ cm} = 37{,}5 \text{ km}.$$

Comparison:

$$\text{Difference} = 37{,}5 \text{ km} - 30 \text{ km} = 7{,}5 \text{ km}.$$

The actual route is 7,5 km longer than the straight-line distance due to topography and road layout. ■

This exercise shows how measured distances on a map can vary depending on the route considered and how the scale is applied to obtain real distances.

Exercise 13.13 A national park has an approximately circular shape and occupies an area of 78,5 cm^2 on a map with a $1:500,000$ scale. Calculate the real radius of the park in kilometers. ■

Demostración. First, calculate the real area:

$$A_{\text{real}} = n^2 \times A_{\text{map}} = (500,000)^2 \times 78{,}5 \text{ cm}^2 = 1{,}9625 \times 10^{14} \text{ cm}^2.$$

Convert the area to square kilometers:

$$1{,}9625 \times 10^{14} \text{ cm}^2 = 1{,}9625 \times 10^{14} \times \left(\frac{1 \text{ m}}{100 \text{ cm}}\right)^2 \times \left(\frac{1 \text{ km}}{1,000 \text{ m}}\right)^2 = 1,962{,}5 \text{ km}^2.$$

Since the park is approximately circular, the area is:

$$A_{\text{real}} = \pi R^2 \implies R = \sqrt{\frac{A_{\text{real}}}{\pi}} = \sqrt{\frac{1,962{,}5}{\pi}} \approx \sqrt{625} = 25 \text{ km}.$$

Thus, the real radius of the park is approximately 25 km. ■

This exercise integrates area calculation and the application of proportions in cartography to determine real measurements.

R Accuracy in cartographic measurements depends on the precision of the scale, the projection used, and the quality of the map. It is essential to consider all these factors when performing calculations and analyses.

In summary, proportions in cartography are essential for correctly interpreting and using maps. Understanding how scales and projections affect distances and areas allows for accurate measurements and informed decisions based on geographic data.

13.4 Solved Exercises

Exercise 13.14 Given a triangle $\triangle ABC$ with sides $AB = 8$ cm, $AC = 6$ cm, and a line DE parallel to BC that intersects sides AB and AC at points D and E, respectively. If $AD = 4$ cm, calculate the length of AE. ▪

Demostración. Since $DE \parallel BC$, we apply the Thales Theorem, which states:

$$\frac{AD}{DB} = \frac{AE}{EC}.$$

Given $AD = 4$ cm and $AB = 8$ cm, we have $DB = 8 - 4 = 4$ cm. This implies $\frac{AD}{DB} = 1$. Therefore, $AE = EC$, and since $AC = 6$ cm, we have $AE = \frac{6}{2} = 3$ cm. ■

Exercise 13.15 In a right triangle $\triangle ABC$ with $\angle C = 90°$, the heights h_a and h_b are drawn from vertices A and B to the hypotenuse AB. If $AB = 10$ cm, $h_a = 4$ cm, and $h_b = 3$ cm, calculate the area of $\triangle ABC$. ▪

Demostración. The area of a triangle is equal to $\frac{1}{2} \times \text{base} \times \text{height}$. Using the hypotenuse $AB = 10$ cm as the base and the corresponding height $h_a = 4$ cm, the area is:

$$\text{Area} = \frac{1}{2} \times 10 \times 4 = 20 \text{ cm}^2.$$

Verification using the height from the other vertex: Area $= \frac{1}{2} \times 10 \times 3 = 20$ cm^2, confirming the area is correct. ■

Exercise 13.16 On a map with a $1:100,000$ scale, the distance between two cities is 5 cm. Calculate the real distance between the cities in kilometers. ▪

Demostración. Given the scale $1:100,000$, each centimeter on the map represents $100,000$ cm in reality. Therefore, the real distance is:

$$\text{Real distance} = 5 \times 100,000 = 500,000 \text{ cm}.$$

Convert to kilometers:

$$500,000 \text{ cm} = 5,000 \text{ m} = 5 \text{ km}.$$

The real distance between the cities is 5 km. ■

Exercise 13.17 A rectangle has a base of 8 cm and a height of 5 cm. If it is scaled by a factor of 1,5, calculate the new area of the rectangle. ▪

Demostración. The new base will be $8 \times 1{,}5 = 12$ cm, and the new height will be $5 \times 1{,}5 = 7{,}5$ cm. Therefore, the new area is:

$$\text{New area} = 12 \times 7{,}5 = 90 \text{ cm}^2.$$

■

Exercise 13.18 In a triangle similar to the original triangle $\triangle ABC$, the side corresponding to $AB = 6$ cm measures 12 cm. If the area of the original triangle is 15 cm^2, calculate the area of the similar triangle. ■

Demostración. The similarity ratio between the triangles is $\frac{12}{6} = 2$. Since the areas of similar triangles are proportional to the square of the similarity ratio, the area of the similar triangle is:

$$\text{Similar area} = 2^2 \times 15 = 4 \times 15 = 60 \text{ cm}^2.$$

■

13.5 Proposed Exercises

13.5.1 Thales' Theorem: Applications in Similar Figures

Exercise 13.19 In a triangle $\triangle ABC$, a line parallel to side BC intersects sides AB and AC at points D and E, respectively. If $AB = 10$ cm, $AC = 8$ cm, and $AD = 4$ cm, calculate the length of AE. ■

Exercise 13.20 In a triangle $\triangle XYZ$, a line parallel to side XY intersects sides XZ and YZ at points P and Q. If $XP = 3$ cm, $PZ = 6$ cm, and $YQ = 5$ cm, find the length of QZ. ■

Exercise 13.21 Given a triangle $\triangle ABC$ with $AB = 15$ cm and $AC = 10$ cm, a line DE parallel to BC intersects AB at D and AC at E, such that $AD = 6$ cm. Calculate the length of AE. ■

Exercise 13.22 In a triangle $\triangle DEF$, a line parallel to EF intersects sides DE and DF at points G and H, respectively. If $DE = 20$ cm, $DF = 15$ cm, and $DG = 8$ cm, find the length of DH. ■

Exercise 13.23 Given a triangle $\triangle ABC$ with $AB = 12$ cm and $AC = 9$ cm, a line parallel to BC intersects AB at D and AC at E. If $AD = 4$ cm, find the length of AE. ■

13.5.2 Similarity Criteria for Triangles: AA, SAS, SSS

Exercise 13.24 Prove that two triangles are similar if they satisfy the AA criterion, i.e., if they have two corresponding angles equal. Use examples with angle measurements. ■

Exercise 13.25 In a triangle $\triangle XYZ$, the sides $XY = 10$ cm, $XZ = 15$ cm, and the angle between them is 60°. In another triangle $\triangle PQR$, the sides $PQ = 5$ cm and $PR = 7{,}5$ cm, and the angle between them is 60°. Verify if the triangles are similar using the SAS criterion. ■

Exercise 13.26 Given a triangle $\triangle ABC$ with sides $AB = 8$ cm, $AC = 6$ cm, and $BC = 10$ cm, and a triangle $\triangle DEF$ with sides $DE = 4$ cm, $DF = 3$ cm, and $EF = 5$ cm, demonstrate that they are similar using the SSS criterion. ■

Exercise 13.27 In a triangle $\triangle KLM$, the angle $\angle KLM$ measures 50°, and the angle $\angle KML$ measures 70°. In another triangle $\triangle PQR$, the corresponding angles measure $\angle PQR = 50°$ and $\angle PRQ = 70°$. Verify if the triangles are similar by the AA criterion. ■

Exercise 13.28 Given a triangle $\triangle GHI$ with sides $GH = 9$ cm, $HI = 12$ cm, and $GI = 15$ cm, and another triangle $\triangle JKL$ with sides $JK = 6$ cm, $KL = 8$ cm, and $JL = 10$ cm. Check if they are similar using the SSS criterion. ■

13.5.3 Application Problems: Scales and Maps

Exercise 13.29 On a map with a scale of $1 : 50,000$, the distance between two points is 7 cm. Calculate the actual distance between these points in kilometers. ■

Exercise 13.30 A house plan has a scale of $1 : 100$. If a room on the plan measures 4 cm in length and 3 cm in width, find the actual dimensions of the room in meters. ■

Exercise 13.31 On a map with a scale of $1 : 200,000$, a nature reserve has an area of 5 cm^2. Calculate the actual area of the reserve in square kilometers. ■

Exercise 13.32 A model of a building is constructed at a scale of $1 : 75$. If the height of the model is 1,2 meters, determine the actual height of the building. ■

Exercise 13.33 On a map with a scale of $1 : 100,000$, the length of a river is measured as 12 cm. Calculate the actual length of the river in kilometers. ■

14. Metric Relations in Triangles

14.1 Pythagorean Theorem: Application in Right Triangles

14.1.1 Application in Physical Problems

The **Pythagorean Theorem** is a fundamental tool in physics to solve problems involving calculations of distances, displacements, velocities, and forces in different directions. Its application simplifies the analysis of physical systems by decomposing vectors into perpendicular components and calculating resultants.

Definition 14.1.1 In a **right triangle**, the *Pythagorean Theorem* states that the square of the length of the hypotenuse is equal to the sum of the squares of the lengths of the legs. Mathematically:

$$c^2 = a^2 + b^2,$$

where c is the hypotenuse and a, b are the legs.

This theorem is widely used in physics to calculate resultant magnitudes when the components are perpendicular to each other.

Theorem 14.1.1 The magnitude of the resultant of two perpendicular vectors is equal to the square root of the sum of the squares of the magnitudes of the individual vectors:

$$|\vec{R}| = \sqrt{|\vec{A}|^2 + |\vec{B}|^2}.$$

Demostración. Let $\vec{A}$ and $\vec{B}$ be two perpendicular vectors. The magnitude of the resultant vector $\vec{R} = \vec{A} + \vec{B}$ is found by applying the Pythagorean Theorem to the right triangle formed by $\vec{A}$, $\vec{B}$, and $\vec{R}$. Thus,

$$|\vec{R}| = \sqrt{|\vec{A}|^2 + |\vec{B}|^2}.$$

■

This result is essential in vector addition in physics, particularly in problems involving kinematics and dynamics.

■ **Example 14.1** A boat sails 6 km east and then 8 km north. What is its direct distance from the starting point, and in what direction is it relative to the east?

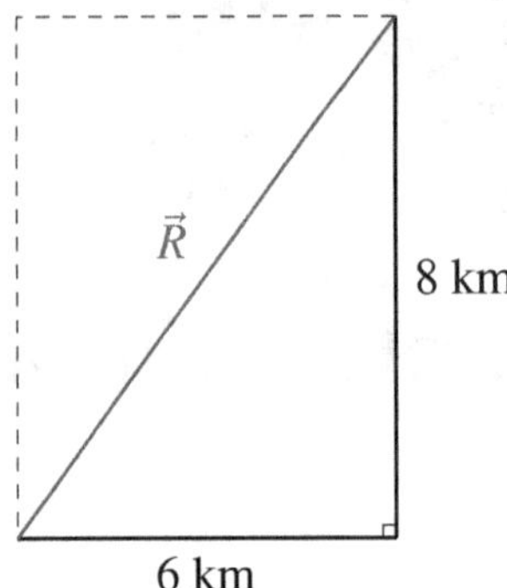

Figura 14.1.1: *Boat displacement represented as perpendicular vectors and its resultant.*

■

Using the Pythagorean Theorem, the direct distance is calculated as:

$$|\vec{R}| = \sqrt{(6\text{ km})^2 + (8\text{ km})^2} = \sqrt{36+64} = \sqrt{100} = 10\text{ km}.$$

To determine the direction, calculate the angle θ relative to the east using the tangent function:

$$\tan\theta = \frac{\text{north component}}{\text{east component}} = \frac{8}{6} = \frac{4}{3} \Longrightarrow \theta = \arctan\left(\frac{4}{3}\right) \approx 53{,}13^\circ.$$

The boat is 10 km away from the starting point at a direction of 53,13° north of east.
This example demonstrates how the Pythagorean Theorem is applied to calculate resultant displacements and directions in navigation problems.

Corollary 14.1.2 In kinematics, the resultant velocity of an object with perpendicular velocities v_x and v_y is:

$$v = \sqrt{v_x^2 + v_y^2}.$$

Demostración. Considering the velocities v_x and v_y as perpendicular components of the total velocity, a right triangle is formed where the hypotenuse is the resultant velocity v. By applying the Pythagorean Theorem:

$$v^2 = v_x^2 + v_y^2.$$

■

This corollary is fundamental for calculating the total velocity in two-dimensional motion.

■ **Example 14.2** An airplane flies at a speed of 250 km/h northward and experiences a crosswind of 100 km/h westward. Determine the resultant velocity and direction of the airplane.

■

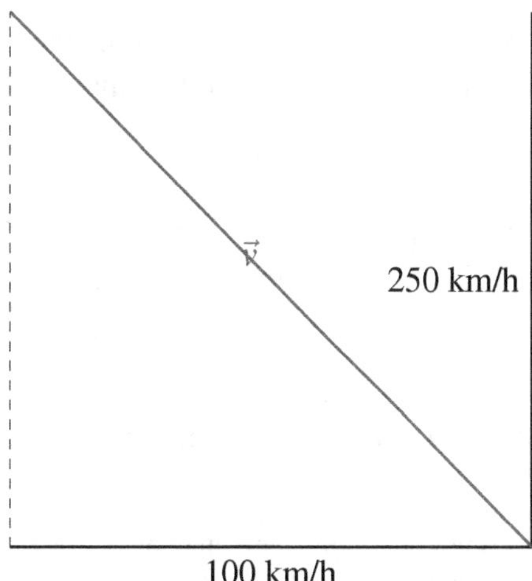

Figura 14.1.2: *Airplane velocity and effect of crosswind.*

The resultant velocity is calculated as:

$$v = \sqrt{(250 \text{ km/h})^2 + (100 \text{ km/h})^2} = \sqrt{62,500 + 10,000} = \sqrt{72,500} \approx 269{,}26 \text{ km/h}.$$

For the direction, calculate the angle θ relative to north:

$$\tan\theta = \frac{\text{west component}}{\text{north component}} = \frac{100}{250} = 0{,}4 \implies \theta = \arctan(0{,}4) \approx 21{,}80^\circ.$$

The airplane moves at approximately 269.26 km/h in a direction of 21,80° west of north.
This case illustrates the application of the Pythagorean Theorem in correcting course due to crosswinds, essential in aerial navigation.

Lema 14.1.1 In a three-dimensional coordinate system, the magnitude of a vector $\vec{V} = (V_x, V_y, V_z)$ is:

$$|\vec{V}| = \sqrt{V_x^2 + V_y^2 + V_z^2}.$$

Demostración. The magnitude of the vector is obtained by extending the Pythagorean Theorem to three-dimensional space. First, calculate the magnitude in the xy plane:

$$R_{xy} = \sqrt{V_x^2 + V_y^2}.$$

Then, apply the Pythagorean Theorem again considering R_{xy} and V_z:

$$|\vec{V}| = \sqrt{R_{xy}^2 + V_z^2} = \sqrt{V_x^2 + V_y^2 + V_z^2}.$$

■

This result is essential for calculating magnitudes of vectors in physics, such as forces or electric fields in three dimensions.

Exercise 14.1 An object moves 5 m east, then 12 m north, and finally 9 m vertically upward. What is the direct distance from the starting point to the final point?

Exercise 14.2 Two forces act on a point in perpendicular directions: one of 60 N southward and the other of 80 N eastward. Calculate the magnitude and direction of the resultant force.

To solve these exercises, apply the Pythagorean Theorem in two and three dimensions, respectively, calculating the resultant magnitudes and using trigonometric functions to determine directions.

(R) The Pythagorean Theorem is not only applicable in geometry and classical physics but is also fundamental in areas such as relativity theory, vector analysis, and Euclidean spaces of higher dimensions.

In summary, the Pythagorean Theorem is an indispensable tool for solving physical problems involving vectors and displacements in multiple dimensions, facilitating the analysis and understanding of complex systems.

14.1.2 Distance Calculation in Space.

Distance calculation in space is fundamental in analytic geometry and various applications in physics and engineering. It allows determining the separation between points in a three-dimensional space, which is essential to understand and solve problems in more complex contexts than the two-dimensional plane.

To begin, it is necessary to extend the concept of distance in the plane to three-dimensional space.

Definition 14.1.2 The **distance** between two points $P(x_1,y_1,z_1)$ and $Q(x_2,y_2,z_2)$ in three-dimensional space is the non-negative real number given by:

$$d(P,Q) = \sqrt{(x_2-x_1)^2+(y_2-y_1)^2+(z_2-z_1)^2}.$$

This formula is a natural extension of the Pythagorean Theorem to three-dimensional space and enables the calculation of the direct distance between any two points in space.

Theorem 14.1.3 The three-dimensional Euclidean space $\mathbb{R}^3$ is a metric space with the distance defined above, satisfying the properties of non-negativity, identity, symmetry, and the triangle inequality.

Demostración. To prove that $\mathbb{R}^3$ is a metric space with the distance d, we must verify the following properties for any points P, Q, and R in $\mathbb{R}^3$:

1. **Non-negativity:** $d(P,Q) \geq 0$.
   By definition, $d(P,Q)$ is the square root of a sum of squares, which is always non-negative.
2. **Identity of indiscernibles:** $d(P,Q) = 0$ if and only if $P = Q$.
   If $P = Q$, then $x_1 = x_2$, $y_1 = y_2$, and $z_1 = z_2$, so $d(P,Q) = 0$. Conversely, if $d(P,Q) = 0$, the sum of squares is zero, implying each difference is zero, hence $P = Q$.
3. **Symmetry:** $d(P,Q) = d(Q,P)$.
   The distance formula is symmetric with respect to P and Q, as $(x_2-x_1)^2 = (x_1-x_2)^2$, and similarly for the other coordinates.

4. **Triangle inequality:** $d(P,R) \leq d(P,Q) + d(Q,R)$.
   This property derives from the Cauchy-Schwarz inequality and applying the Pythagorean Theorem in three-dimensional space.

■

This theorem confirms that three-dimensional space with Euclidean distance is a metric space, allowing the use of tools and properties of such spaces in calculations and proofs.

Lema 14.1.2 The distance between two points in space is invariant under translations and rotations. That is, the distance does not change if the coordinate system is translated or rotated.

Demostración. Translations and rotations in space correspond to isometric transformations that preserve distances. Mathematically, if a transformation T is applied such that $T(P) = P'$ and $T(Q) = Q'$, then:

$$d(P',Q') = d(T(P),T(Q)) = d(P,Q).$$

This is because isometric transformations conserve the distances between points. ■

This lemma is crucial in physical applications, where positions can be described in different reference systems without altering the real distances between objects.

■ **Example 14.3** Calculate the distance between points $A(1,2,3)$ and $B(4,6,8)$.

■

Demostración. We apply the distance formula in space:

$$d(A,B) = \sqrt{(4-1)^2 + (6-2)^2 + (8-3)^2} = \sqrt{3^2 + 4^2 + 5^2} = \sqrt{9+16+25} = \sqrt{50}.$$

Simplifying:

$$\sqrt{50} = \sqrt{25 \times 2} = 5\sqrt{2}.$$

Therefore, the distance between A and B is $5\sqrt{2}$ units.

■

This example illustrates the direct application of the distance formula in space to calculate the separation between two given points.

Theorem 14.1.4 The minimum distance from a point $P(x_0, y_0, z_0)$ to a plane α given by the equation $Ax + By + Cz + D = 0$ is:

$$d = \left| \frac{Ax_0 + By_0 + Cz_0 + D}{\sqrt{A^2 + B^2 + C^2}} \right|.$$

Demostración. The distance from a point $P(x_0, y_0, z_0)$ to a plane $Ax + By + Cz + D = 0$ is the length of the orthogonal projection of the vector $\vec{v} = (x_0, y_0, z_0)$ onto the normal vector $\vec{n} = (A,B,C)$, which is perpendicular to the plane. The distance formula is:

$$d = \frac{|\vec{v} \cdot \vec{n} + D|}{|\vec{n}|} = \frac{|Ax_0 + By_0 + Cz_0 + D|}{\sqrt{A^2 + B^2 + C^2}}.$$

■

This formula is essential for solving problems where determining the relative position of a point with respect to a plane is required.

■ **Example 14.4** Find the distance from the point $P(2,-1,3)$ to the plane $\alpha : 2x - y + 2z - 5 = 0$. ■

Demostración. We apply the formula:

$$d = \left| \frac{2(2) - (-1) + 2(3) - 5}{\sqrt{(2)^2 + (-1)^2 + (2)^2}} \right| = \left| \frac{4+1+6-5}{\sqrt{4+1+4}} \right| = \left| \frac{6}{\sqrt{9}} \right| = \left| \frac{6}{3} \right| = 2.$$

Therefore, the distance from point P to plane α is 2 units. ■

This example shows how to calculate the distance from a point to a plane using the derived formula.

Corollary 14.1.5 The distance between two parallel planes $\alpha_1 : Ax + By + Cz + D_1 = 0$ and $\alpha_2 : Ax + By + Cz + D_2 = 0$ is:

$$d = \frac{|D_2 - D_1|}{\sqrt{A^2 + B^2 + C^2}}.$$

Demostración. Since the planes α_1 and α_2 are parallel, they share the same normal vector $\vec{n} = (A, B, C)$. The distance between them is the distance from any point on the plane α_1 to the plane α_2. Let $P(x_0, y_0, z_0)$ be a point on α_1, then $Ax_0 + By_0 + Cz_0 + D_1 = 0$. The distance from P to α_2 is:

$$d = \frac{|Ax_0 + By_0 + Cz_0 + D_2|}{\sqrt{A^2 + B^2 + C^2}} = \frac{|D_2 - D_1|}{\sqrt{A^2 + B^2 + C^2}}.$$

■

Demostración. Since the planes are parallel, they share the same normal vector $\vec{n} = (A, B, C)$. The distance between the planes can be calculated by taking any point on one plane and applying the point-to-plane distance formula with respect to the other plane. Simplifying yields the stated formula. ■

This formula is useful in geometry and physics to determine separations between planes in space.

Lema 14.1.3 The distance between two skew lines in space is equal to the length of the common perpendicular segment between the two lines.

Demostración. For two lines that do not intersect and are not parallel (skew lines), there exists a unique common perpendicular segment between them. The length of this segment represents the minimum distance between the lines. This segment can be found using the direction vectors of the lines and computing the cross product to obtain a perpendicular vector. ■

This result is relevant when studying the relative positions of lines in three-dimensional space.

■ **Example 14.5** Calculate the distance between the lines r_1 and r_2, given by:

$$r_1 : \frac{x-1}{2} = \frac{y+1}{-1} = \frac{z}{3}, \quad r_2 : \frac{x}{1} = \frac{y-2}{2} = \frac{z-1}{-2}.$$

■

Demostración. First, find points P_1 on r_1 and P_2 on r_2. Let $t = 0$ for both lines:
For r_1 when $t = 0$:

$$x = 1+2(0) = 1, \quad y = -1-1(0) = -1, \quad z = 0+3(0) = 0, \quad P_1(1,-1,0).$$

For r_2 when $s = 0$:

$$x = 0+1(0) = 0, \quad y = 2+2(0) = 2, \quad z = 1-2(0) = 1, \quad P_2(0,2,1).$$

The direction vectors are:

$$\vec{u} = (2,-1,3), \quad \vec{v} = (1,2,-2).$$

The vector $P_1\vec{P}_2 = P_2 - P_1 = (-1,3,1)$.
The distance between the lines is:

$$d = \frac{|P_1\vec{P}_2 \cdot (\vec{u} \times \vec{v})|}{|\vec{u} \times \vec{v}|}.$$

Calculate the cross product:

$$\vec{u} \times \vec{v} = \begin{vmatrix} \vec{i} & \vec{j} & \vec{k} \\ 2 & -1 & 3 \\ 1 & 2 & -2 \end{vmatrix}$$

$$= (-1\cdot-2-3\cdot2)\vec{i} - (2\cdot-2-3\cdot1)\vec{j} + (2\cdot2-(-1)\cdot1)\vec{k} = (-2-6)\vec{i} - (-4-3)\vec{j} + (4+1)\vec{k} = (-8,7,5).$$

The norm of the cross product is:

$$|\vec{u} \times \vec{v}| = \sqrt{(-8)^2+7^2+5^2} = \sqrt{64+49+25} = \sqrt{138}.$$

Calculate the numerator:

$$|P_1\vec{P}_2 \cdot (\vec{u} \times \vec{v})| = |(-1,3,1)\cdot(-8,7,5)| = |-1(-8)+3(7)+1(5)| = |8+21+5| = |34| = 34.$$

Finally, the distance is:

$$d = \frac{34}{\sqrt{138}} \approx 2{,}891.$$

■

This example demonstrates the procedure to calculate the distance between two skew lines in space, applying vector algebra concepts.

Exercise 14.3 Find the distance from the point $Q(3,-2,5)$ to the line r passing through the point $P(1,0,2)$ with direction vector $\vec{d}=(2,-1,2)$.

Exercise 14.4 Determine the distance between the parallel planes $\alpha_1 : x+2y-2z+5=0$ and $\alpha_2 : x+2y-2z-3=0$.

These exercises allow practice in distance calculations in different three-dimensional contexts, consolidating the concepts and techniques learned.

(R) Distance calculation in space is an essential tool in many areas of mathematics and physics, including analytic geometry, vector analysis, and mechanics. A solid understanding of these concepts facilitates the study of more complex structures and the resolution of advanced problems.

In conclusion, distance calculation in space provides the foundation to analyze and understand three-dimensional geometry, enabling the resolution of problems ranging from simple point separations to analyzing relative positions of more complex geometric figures such as lines and planes.

14.2 Altitudes, Medians, and Angle Bisectors: Definitions and Properties.

14.2.1 Application in Triangle Construction.

The construction of triangles using their notable elements, such as altitudes, medians, and angle bisectors, is fundamental in geometry and has multiple applications in design problems and mathematical reasoning. In this section, we explore how these elements can be used to construct triangles and analyze their properties through definitions, theorems, and examples.

Definition 14.2.1 In a triangle, an **altitude** is the perpendicular segment drawn from a vertex to the opposite side (or its extension). The point where the altitude intersects the opposite side is called the **foot of the altitude**.

Definition 14.2.2 A **median** of a triangle is the segment that connects a vertex with the midpoint of the opposite side.

Definition 14.2.3 An **angle bisector** in a triangle is the ray that divides an angle into two congruent angles, starting from the vertex and reaching the opposite side.

These notable elements play a crucial role in the construction and analysis of triangles, as they establish essential metric relationships and geometric properties.

Theorem 14.2.1 In any triangle, the three medians intersect at a point called the **centroid** or **center of gravity**, which divides each median in a 2 : 1 ratio, measured from the vertex.

Demostración. Let $\triangle ABC$ have medians AM, BN, and CP intersecting at point G. By the properties of medians, G divides each median in the ratio $AG : GM = 2 : 1$. The proof involves using position vectors and verifying that the medians concur at G and satisfy the segment ratio. ■

The centroid is a significant point in triangle construction as it acts as the center of mass if the triangle is considered a uniform lamina.

■ **Example 14.6** Construct a triangle given the midpoints of its sides.

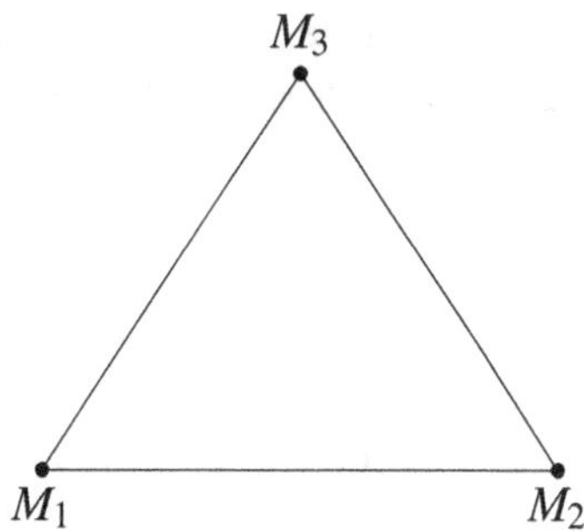

Figura 14.2.1: *Given midpoints M_1, M_2, M_3.*

Demostración. To construct the original triangle, we use the theorem stating that the midpoints of the sides of a triangle form a triangle called the **medial triangle**, which is similar to the original triangle and homothetic with a ratio of $1/2$.
The steps are:
1. Given the triangle formed by the midpoints M_1, M_2, M_3, determine the center of homothety, which is the centroid of the original triangle. 2. From each M_i, draw a line passing through the centroid G and extend it twice the distance GM_i to find the vertices A, B, C of the original triangle.
Alternatively, using parallel lines:
1. Draw lines parallel to the sides of the triangle $M_1M_2M_3$ from the corresponding midpoints. 2. The intersections of these lines will be the vertices A, B, C of the original triangle.

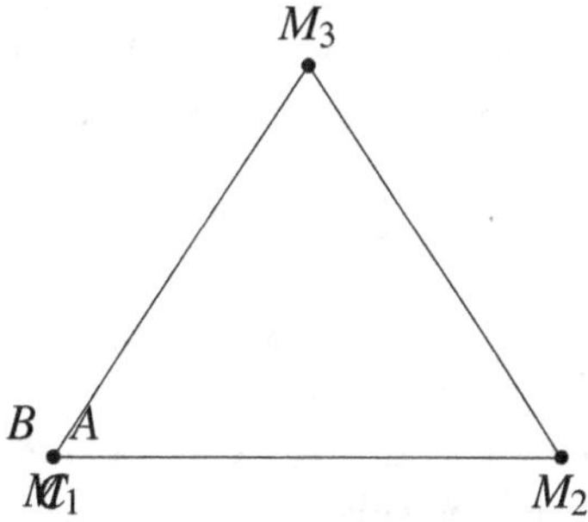

Figura 14.2.2: *Construction of the original triangle from the midpoints M_1, M_2, M_3.*

■

This example demonstrates how a triangle can be reconstructed from certain elements by applying properties of medians and midpoints.

Theorem 14.2.2 In a triangle, the three altitudes intersect at a point called the **orthocenter**.

Demostración. The proof is based on showing that the three altitudes are concurrent. This can be done using analytic geometry by assigning coordinates to the vertices and demonstrating that the equations of the altitudes intersect at a common point.
Alternatively, the circumcircle can be used to prove that the orthocenter is the isogonal conjugate of the centroid. ■

The orthocenter is another notable point that is useful in triangle constructions, especially in problems where the altitudes or their properties are known.

■ **Example 14.7** Given a triangle with side lengths $a = 13$ cm, $b = 14$ cm, and the altitude corresponding to side a equal to 12 cm, construct the triangle and determine the length of the altitude corresponding to side b.

■

Demostración. First, construct the triangle:
1. Draw the side $a = 13$ cm. 2. From one end of a, draw a perpendicular line and mark the altitude $h_a = 12$ cm. 3. Using the obtained point and the other end of a, draw an arc with radius $b = 14$ cm that intersects the perpendicular line at point C, forming the triangle $\triangle ABC$.
Now, to find the altitude h_b corresponding to side b, use the relationship between areas:
The area of the triangle is:

$$\text{Area} = \frac{ah_a}{2} = \frac{13 \times 12}{2} = 78 \text{ cm}^2.$$

Also, the area is:

$$\text{Area} = \frac{bh_b}{2} \implies h_b = \frac{2 \times \text{Area}}{b} = \frac{2 \times 78}{14} = \frac{156}{14} = 11{,}14 \text{ cm}.$$

■

This example illustrates how to use altitudes and area properties to construct a triangle and determine unknown measurements.

Theorem 14.2.3 In any triangle, the internal angle bisectors intersect at a point called the **incenter**, which is the center of the inscribed circle of the triangle.

Demostración. The internal angle bisectors of a triangle are concurrent. This can be demonstrated using the concurrency criterion for bisectors and the properties of adjacent angles. The incenter is equidistant from the three sides of the triangle, enabling the inscribed circle to be drawn. ■

The incenter is fundamental in constructions involving the inscribed circle and in problems that require finding points equidistant from the sides.

Lema 14.2.1 The length of the internal angle bisector relative to angle A in a triangle $\triangle ABC$ is given by:

$$l_a = \frac{2bc\cos\frac{A}{2}}{b+c}.$$

Demostración. The formula is derived using the properties of angle bisectors and trigonometric relationships in the triangle. By applying the law of cosines and the definition of the angle bisector, the expression for l_a is obtained. ■

This result is useful for calculating the length of an angle bisector when the sides and angles of the triangle are known.

Exercise 14.5 Construct a triangle given two sides $b = 8$ cm, $c = 6$ cm, and the internal angle bisector of the included angle with length $l_a = 5$ cm.

■

Exercise 14.6 In a triangle, the lengths of the medians are $m_a = 5$ cm, $m_b = 6$ cm, and $m_c = 7$ cm. Is it possible to construct such a triangle? Justify your answer.

To solve these exercises, apply the properties of medians and angle bisectors, as well as the criteria for triangle construction, considering possible restrictions and existence conditions.

(R) Constructing triangles from their notable elements strengthens the understanding of geometric and metric relationships in these figures. It also allows tackling more complex problems in geometry and mathematical analysis by applying advanced concepts and logical reasoning.

In conclusion, altitudes, medians, and angle bisectors are fundamental tools in the construction and analysis of triangles. Understanding their properties and relationships allows solving a wide variety of geometric problems and developing advanced mathematical reasoning skills.

14.2.2 Use of Medians in Geometric Design.

The medians of a triangle are fundamental tools in geometric design, as they allow dividing figures into proportional parts and locating key points such as the centroid. Their study is essential for understanding the intrinsic properties of triangles and their application to more complex problems.

Definition 14.2.4 In a triangle, a **median** is the segment that joins a vertex with the midpoint of the opposite side.

Medians have notable properties that are utilized in geometric design to create balanced and symmetrical constructions.

Theorem 14.2.4 The three medians of a triangle intersect at a unique point called the **centroid**, which divides each median in a ratio of $2:1$, counting from the vertex.

Demostración. Let $\triangle ABC$ have medians AM_a, BM_b, and CM_c, where M_a, M_b, and M_c are the midpoints of the sides opposite A, B, and C, respectively. The medians intersect at the point G (centroid).
Using barycentric coordinates, we can establish that the centroid G has coordinates that are the average of the vertices' coordinates:

$$G = \left(\frac{x_A + x_B + x_C}{3}, \frac{y_A + y_B + y_C}{3}\right).$$

This demonstrates that the medians intersect at a unique point and that this point divides each median in a $2:1$ ratio. ■

This property is fundamental in geometric design, as the centroid acts as the center of balance of the triangle and can be used to create figures with specific symmetries and proportions.

■ **Example 14.8** Design an equilateral triangle and locate its centroid using the medians.

An equilateral triangle $\triangle ABC$ is constructed. The midpoints of the sides are M_a, M_b, and M_c. By drawing the medians from each vertex to the midpoint of the opposite side, they intersect at the centroid G. In an equilateral triangle, due to its symmetry, the centroid coincides with the center of the triangle and is equidistant from the vertices and the sides.

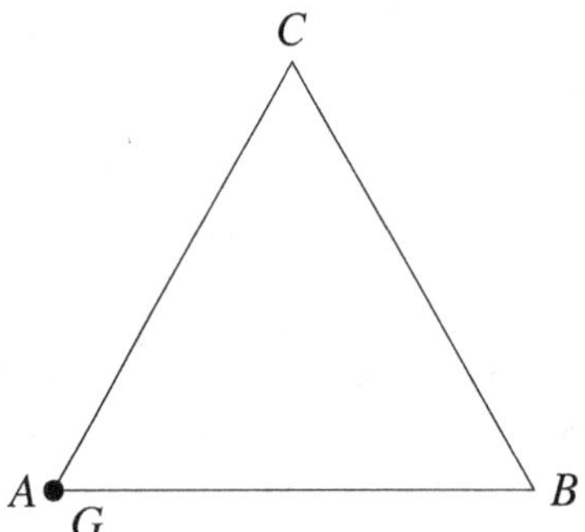

Figura 14.2.3: *Equilateral triangle with its medians and centroid G.*

The use of medians in geometric design allows dividing figures into equivalent areas, which is especially useful in industrial and architectural design.

Lema 14.2.2 In a triangle, the centroid divides each median in a $2:1$ ratio, with the longer segment being between the vertex and the centroid.

Demostración. Let $\triangle ABC$ have a median AM_a and centroid G. Then:

$$\frac{AG}{GM_a} = 2.$$

This is demonstrated using position vectors. Taking the origin at A, we have:

$$\vec{AG} = \frac{1}{3}(\vec{AB} + \vec{AC}).$$

The midpoint M_a has a position vector:

$$\vec{AM_a} = \frac{1}{2}(\vec{AB} + \vec{AC}).$$

Thus:

$$\vec{GM_a} = \vec{AM_a} - \vec{AG} = \left(\frac{1}{2} - \frac{1}{3}\right)(\vec{AB} + \vec{AC}) = \frac{1}{6}(\vec{AB} + \vec{AC}) = \frac{1}{2}\vec{AG}.$$

Therefore, $AG = 2GM_a$, proving the $2:1$ ratio. ■

Corollary 14.2.5 The centroid of a triangle divides the triangle into six smaller triangles of equal area.

Demostración. The three medians divide the original triangle into six smaller triangles. Each of these triangles shares the same vertex G and has the same base (a portion of the side of the original triangle) and the same height relative to that base. Thus, all have equal area. ■

This property is utilized in geometric design to create patterns and subdivisions with equivalent areas, facilitating the uniform distribution of materials or spaces.

■ **Example 14.9** Design a geometric pattern in the shape of a triangle divided into six regions of equal area using the medians.

■

By drawing the medians in the triangle and marking the centroid G, the triangle is divided into six smaller triangles of equal area. These triangles can be used to create a geometric pattern or to distribute design elements uniformly.

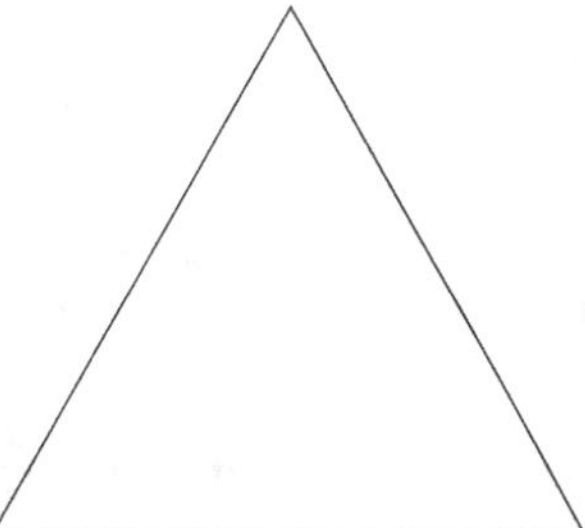

Figura 14.2.4: *Triangle divided into six equal areas using medians and centroid.*

Exercise 14.7 In a triangle $\triangle ABC$, the medians measure $m_a = 9$ cm, $m_b = 12$ cm, and $m_c = 15$ cm. Calculate the area of the triangle using the lengths of the medians. ■

Use the formula for the area of a triangle in terms of its medians:

$$\text{Area} = \frac{4}{3}\sqrt{s_m(s_m - m_a)(s_m - m_b)(s_m - m_c)},$$

where $s_m = \dfrac{m_a + m_b + m_c}{2}$ is the semi-perimeter of the medians.

Exercise 14.8 Design a triangle where one of the medians is perpendicular to a side of the triangle. Prove that the resulting triangle is isosceles. ■

[of Exercise 1] First, calculate the semi-perimeter of the medians:

$$s_m = \frac{9 + 12 + 15}{2} = \frac{36}{2} = 18 \text{ cm}.$$

Then, apply the formula for the area in terms of the medians:

$$\text{Area} = \frac{4}{3}\sqrt{18(18-9)(18-12)(18-15)} = \frac{4}{3}\sqrt{18 \times 9 \times 6 \times 3}.$$

Calculate the radicand:

$$18 \times 9 \times 6 \times 3 = 18 \times 9 \times 18 = 18^2 \times 9 = 324 \times 9 = 2916.$$

Thus:

$$\text{Area} = \frac{4}{3}\sqrt{2916} = \frac{4}{3} \times 54 = \frac{4 \times 54}{3} = 72 \text{ cm}^2.$$

Therefore, the area of the triangle is 72 cm^2.

(R) The formula for the area in terms of the medians is a powerful tool in geometry, especially when the side lengths are not directly known.

The use of medians in geometric design not only facilitates the creation of balanced figures but also enables solving complex problems where internal properties of the triangle are required without relying solely on its sides or angles.
In more advanced applications, medians can be used in the study of geometric transformations, optimization of areas and volumes, and solving problems in vector spaces.

14.3 Area Calculation: Using Formulas for Triangles and Quadrilaterals

14.3.1 Heron's Formula for Triangles

The **Heron's Formula** is a powerful tool in geometry that allows the calculation of a triangle's area when the lengths of its three sides are known, without requiring the height. This formula is particularly useful in scenarios where determining the height is challenging or impossible.

Definition 14.3.1 Let $\triangle ABC$ be a triangle with sides of lengths a, b, and c. The **semiperimeter** s of the triangle is defined as:

$$s = \frac{a+b+c}{2}.$$

This semiperimeter is half the perimeter of the triangle and plays a crucial role in Heron's Formula.

Theorem 14.3.1 — **Heron's Formula.** The area A of a triangle with sides of lengths a, b, and c is:

$$A = \sqrt{s(s-a)(s-b)(s-c)},$$

where s is the semiperimeter of the triangle.

Demostración. Let $s = \frac{a+b+c}{2}$ be the semiperimeter of the triangle. The area A of a triangle can be expressed in terms of its sides and semiperimeter using the identity:

$$A = \sqrt{s(s-a)(s-b)(s-c)}.$$

This formula is derived from the trigonometric identity for the area of a triangle and through algebraic manipulations using the semiperimeter. ■

This formula enables the calculation of the area of any triangle when its three sides are known, without needing to compute angles or heights.

■ **Example 14.10** Calculate the area of a triangle with sides measuring $a = 13$ cm, $b = 14$ cm, and $c = 15$ cm.

■

First, compute the semiperimeter:

$$s = \frac{a+b+c}{2} = \frac{13+14+15}{2} = \frac{42}{2} = 21 \text{ cm}.$$

Next, apply Heron's Formula:

$$\begin{aligned} A &= \sqrt{s(s-a)(s-b)(s-c)} \\ &= \sqrt{21(21-13)(21-14)(21-15)} \\ &= \sqrt{21 \times 8 \times 7 \times 6} \\ &= \sqrt{21 \times 8 \times 7 \times 6}. \end{aligned}$$

Compute the product inside the square root:

$$21 \times 8 \times 7 \times 6 = 7056.$$

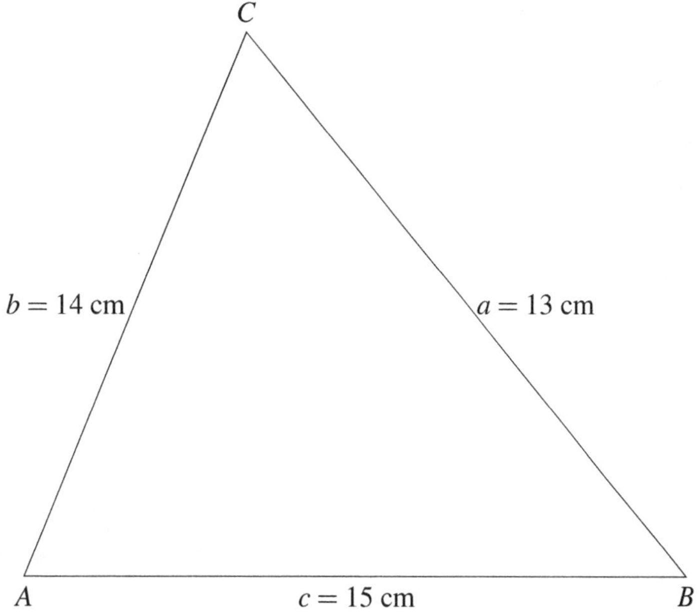

Figura 14.3.1: *Triangle with sides $a = 13$ cm, $b = 14$ cm, and $c = 15$ cm.*

Thus:

$$A = \sqrt{7056} = 84 \text{ cm}^2.$$

The area of the triangle is 84 cm^2.

This example demonstrates the direct application of Heron's Formula to calculate a triangle's area when its three sides are known.

Heron's Formula is particularly useful when the triangle's height is unavailable or difficult to calculate directly.

Corollary 14.3.2 For an equilateral triangle with side l, the area is:

$$A = \frac{\sqrt{3}}{4} l^2.$$

Demostración. In an equilateral triangle, $a = b = c = l$. The semiperimeter is:

$$s = \frac{3l}{2}.$$

Applying Heron's Formula:

$$
\begin{aligned}
A &= \sqrt{s(s-a)^3} \\
&= \sqrt{\frac{3l}{2}\left(\frac{3l}{2}-l\right)^3} \\
&= \sqrt{\frac{3l}{2}\left(\frac{l}{2}\right)^3} \\
&= \sqrt{\frac{3l}{2}\times\frac{l^3}{8}} \\
&= \sqrt{\frac{3l^4}{16}} \\
&= \frac{\sqrt{3}l^2}{4}.
\end{aligned}
$$

■

This corollary provides a straightforward formula for calculating the area of an equilateral triangle.

■ **Example 14.11** Calculate the area of an equilateral triangle with side $l = 10$ cm using Heron's Formula.

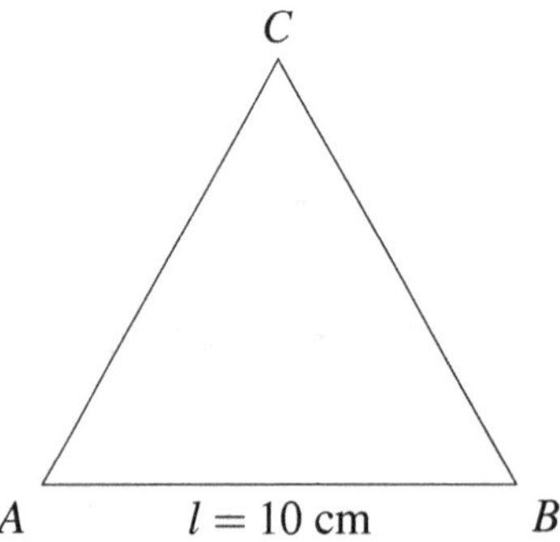

Figura 14.3.2: *Equilateral triangle with side $l = 10$ cm.*

■

Using the previous corollary:

$$A = \frac{\sqrt{3}}{4}l^2 = \frac{\sqrt{3}}{4}(10\text{ cm})^2 = \frac{\sqrt{3}}{4}\times 100\text{ cm}^2 = 25\sqrt{3}\text{ cm}^2 \approx 43{,}30\text{ cm}^2.$$

This result aligns with the one obtained by directly applying Heron's Formula to an equilateral triangle.

Exercise 14.9 A triangle has sides of lengths $a = 7$ cm, $b = 8$ cm, and $c = 9$ cm. Calculate the area of the triangle using Heron's Formula. ■

Exercise 14.10 Prove that for any right triangle with legs of lengths a and b, and hypotenuse c, the area calculated using Heron's Formula matches the area calculated by $A = \frac{ab}{2}$. ■

These exercises allow for applying and deepening the understanding of Heron's Formula and its relationship with other expressions for a triangle's area.

R Heron's Formula is also useful in other fields, such as surveying, to calculate the areas of irregular terrains by dividing them into triangles.

In conclusion, Heron's Formula is an essential tool in geometry that facilitates the calculation of a triangle's area from its sides, without needing the height or angles, with practical applications in diverse fields of mathematics and engineering.

14.3.2 Irregular Area Calculation

The calculation of irregular areas is a fundamental part of geometry and mathematical analysis, especially when dealing with shapes that do not have a direct formula for area calculation. In this section, we will explore various techniques and methods to determine the area of irregular shapes using advanced concepts and mathematical tools.

Definition 14.3.2 An **irregular shape** is a geometric figure that does not conform to standard geometric forms (such as regular polygons or circles) and therefore lacks a straightforward formula for calculating its area. These shapes may have complex forms or curved edges, requiring more advanced methods to determine their area.

One of the most common techniques for calculating the area of irregular shapes is decomposition into simpler figures.

Theorem 14.3.3 The area of an irregular shape can be calculated by dividing the shape into simpler parts, whose areas can be computed directly, and then summing or subtracting these areas as needed.

Demostración. By dividing the irregular shape into simpler figures (such as triangles, rectangles, or circles), we can use known formulas to calculate the area of each part. The sum of the areas of all parts gives the total area of the original shape. This method is based on the additivity principle of area, which states that the total area equals the sum of the non-overlapping parts' areas. ■

R Decomposition is particularly useful in planar geometry and practical problems such as surveying and architectural design, where shapes may have complex forms that do not fit standard formulas.

■ **Example 14.12** Calculate the area of an irregular shape formed by a rectangle with a base of 8 m and a height of 5 m, with a semicircle of radius 2,5 m attached to one of its shorter sides. ■

To calculate the total area of the shape, we divide it into two parts: the rectangle and the attached semicircle.

Area of the rectangle:

$$A_{\text{rectangle}} = \text{base} \times \text{height} = 8\,\text{m} \times 5\,\text{m} = 40\,\text{m}^2.$$

Area of the semicircle:

$$A_{\text{semicircle}} = \frac{1}{2}\pi r^2 = \frac{1}{2}\pi(2{,}5\,\text{m})^2 = \frac{1}{2}\pi \times 6{,}25\,\text{m}^2 = 3{,}125\pi\,\text{m}^2.$$

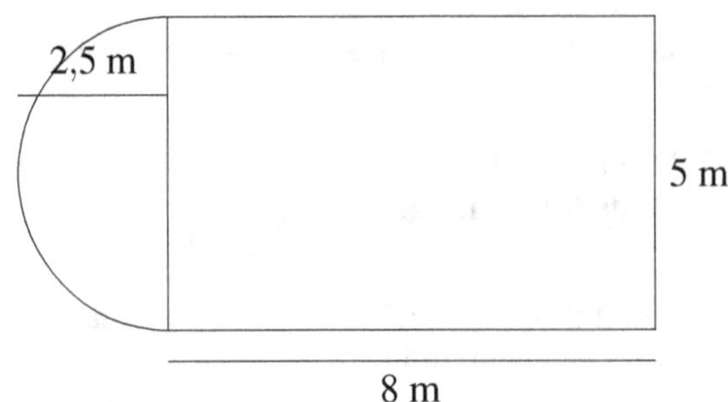

Figura 14.3.3: *Shape composed of a rectangle and a semicircle attached to one of its shorter sides.*

Total area:

$$A_{\text{total}} = A_{\text{rectangle}} + A_{\text{semicircle}} = 40\,\text{m}^2 + 3{,}125\pi\,\text{m}^2 \approx 40\,\text{m}^2 + 9{,}82\,\text{m}^2 = 49{,}82\,\text{m}^2.$$

Thus, the total area of the shape is approximately 49,82 m^2.
Another fundamental technique for calculating irregular areas is the use of definite integrals, especially for areas under curves or between functions.

Theorem 14.3.4 The area under a continuous and positive curve $y = f(x)$ on the interval $[a,b]$ can be calculated using the definite integral:

$$A = \int_a^b f(x)\,dx.$$

Demostración. We divide the interval $[a,b]$ into n subintervals of width $\Delta x = \frac{b-a}{n}$. In each subinterval, we take a point x_i^* and form a rectangle with height $f(x_i^*)$ and base Δx, whose area is approximately $f(x_i^*)\Delta x$. The sum of the areas of these rectangles approximates the area under the curve:

$$\sum_{i=1}^{n} f(x_i^*)\Delta x.$$

Taking the limit as $n \to \infty$, this sum becomes the definite integral:

$$A = \int_a^b f(x)\,dx.$$

■

■ **Example 14.13** Calculate the area enclosed by the parabola $y = x^2$ and the lines $x = 0$ and $x = 2$. ■

Demostración. The area A under the curve $y = x^2$ from $x = 0$ to $x = 2$ is:

$$A = \int_0^2 x^2\,dx = \left.\frac{x^3}{3}\right|_0^2 = \frac{(2)^3}{3} - \frac{(0)^3}{3} = \frac{8}{3} - 0 = \frac{8}{3}.$$

Thus, the area is $\frac{8}{3}$ square units. ■

Moreover, we can calculate the area between two curves.

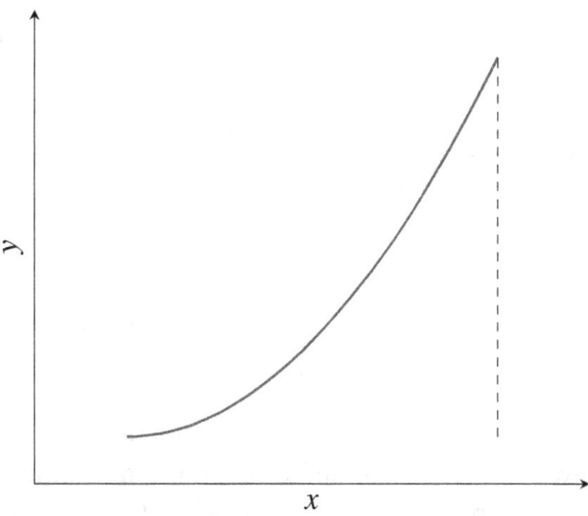

Figura 14.3.4: *Area under the curve* $y = x^2$ *between* $x = 0$ *and* $x = 2$.

Corollary 14.3.5 The area A between two continuous curves $y = f(x)$ and $y = g(x)$ on the interval $[a,b]$ is given by:

$$A = \int_a^b |f(x) - g(x)|\,dx.$$

Demostración. The area between the two curves is computed as the integral of the difference of the functions, representing the height between them, over the interval $[a,b]$. If $f(x) \geq g(x)$ throughout the interval, the integral simplifies to:

$$A = \int_a^b [f(x) - g(x)]\,dx.$$

■

■ **Example 14.14** Calculate the area enclosed between the curves $y = x$ and $y = x^2$ from $x = 0$ to $x = 1$.

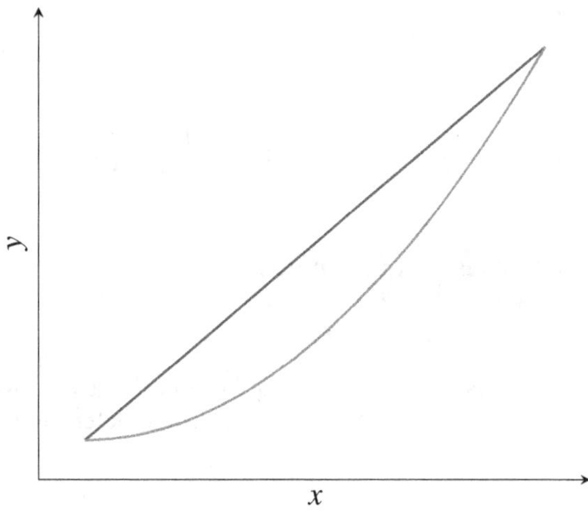

Figura 14.3.5: *Area between the curves* $y = x$ *and* $y = x^2$ *on* $[0,1]$.

■

Demostración. We observe that on the interval $[0,1]$, $y=x$ is greater than $y=x^2$.
The area A is:

$$A=\int_0^1 (x-x^2)\,dx=\left(\frac{x^2}{2}-\frac{x^3}{3}\right)\bigg|_0^1=\left(\frac{1}{2}-\frac{1}{3}\right)-(0-0)=\frac{1}{2}-\frac{1}{3}=\frac{3}{6}-\frac{2}{6}=\frac{1}{6}.$$

Thus, the area between the curves is $\frac{1}{6}$ square units. ■

Lema 14.3.1 If a planar figure is bounded by continuous functions on a closed interval, the area of the figure can be calculated using the definite integral of the functions describing its boundaries.

Demostración. The definite integral sums infinitesimal elements of area along an interval. If the figure's boundaries can be expressed as continuous functions, the definite integral of the difference of these functions over the corresponding interval will yield the total area of the figure. ■

Exercise 14.11 Calculate the area enclosed between the curves $y=\sqrt{x}$ and $y=x$ from $x=0$ to $x=1$. ■

First, determine the interval where $y=\sqrt{x}$ is greater than $y=x$, and then integrate the difference of the functions over that interval.

Exercise 14.12 A planar region is bounded by the function $y=\sin x$ between $x=0$ and $x=\pi$. Calculate the area under the curve. ■

of Exercise 1. First, find the points of intersection between the curves $y=\sqrt{x}$ and $y=x$.
Equate the functions:

$$\sqrt{x}=x \implies x=x^2 \implies x^2-x=0 \implies x(x-1)=0.$$

Thus, the curves intersect at $x=0$ and $x=1$.
On the interval $[0,1]$, for $x\in(0,1)$, $\sqrt{x}>x$.
The area A is:

$$A=\int_0^1 (\sqrt{x}-x)\,dx.$$

Compute the integral:

$$A=\int_0^1 (x^{1/2}-x)\,dx=\left(\frac{x^{3/2}}{(3/2)}-\frac{x^2}{2}\right)\bigg|_0^1=\left(\frac{2}{3}-\frac{1}{2}\right)-(0-0)=\frac{2}{3}-\frac{1}{2}=\frac{4}{6}-\frac{3}{6}=\frac{1}{6}.$$

Thus, the area between the curves is $\frac{1}{6}$ square units. ■

(R) The use of definite integrals to calculate irregular areas is a fundamental application of calculus and is widely used in engineering and physics to determine areas and volumes of complex figures.

In summary, the calculation of irregular areas requires more advanced methods, such as decomposition into simple figures and the use of definite integrals. These techniques enable tackling a wide variety of problems and are essential in mathematical analysis and practical applications across various disciplines.

14.4 Solved Exercises

Exercise 14.13 Given a right triangle with legs of length $a = 6$ cm and $b = 8$ cm, calculate the length of the hypotenuse. ■

Demostración. We apply the **Pythagorean Theorem**, which states that in a right triangle, the square of the hypotenuse c is equal to the sum of the squares of the legs:

$$c^2 = a^2 + b^2.$$

Substituting the given values:

$$c^2 = 6^2 + 8^2 = 36 + 64 = 100.$$

Taking the square root of both sides:

$$c = \sqrt{100} = 10 \text{ cm}.$$

Therefore, the length of the hypotenuse is $c = 10$ cm. ■

Exercise 14.14 Find the distance between the points $A(1,2,3)$ and $B(4,6,8)$ in three-dimensional space. ■

Demostración. We use the formula for the distance between two points in three-dimensional space:

$$d = \sqrt{(x_2 - x_1)^2 + (y_2 - y_1)^2 + (z_2 - z_1)^2}.$$

Substituting the coordinates of A and B:

$$d = \sqrt{(4-1)^2 + (6-2)^2 + (8-3)^2} = \sqrt{3^2 + 4^2 + 5^2} = \sqrt{9 + 16 + 25} = \sqrt{50}.$$

Simplifying:

$$d = 5\sqrt{2}.$$

Therefore, the distance between the points A and B is $5\sqrt{2}$. ■

Exercise 14.15 In a triangle, the sides are $a = 7$ cm, $b = 24$ cm, and $c = 25$ cm. Verify if the triangle is a right triangle. ■

Demostración. To verify if the triangle is a right triangle, we check if the Pythagorean Theorem holds with c as the hypotenuse:

$$c^2 = a^2 + b^2.$$

Calculating each term:

$$c^2 = 25^2 = 625,$$

$$a^2 + b^2 = 7^2 + 24^2 = 49 + 576 = 625.$$

Since $c^2 = a^2 + b^2$, the triangle is a right triangle. ■

Exercise 14.16 Determine the area of a triangle with sides $a = 13$ cm, $b = 14$ cm, and $c = 15$ cm using Heron's Formula. ■

Demostración. First, calculate the semiperimeter:

$$s = \frac{a+b+c}{2} = \frac{13+14+15}{2} = \frac{42}{2} = 21 \text{ cm}.$$

Apply Heron's Formula for the area:

$$A = \sqrt{s(s-a)(s-b)(s-c)}.$$

Substituting the values:

$$A = \sqrt{21(21-13)(21-14)(21-15)} = \sqrt{21 \times 8 \times 7 \times 6} = \sqrt{7056}.$$

Thus:

$$A = 84 \text{ cm}^2.$$

The area of the triangle is 84 cm². ■

Exercise 14.17 Calculate the distance from the point $P(2,-1,3)$ to the plane $\alpha : 2x - y + 2z - 5 = 0$. ■

Demostración. The formula for the distance d from a point $P(x_0, y_0, z_0)$ to a plane $Ax + By + Cz + D = 0$ is:

$$d = \left| \frac{Ax_0 + By_0 + Cz_0 + D}{\sqrt{A^2 + B^2 + C^2}} \right|.$$

Substituting the values of P and the plane's coefficients:

$$d = \left| \frac{2(2) - (-1) + 2(3) - 5}{\sqrt{2^2 + (-1)^2 + 2^2}} \right| = \left| \frac{4+1+6-5}{\sqrt{4+1+4}} \right| = \left| \frac{6}{\sqrt{9}} \right| = \left| \frac{6}{3} \right| = 2.$$

Therefore, the distance from the point P to the plane α is 2 units. ■

14.5 Proposed Exercises

14.5.1 Pythagorean Theorem: Applications in Right Triangles

Exercise 14.18 Given a right triangle with legs of length $a = 5$ cm and $b = 12$ cm, calculate the length of the hypotenuse. ■

Exercise 14.19 In a right triangle, the hypotenuse measures 13 cm, and one of the legs measures 5 cm. Calculate the length of the other leg. ■

Exercise 14.20 A climber ascends a hill that forms a right angle with the ground. If they move 6 meters horizontally and 8 meters vertically, what is the straight-line distance they traveled? ■

Exercise 14.21 A 15-meter pole casts a 9-meter shadow on the ground, forming a right triangle with the pole and the ground. Calculate the distance from the top of the pole to the end of the shadow. ■

Exercise 14.22 In a right triangle, the hypotenuse is 10 cm, and one leg is 6 cm. Find the length of the other leg and verify if it satisfies the Pythagorean Theorem. ■

14.5.2 Altitudes, Medians, and Angle Bisectors: Definitions and Properties

Exercise 14.23 Draw an equilateral triangle with side $l = 10$ cm and draw its three altitudes. What is the length of each altitude? ■

Exercise 14.24 In an isosceles triangle, the base measures 12 cm, and each of the equal sides measures 10 cm. Calculate the length of the median corresponding to the base. ■

Exercise 14.25 Draw a triangle with sides $a = 8$ cm, $b = 10$ cm, and $c = 6$ cm. Draw the three angle bisectors and determine the point where they intersect. ■

Exercise 14.26 A triangle has sides measuring 7 cm, 8 cm, and 9 cm. Calculate the length of the median corresponding to the side measuring 9 cm. ■

Exercise 14.27 In a right triangle with legs measuring 9 cm and 12 cm, calculate the length of the altitude relative to the hypotenuse. ■

14.5.3 Area Calculation: Using Formulas for Triangles and Quadrilaterals

Exercise 14.28 Calculate the area of a triangle with sides $a = 7$ cm, $b = 8$ cm, and $c = 9$ cm using Heron's Formula. ■

Exercise 14.29 Draw a square with a side length of 5 cm and an equilateral triangle with a side length of 5 cm. Compare their areas. ■

Exercise 14.30 A rectangle has an area of 48 cm^2 and a base of 8 cm. Calculate its height. ■

Exercise 14.31 Calculate the area of an isosceles trapezoid whose larger base measures 10 cm, smaller base 6 cm, and height 4 cm. ■

Exercise 14.32 Find the area of an isosceles triangle with a base of 10 cm and a height of 12 cm. ■

15. Planar Regions and Spatial Location

15.1 Perimeters and Areas of Plane Figures: Circles, Triangles, and Polygons

15.1.1 Area Calculation for Regular Polygons

The study of regular polygons is fundamental in geometry, as these figures possess symmetrical properties that facilitate the calculation of their areas and perimeters. In this section, we explore advanced methods to determine the area of regular polygons, leveraging concepts such as the apothem, radius, and trigonometric relationships.

Definition 15.1.1 A **regular polygon** is a convex polygon in which all sides have the same length and all interior angles are equal. Common examples include the equilateral triangle, square, and regular pentagon.

The symmetry of regular polygons allows them to be divided into congruent triangles, which is key for area calculation.

Theorem 15.1.1 The area A of a regular polygon with n sides, each of length l, is:

$$A = \frac{nla}{2},$$

where a is the **apothem** of the polygon, i.e., the distance from the center of the polygon to any of its sides.

Demostración. Divide the regular polygon into n congruent triangles, each with base l and height a. The area of one of these triangles is $\frac{la}{2}$. Since there are n triangles, the total area of the polygon is:

$$A = n \cdot \frac{la}{2} = \frac{nla}{2}.$$

■

To calculate the apothem, we use trigonometric relationships based on the central angles of the polygon.

Lema 15.1.1 In a regular polygon with n sides, the apothem a is related to the radius R of the circumscribed circle and the central angle θ by:

$$a = R\cos\left(\frac{\theta}{2}\right),$$

where $\theta = \dfrac{2\pi}{n}$.

Demostración. The central angle θ is the angle subtended by each side of the polygon at the center. Consider one of the isosceles triangles formed by connecting the center to two adjacent vertices. The apothem is the height of this triangle and can be expressed in terms of the radius and angle:

$$a = R\cos\left(\frac{\theta}{2}\right).$$

■

Using these relationships, we can derive formulas for area calculation based on known elements.

■ **Example 15.1** Calculate the area of a regular pentagon with side $l = 6$ cm. ■

Demostración. First, determine the number of sides: $n = 5$.
The central angle is:

$$\theta = \frac{2\pi}{n} = \frac{2\pi}{5}.$$

The radius of the circumscribed circle R is calculated using the relationship between the side and the radius:

$$l = 2R\sin\left(\frac{\theta}{2}\right) \implies R = \frac{l}{2\sin\left(\frac{\theta}{2}\right)}.$$

Calculate:

$$\frac{\theta}{2} = \frac{\pi}{5} \approx 36^\circ,$$

$$\sin\left(\frac{\theta}{2}\right) = \sin(36^\circ) \approx 0{,}5878,$$

$$R = \frac{6 \text{ cm}}{2 \times 0{,}5878} \approx \frac{6}{1{,}1756} \approx 5{,}105 \text{ cm}.$$

Now, calculate the apothem:

$$a = R\cos\left(\frac{\theta}{2}\right) = 5{,}105 \times \cos(36^\circ) \approx 5{,}105 \times 0{,}8090 \approx 4{,}131 \text{ cm}.$$

Finally, the area is:

$$A = \frac{nla}{2} = \frac{5 \times 6 \times 4{,}131}{2} = \frac{123{,}93}{2} \approx 61{,}97 \text{ cm}^2.$$

■

This example illustrates how to apply trigonometric relationships and the properties of regular polygons to calculate areas.

Theorem 15.1.2 The area A of a regular polygon with n sides inscribed in a circle of radius R is:

$$A = \frac{nR^2 \sin\theta}{2},$$

where $\theta = \frac{2\pi}{n}$.

Demostración. Each isosceles triangle formed has an area:

$$A_{\text{triangle}} = \frac{R^2 \sin\theta}{2}.$$

Multiplying by the number of sides:

$$A = n \times A_{\text{triangle}} = n \times \frac{R^2 \sin\theta}{2} = \frac{nR^2 \sin\theta}{2}.$$

■

This theorem is useful when the radius of the circumscribed circle of the polygon is known.

Corollary 15.1.3 When $n \to \infty$, the regular polygon approaches a circle, and the area converges to $A = \pi R^2$.

Demostración. As $n \to \infty$, $\theta \to 0$, and $\sin\theta \approx \theta$. Thus:

$$A \approx \frac{nR^2\theta}{2} = \frac{nR^2\left(\frac{2\pi}{n}\right)}{2} = \pi R^2.$$

■

This corollary illustrates the connection between regular polygons and the circle, highlighting the importance of these concepts in area calculations.

Exercise 15.1 Calculate the area of a regular hexagon inscribed in a circle of radius $R = 10$ cm.

■

Demostración. For a regular hexagon, $n = 6$ and $\theta = \frac{2\pi}{6} = \frac{\pi}{3}$.
Using the theorem:

$$A = \frac{nR^2 \sin\theta}{2} = \frac{6 \times 10^2 \times \sin\left(\frac{\pi}{3}\right)}{2} = \frac{6 \times 100 \times \frac{\sqrt{3}}{2}}{2} = \frac{600 \times \frac{\sqrt{3}}{2}}{2} = \frac{300\sqrt{3}}{2} = 150\sqrt{3}\ \text{cm}^2.$$

Therefore, the area of the hexagon is $150\sqrt{3}$ cm^2. ■

Exercise 15.2 Demonstrate that the area of a regular polygon with n sides and side length l can be expressed as:

$$A = \frac{nl^2}{4\tan\left(\frac{\pi}{n}\right)}.$$

■

Demostración. Consider a regular polygon with n sides of length l. The apothem a can be expressed in terms of the side l:

$$a = \frac{l}{2\tan\left(\frac{\pi}{n}\right)}.$$

The area is:

$$A = \frac{nla}{2} = \frac{nl\left(\frac{l}{2\tan\left(\frac{\pi}{n}\right)}\right)}{2} = \frac{nl^2}{4\tan\left(\frac{\pi}{n}\right)}.$$

■

To solve these exercises, it is necessary to apply the previously established properties and theorems, using trigonometric identities and advanced geometric reasoning.

Ⓡ A deep understanding of the properties of regular polygons and their relationship with trigonometric functions is essential for tackling complex problems in geometry and design.

Graphs and visual representations are valuable tools for better understanding these concepts.

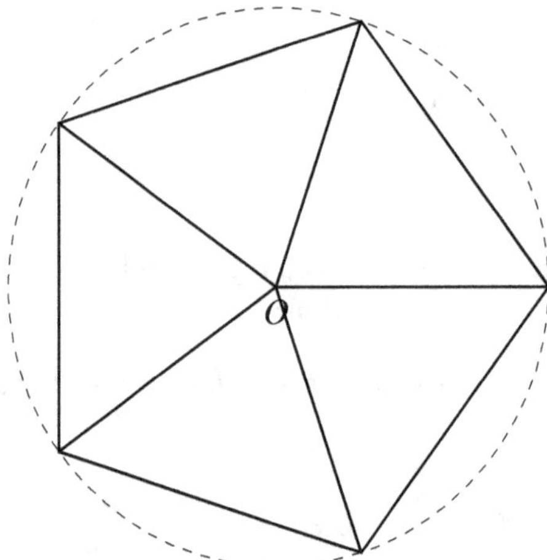

Figura 15.1.1: *Regular pentagon inscribed in a circle of radius R.*

Figure 15.1.1 shows a regular pentagon inscribed in a circle, where the isosceles triangles formed and the relationship between the radius, the apothem, and the polygon's sides can be observed. In conclusion, calculating areas in regular polygons relies on decomposition into congruent triangles and the use of trigonometric relationships. These techniques are fundamental in advanced geometry and have applications in diverse fields such as architecture, design, and engineering.

15.1.2 Applications in Architectural Design Problems

Mathematics, particularly geometry, plays a fundamental role in architectural design. Concepts such as proportion, symmetry, scale, and planar and spatial geometry are essential tools for architects when conceptualizing and materializing their projects. In this section, we explore how mathematical concepts are applied to real-world architectural design problems, emphasizing the mathematical reasoning behind design decisions.

Definition 15.1.2 An **architectural module** is a unit of measurement or repetitive element used to establish a coherent structure of dimensions and proportions in an architectural design.

The use of modules enables architects to create harmonious and functional designs, facilitating the distribution of spaces and standardization of construction elements.

Theorem 15.1.4 The **Golden Ratio Principle** states that a ratio is golden if the ratio between the whole and the larger part is equal to the ratio between the larger part and the smaller part, that is, if $\frac{a+b}{a} = \frac{a}{b} = \phi$, where $\phi \approx 1{,}618$ is the golden number.

Demostración. Let a be the larger length and b the smaller length of a segment divided such that the ratio between the whole and the larger part equals the ratio between the larger part and the smaller part. Then:

$$\frac{a+b}{a} = \frac{a}{b} \implies \frac{a+b}{a} - \frac{a}{b} = 0.$$

Multiplying both sides by ab:

$$b(a+b) - a^2 = 0 \implies ab + b^2 - a^2 = 0.$$

Dividing all terms by b^2:

$$\frac{a}{b} + 1 - \left(\frac{a}{b}\right)^2 = 0.$$

Let $x = \frac{a}{b}$. Then:

$$x^2 - x - 1 = 0.$$

Solving this quadratic equation:

$$x = \frac{1 \pm \sqrt{5}}{2}.$$

Taking the positive solution, as lengths are positive:

$$x = \frac{1+\sqrt{5}}{2} = \phi \approx 1{,}618.$$

■

This principle is widely used in architecture to achieve aesthetically pleasing proportions.

■ **Example 15.2** An architect wants to design a rectangular facade of a building using the golden ratio, where the height h and width w of the facade satisfy $w = \phi h$. If the planned height is 10 meters, determine the width of the facade.

■

Using the relation $w = \phi h$ with $h = 10$ m:

$$w = \phi \times 10 \text{ m} \approx 1{,}618 \times 10 \text{ m} = 16{,}18 \text{ m}.$$

Therefore, the width of the facade should be approximately 16,18 meters to satisfy the golden ratio. This example demonstrates how mathematical proportions are applied in the design of architectural elements to achieve visual harmony.

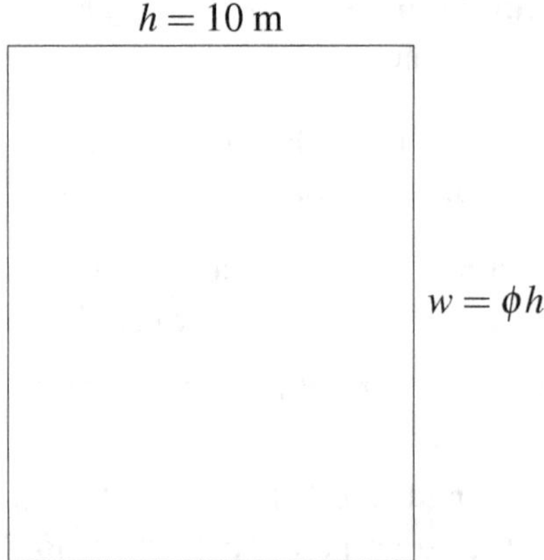

Figura 15.1.2: *Rectangular facade with golden ratio.*

Lema 15.1.2 In a regular polygon used in architectural design, the **structural density** is inversely proportional to the number of sides of the polygon, considering polygons inscribed in a fixed circle.

Demostración. Consider regular polygons inscribed in a circle of radius R. The area A_n of a regular polygon with n sides is:

$$A_n = \frac{nR^2 \sin\left(\frac{2\pi}{n}\right)}{2}.$$

As n increases, $\sin\left(\frac{2\pi}{n}\right) \approx \frac{2\pi}{n}$ for large values of n, so:

$$A_n \approx \frac{nR^2\left(\frac{2\pi}{n}\right)}{2} = \pi R^2.$$

The structural density, defined as the ratio between the perimeter and the area, decreases as n increases, since the perimeter grows linearly with n, but the area approaches a constant limit. ∎

This lemma is relevant when deciding the shape of polygonal structures in architecture, such as building layouts or decorative elements.

Theorem 15.1.5 In architectural design, the use of **regular tessellations** allows complete coverage of a plane without overlaps or gaps using congruent regular polygons. Only three regular tessellations are possible: with equilateral triangles, squares, or regular hexagons.

Demostración. For regular polygons to tessellate the plane, the interior angles must sum to 360° around each vertex. The interior angle of a regular polygon with n sides is:

$$\alpha = \frac{(n-2)\times 180^\circ}{n}.$$

The number of polygons meeting at a vertex is k, where:

$$k \times \alpha = 360^\circ.$$

Testing for $n = 3$ (equilateral triangle):

$$\alpha = \frac{(3-2)\times 180^\circ}{3} = 60^\circ, \quad k = \frac{360^\circ}{60^\circ} = 6.$$

For $n = 4$ (square):

$$\alpha = 90^\circ, \quad k = \frac{360^\circ}{90^\circ} = 4.$$

For $n = 6$ (regular hexagon):

$$\alpha = 120^\circ, \quad k = \frac{360^\circ}{120^\circ} = 3.$$

No other integer values of n and k satisfy $k\alpha = 360^\circ$ with α corresponding to a regular polygon. Therefore, only these three regular polygons can tessellate the plane. ■

Tessellations are used in architecture to design pavements, claddings, and decorative elements that cover flat surfaces.

■ **Example 15.3** An architect wants to design a floor using tessellation with regular hexagons of side $l = 50$ cm. Calculate the area of each hexagon and determine how many hexagons are needed to cover a rectangular room 10 m wide by 15 m long.

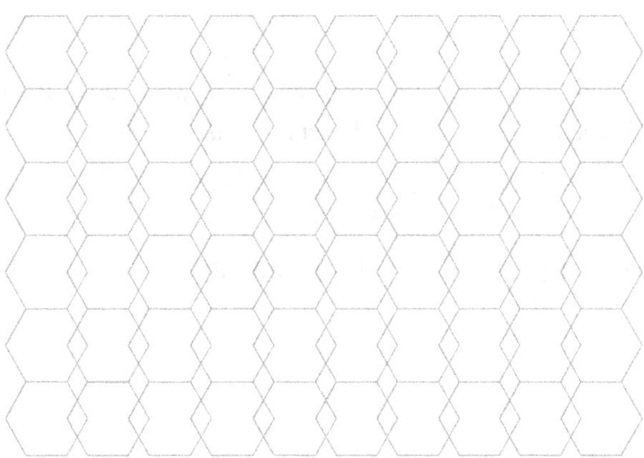

Figura 15.1.3: *Tessellation of regular hexagons on a floor.*

■

Area of a regular hexagon:
For a regular hexagon with side l, the area A is:

$$A = \frac{3\sqrt{3}}{2}l^2 = \frac{3\sqrt{3}}{2}(0{,}5\ \text{m})^2 = \frac{3\sqrt{3}}{2}(0{,}25\ \text{m}^2) = \frac{3\sqrt{3}\times 0{,}25}{2}\ \text{m}^2 = \frac{0{,}75\sqrt{3}}{2}\ \text{m}^2 \approx 0{,}6495\ \text{m}^2.$$

Area of the room:

$$A_{\text{room}} = 10\ \text{m} \times 15\ \text{m} = 150\ \text{m}^2.$$

Number of hexagons needed:

$$N = \frac{A_{\text{room}}}{A_{\text{hexagon}}} = \frac{150}{0{,}6495} \approx 231.$$

Therefore, approximately 231 hexagons are needed to cover the floor of the room.
This example demonstrates the practical application of geometric calculations in architectural design and planning.

Exercise 15.3 An architect is designing a geodesic dome based on a regular icosahedron. If the radius of the circumscribed sphere of the icosahedron is 10 meters, calculate the edge length of the icosahedron and the total area of the faces that will form the dome.

Use the properties of the regular icosahedron, which has 20 equilateral triangular faces. The relationship between the radius R of the circumscribed sphere and the edge l is:

$$l = R \times \frac{2}{\phi}\sqrt{\frac{5-\sqrt{5}}{5}},$$

where ϕ is the golden ratio.
The area of one face is:

$$A_{\text{face}} = \frac{\sqrt{3}}{4}l^2.$$

The total area is:

$$A_{\text{total}} = 20 \times A_{\text{face}}.$$

Exercise 15.4 In the design of a garden, a pavilion is planned with the shape of a regular dodecagon (12 sides). If the apothem of the dodecagon is 5 meters, determine the perimeter and the area of the pavilion.

[of Exercise 1] **Calculation of the edge l of the icosahedron:**

The golden ratio is $\phi = \frac{1+\sqrt{5}}{2} \approx 1{,}618$.
We calculate:

$$l = 10\text{ m} \times \frac{2}{1{,}618}\sqrt{\frac{5-\sqrt{5}}{5}}.$$

First, calculate $\frac{5-\sqrt{5}}{5}$:

$$\frac{5-\sqrt{5}}{5} = \frac{5-2{,}236}{5} = \frac{2{,}764}{5} = 0{,}5528.$$

The square root:

$$\sqrt{0{,}5528} \approx 0{,}7435.$$

Now, calculate l:

$$l = 10\text{ m} \times \frac{2}{1{,}618} \times 0{,}7435 \approx 10\text{ m} \times 1{,}2361 \times 0{,}7435 \approx 10\text{ m} \times 0{,}9185 \approx 9{,}185\text{ m}.$$

Area of one triangular face:

$$A_{\text{face}} = \frac{\sqrt{3}}{4}l^2 = \frac{\sqrt{3}}{4}(9{,}185\text{ m})^2 \approx \frac{1{,}732}{4} \times 84{,}38\text{ m}^2 \approx 0{,}433 \times 84{,}38\text{ m}^2 \approx 36{,}53\text{ m}^2.$$

Total area of the faces:

$$A_{\text{total}} = 20 \times A_{\text{face}} = 20 \times 36{,}53\ \text{m}^2 \approx 730{,}6\ \text{m}^2.$$

Thus, the total area of the faces forming the dome is approximately 730,6 m^2.

R The application of Platonic solids, such as the icosahedron, in architecture allows the creation of strong and aesthetically striking structures, leveraging the geometric properties of these polyhedra.

In architectural design, mathematical reasoning is essential for solving complex problems and ensuring structures are functional, safe, and visually appealing. The presented examples and exercises illustrate how mathematics provides fundamental tools for creating and analyzing advanced architectural designs.

15.2 Coordinate System: Point Location, Distances, and Slopes

15.2.1 Calculating Distances in the Cartesian Plane.

The Cartesian plane is a fundamental tool in analytic geometry, allowing the representation of points and figures through coordinates. One of the most essential operations is the calculation of the distance between two points. In this section, we explore the formulas and properties related to distances in the Cartesian plane, providing a solid foundation for more advanced applications in geometry and mathematical analysis.

Definition 15.2.1 The **Euclidean distance** between two points $P(x_1,y_1)$ and $Q(x_2,y_2)$ in the Cartesian plane is the non-negative real number given by:

$$d(P,Q) = \sqrt{(x_2-x_1)^2+(y_2-y_1)^2}.$$

This formula is a direct application of the Pythagorean Theorem in the plane and is fundamental for measuring the separation between points in $\mathbb{R}^2$.

Theorem 15.2.1 The function $d : \mathbb{R}^2 \times \mathbb{R}^2 \to \mathbb{R}$ defined by $d(P,Q)$ is a **metric** in the Cartesian plane. That is, for all points $P,Q,R \in \mathbb{R}^2$, the following properties hold:

1. **Non-negativity**: $d(P,Q) \geq 0$, and $d(P,Q) = 0$ if and only if $P = Q$.
2. **Symmetry**: $d(P,Q) = d(Q,P)$.
3. **Triangle inequality**: $d(P,R) \leq d(P,Q) + d(Q,R)$.

Demostración. We will prove each property separately:

1. **Non-negativity**: Since $(x_2-x_1)^2 \geq 0$ and $(y_2-y_1)^2 \geq 0$, their sum is also non-negative, and the square root of a non-negative number is non-negative. If $d(P,Q) = 0$, then $(x_2-x_1)^2 + (y_2-y_1)^2 = 0$, implying $x_2 = x_1$ and $y_2 = y_1$, that is, $P = Q$.
2. **Symmetry**:

$$d(P,Q) = \sqrt{(x_2-x_1)^2+(y_2-y_1)^2} = \sqrt{(x_1-x_2)^2+(y_1-y_2)^2} = d(Q,P).$$

3. **Triangle inequality**:

$$d(P,R) = \sqrt{(x_3-x_1)^2+(y_3-y_1)^2}.$$

Applying Minkowski's inequality or the triangle inequality theorem in $\mathbb{R}^2$, we can demonstrate that:

$$d(P,R) \leq d(P,Q) + d(Q,R).$$

■

This theorem establishes that the Cartesian plane with Euclidean distance is a *metric space*, enabling the use of advanced tools from mathematical analysis.

Ⓡ The Euclidean metric is **invariant under isometric transformations**, meaning that distances between points remain unchanged under translations and rotations in the plane.

■ **Example 15.4** Calculate the distance between the points $A(1,3)$ and $B(4,7)$ in the Cartesian plane.

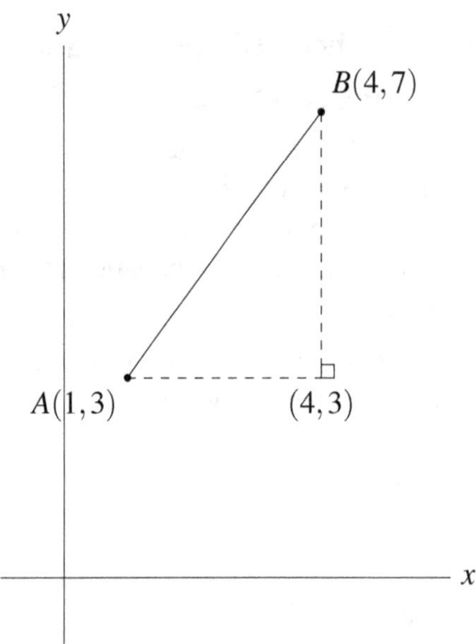

Figura 15.2.1: *Distance between points A and B in the Cartesian plane.*

■

Demostración. Using the distance formula:

$$d(A,B) = \sqrt{(4-1)^2 + (7-3)^2} = \sqrt{(3)^2 + (4)^2} = \sqrt{9+16} = \sqrt{25} = 5.$$

Thus, the distance between A and B is 5 units. ■

This example demonstrates how to directly apply the distance formula to calculate the separation between two given points.

Lema 15.2.1 The **midpoint** M of the segment connecting the points $P(x_1,y_1)$ and $Q(x_2,y_2)$ has coordinates:

$$M\left(\frac{x_1+x_2}{2}, \frac{y_1+y_2}{2}\right).$$

Demostración. The midpoint M is the point that divides the segment PQ into two equal parts. By definition, its coordinates are the averages of the corresponding coordinates of P and Q. ■

■ **Example 15.5** Determine the coordinates of the midpoint of the segment connecting the points $C(-2,5)$ and $D(6,-3)$. ■

Demostración. Using the midpoint formula:

$$M\left(\frac{-2+6}{2},\frac{5+(-3)}{2}\right)=M\left(\frac{4}{2},\frac{2}{2}\right)=M(2,1).$$

Thus, the midpoint is $M(2,1)$. ■

The concept of the midpoint is essential in analytic geometry and has applications in various areas, such as determining perpendicular bisectors and constructing geometric figures.

Theorem 15.2.2 The **minimum distance** from a point $P(x_0,y_0)$ to a line $L: Ax+By+C=0$ is given by:

$$d=\left|\frac{Ax_0+By_0+C}{\sqrt{A^2+B^2}}\right|.$$

Demostración. The distance from point $P(x_0,y_0)$ to the line $Ax+By+C=0$ is the orthogonal projection of the vector from a point on the line to P onto the normal vector $\vec{n}=(A,B)$ of the line. This distance is given by:

$$d=\frac{|Ax_0+By_0+C|}{\sqrt{A^2+B^2}}.$$

■

■ **Example 15.6** Calculate the distance from the point $P(2,-1)$ to the line $L: 3x-4y+12=0$.

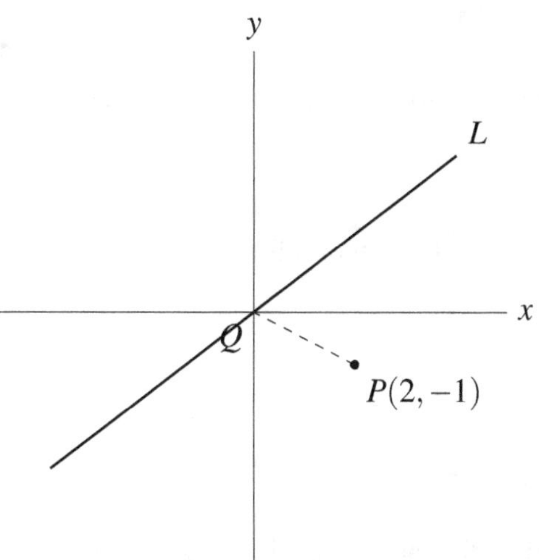

Figura 15.2.2: *Distance from point P to line L.*

■

Demostración. We apply the formula:

$$d = \left|\frac{3(2)-4(-1)+12}{\sqrt{3^2+(-4)^2}}\right| = \left|\frac{6+4+12}{5}\right| = \left|\frac{22}{5}\right| = \frac{22}{5} = 4{,}4.$$

Thus, the distance from point P to line L is 4,4 units. ■

This result is useful in optimization and minimization problems in analytic geometry.

> **Corollary 15.2.3** The **locus** of points equidistant from two fixed points $A(x_1,y_1)$ and $B(x_2,y_2)$ is the **perpendicular bisector** of segment AB, which is the line perpendicular to segment AB at its midpoint.

Demostración. Let M be the midpoint of AB. Any point $P(x,y)$ equidistant from A and B satisfies:

$$d(P,A) = d(P,B).$$

Squaring both sides:

$$(x-x_1)^2+(y-y_1)^2 = (x-x_2)^2+(y-y_2)^2.$$

Simplifying, we obtain the equation of the perpendicular bisector, which is a line perpendicular to AB passing through M. ■

■ **Example 15.7** Find the equation of the perpendicular bisector of the segment connecting the points $E(1,4)$ and $F(5,2)$. ■

Demostración. First, we calculate the midpoint M:

$$M\left(\frac{1+5}{2},\frac{4+2}{2}\right) = M(3,3).$$

The slope of segment EF is:

$$m_{EF} = \frac{2-4}{5-1} = \frac{-2}{4} = -\frac{1}{2}.$$

The slope of the perpendicular bisector is the negative reciprocal:

$$m_{\text{bisector}} = 2.$$

Using the point-slope form of the equation of a line:

$$y-y_0 = m(x-x_0),$$

we get:

$$y-3 = 2(x-3) \implies y = 2x-3.$$

Thus, the equation of the perpendicular bisector is $y = 2x-3$. ■

This example demonstrates how to use the concept of a perpendicular bisector to find equations of lines in the plane.

Exercise 15.5 Determine whether the triangle formed by the points $A(0,0)$, $B(6,0)$, and $C(3,3\sqrt{3})$ is equilateral. ■

Exercise 15.6 Calculate the distance between the points $G(-2,-3)$ and $H(4,1)$, and find the coordinates of the point that divides the segment GH in a $2:1$ ratio internally. ■

of Exercise 1. First, calculate the distances between the vertices:
Side AB**:**

$$d(A,B) = \sqrt{(6-0)^2 + (0-0)^2} = \sqrt{36} = 6.$$

Side AC**:**

$$d(A,C) = \sqrt{(3-0)^2 + (3\sqrt{3}-0)^2} = \sqrt{9+27} = \sqrt{36} = 6.$$

Side BC**:**

$$d(B,C) = \sqrt{(3-6)^2 + (3\sqrt{3}-0)^2} = \sqrt{9+27} = \sqrt{36} = 6.$$

Since all three sides measure 6 units, the triangle is equilateral. ■

(R) Calculating distances in the Cartesian plane is fundamental for determining properties of geometric figures, such as congruence and similarity of triangles, and for solving location and optimization problems in the plane.

In conclusion, mastering distance calculations in the Cartesian plane is essential for advancing in the study of analytic geometry and its applications in higher mathematics. The tools and concepts presented in this section serve as a foundation for exploring more complex topics in spatial geometry, vector analysis, and other branches of mathematics.

15.2.2 Slopes and Angles Between Lines.

In analytic geometry, the slope of a line is an essential measure that describes its inclination and direction in the Cartesian plane. Understanding how to calculate the slope and how it relates to the angle between two lines is fundamental for advanced geometric analysis.

Definition 15.2.2 The **slope** m of a line in the Cartesian plane is a measure of its inclination with respect to the x-axis. If a line passes through two distinct points $P(x_1,y_1)$ and $Q(x_2,y_2)$, its slope is given by:

$$m = \frac{y_2 - y_1}{x_2 - x_1}, \quad \text{with } x_2 \neq x_1.$$

This definition allows us to quantify the direction and inclination of a line, providing a basis for analyzing its behavior and relationship with other lines in the plane.

Theorem 15.2.4 The **angle** θ between two lines with slopes m_1 and m_2 is determined by:

$$\tan\theta = \left|\frac{m_2 - m_1}{1 + m_1 m_2}\right|,$$

provided that $1+m_1m_2 \neq 0$.

Demostración. Let θ be the angle between two lines L_1 and L_2 with slopes m_1 and m_2, respectively. The angles these lines form with the x-axis are $\alpha = \arctan m_1$ and $\beta = \arctan m_2$. Therefore, the angle between the lines is $\theta = |\beta - \alpha|$.
Using the tangent difference identity:

$$\tan(\beta - \alpha) = \frac{\tan\beta - \tan\alpha}{1+\tan\alpha\tan\beta} = \frac{m_2 - m_1}{1+m_1m_2}.$$

Taking the absolute value, since the angle between lines is always positive:

$$\tan\theta = \left|\frac{m_2 - m_1}{1+m_1m_2}\right|.$$

■

This result is fundamental for calculating the angle between two lines when their slopes are known, which is essential in various mathematical and physical applications.

Corollary 15.2.5 Two lines are **parallel** if and only if their slopes are equal:

$$m_1 = m_2.$$

Demostración. If the slopes are equal, $m_1 = m_2$, then:

$$\tan\theta = \left|\frac{m_2 - m_1}{1+m_1^2}\right| = \left|\frac{0}{1+m_1^2}\right| = 0.$$

Thus, the angle between the lines is $\theta = 0°$, indicating that the lines are parallel. ■

Corollary 15.2.6 Two lines are **perpendicular** if and only if the product of their slopes is -1:

$$m_1m_2 = -1.$$

Demostración. If the lines are perpendicular, the angle between them is $\theta = 90°$. Therefore:

$$\tan\theta = \tan 90° = \infty.$$

This implies that the denominator in the formula for $\tan\theta$ is zero:

$$1+m_1m_2 = 0 \implies m_1m_2 = -1.$$

■

These properties are essential for identifying perpendicularity and parallelism relationships between lines in the Cartesian plane.

■ **Example 15.8** Determine the angle between the lines L_1 and L_2 given by the equations:

$$L_1 : y = 2x+3, \quad L_2 : y = -\frac{1}{2}x+1.$$

■

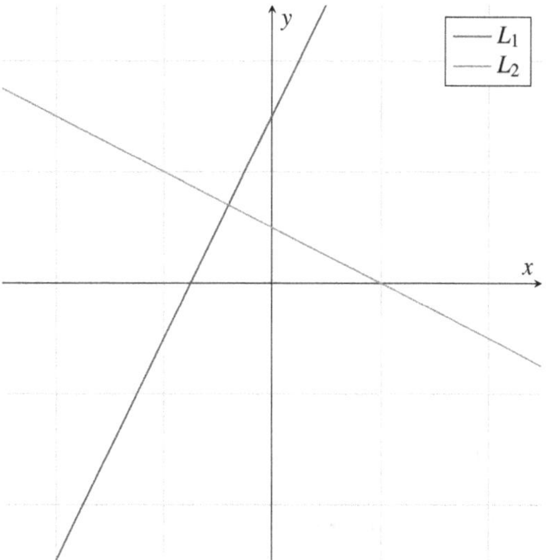

Figura 15.2.3: *Graph of lines L_1 and L_2.*

Demostración. The slopes of the lines are $m_1 = 2$ and $m_2 = -\frac{1}{2}$.
We calculate the angle θ between the lines using the formula:

$$\tan\theta = \left|\frac{m_2 - m_1}{1 + m_1 m_2}\right| = \left|\frac{-\frac{1}{2} - 2}{1 + 2\left(-\frac{1}{2}\right)}\right| = \left|\frac{-\frac{5}{2}}{1-1}\right|.$$

The denominator is zero, indicating that $\tan\theta$ is undefined and, therefore, $\theta = 90^\circ$.
We conclude that lines L_1 and L_2 are **perpendicular**.

■

This example demonstrates how to determine the angle between two lines and verify their perpendicularity through their slopes.

Lema 15.2.2 The slope of a line is the tangent of the **inclination angle** α it forms with the x-axis:

$$m = \tan\alpha.$$

Demostración. By definition, the slope m is the ratio of the change in y to the change in x between two points on the line. This ratio is precisely the tangent of the angle α between the line and the x-axis:

$$m = \frac{\Delta y}{\Delta x} = \tan\alpha.$$

■

This relationship is fundamental for connecting the slope of a line with its orientation in the plane.

■ **Example 15.9** Calculate the inclination angle of the line $y = 3x - 4$.

■

Demostración. The slope of the line is $m = 3$.
The inclination angle α is obtained using:

$$\alpha = \arctan m = \arctan 3.$$

Calculating:

$$\alpha \approx \arctan 3 \approx 71{,}57^\circ.$$

Thus, the line forms an angle of approximately $71{,}57^\circ$ with the x-axis. ∎

This example illustrates how to find the inclination angle from the slope of a line.

Theorem 15.2.7 The **angle between two lines** with slopes m_1 and m_2 can also be expressed using the formula:

$$\cos\theta = \frac{1 + m_1 m_2}{\sqrt{1+m_1^2}\sqrt{1+m_2^2}}.$$

Demostración. Consider the direction vectors of the lines:

$$\vec{u} = (1, m_1), \quad \vec{v} = (1, m_2).$$

The cosine of the angle θ between $\vec{u}$ and $\vec{v}$ is:

$$\cos\theta = \frac{\vec{u}\cdot\vec{v}}{|\vec{u}||\vec{v}|} = \frac{1\cdot 1 + m_1 m_2}{\sqrt{1+m_1^2}\sqrt{1+m_2^2}}.$$

∎

This formula is useful when calculating the angle between lines using the cosine instead of the tangent.

Exercise 15.7 Find the angle between the lines $y = x$ and $y = \sqrt{3}x$. ∎

Exercise 15.8 Determine whether the lines passing through points $A(-1,2)$ and $B(3,6)$, and the line passing through points $C(2,5)$ and $D(6,9)$ are parallel, perpendicular, or neither. ∎

of Exercise 1. The slopes are $m_1 = 1$ and $m_2 = \sqrt{3}$.
Using the formula for the angle between lines:

$$\tan\theta = \left|\frac{\sqrt{3}-1}{1+1\times\sqrt{3}}\right| = \left|\frac{\sqrt{3}-1}{1+\sqrt{3}}\right|.$$

Multiply numerator and denominator by the conjugate of the denominator to simplify:

$$\tan\theta = \left|\frac{(\sqrt{3}-1)(1-\sqrt{3})}{(1+\sqrt{3})(1-\sqrt{3})}\right| = \left|\frac{-(\sqrt{3}-1)^2}{1-3}\right| = \left|\frac{-(\sqrt{3}-1)^2}{-2}\right| = \frac{(\sqrt{3}-1)^2}{2}.$$

Calculating:

$$(\sqrt{3}-1)^2 = (\sqrt{3})^2 - 2\sqrt{3} + 1 = 3 - 2\sqrt{3} + 1 = 4 - 2\sqrt{3}.$$

Thus:

$$\tan\theta = \frac{4-2\sqrt{3}}{2} = 2-\sqrt{3}.$$

Calculating θ:

$$\theta = \arctan(2-\sqrt{3}) \approx \arctan(0{,}2679) \approx 15^\circ.$$

Therefore, the angle between the lines is approximately 15°.

■

(R) Understanding slopes and angles between lines is essential in fields such as engineering, physics, and computer graphics, where it is necessary to model and analyze geometric relationships in the plane and space.

Through these definitions, theorems, and examples, we have explored how slopes determine the inclination of lines and how to calculate the angle between them. This knowledge is fundamental for advancing in the study of analytic geometry and its applications in complex mathematical problems.

15.3 Lines and Planes in Space: Parallelism and Perpendicularity.

15.3.1 Applications in 3D Design.

Three-dimensional design is a field that combines advanced mathematical concepts with practical applications in engineering, architecture, and computer graphics. Understanding geometric relationships in three-dimensional space is essential for modeling and manipulating 3D objects accurately and efficiently.

Definition 15.3.1 A **vector in three-dimensional space** is an element of the space $\mathbb{R}^3$, represented as an ordered triplet of real numbers (x,y,z). Geometrically, this vector can be interpreted as a directed segment from the origin to the point (x,y,z) in space.

Vectors are fundamental in 3D design as they describe positions, displacements, and orientations of objects in space.

Proposition 15.3.1 The dot product (or scalar product) of two vectors $\vec{u} = (u_1,u_2,u_3)$ and $\vec{v} = (v_1,v_2,v_3)$ in $\mathbb{R}^3$ is defined as:

$$\vec{u}\cdot\vec{v} = u_1v_1 + u_2v_2 + u_3v_3.$$

The dot product is useful for calculating angles between vectors and determining orthogonality in 3D space.

Theorem 15.3.2 Two vectors $\vec{u}$ and $\vec{v}$ in $\mathbb{R}^3$ are **perpendicular** if and only if their dot product is zero:

$$\vec{u}\cdot\vec{v} = 0 \iff \vec{u} \perp \vec{v}.$$

Demostración. If $\vec{u}\cdot\vec{v}=0$, then the angle θ between $\vec{u}$ and $\vec{v}$ satisfies $\cos\theta=0$, implying $\theta=90°$, i.e., the vectors are perpendicular. Conversely, if $\vec{u}\perp\vec{v}$, then $\theta=90°$, and therefore $\cos\theta=0$, which implies $\vec{u}\cdot\vec{v}=0$. ■

This result is crucial in 3D design to ensure structural elements are correctly aligned and avoid unwanted intersections.

Definition 15.3.2 The **cross product** of two vectors $\vec{u}$ and $\vec{v}$ in $\mathbb{R}^3$ is a vector $\vec{w}=\vec{u}\times\vec{v}$ such that:

$$\vec{w}=(u_2v_3-u_3v_2,\ u_3v_1-u_1v_3,\ u_1v_2-u_2v_1).$$

The cross product results in a vector perpendicular to the plane defined by $\vec{u}$ and $\vec{v}$, with its direction determined by the right-hand rule.

Theorem 15.3.3 The magnitude of the cross product $\vec{w}=\vec{u}\times\vec{v}$ equals the area of the parallelogram defined by $\vec{u}$ and $\vec{v}$:

$$|\vec{w}|=|\vec{u}\times\vec{v}|=|\vec{u}||\vec{v}|\sin\theta,$$

where θ is the angle between $\vec{u}$ and $\vec{v}$.

Demostración. By the definition of the cross product and trigonometric properties, the magnitude of $\vec{w}$ is:

$$|\vec{w}|=\sqrt{(u_2v_3-u_3v_2)^2+(u_3v_1-u_1v_3)^2+(u_1v_2-u_2v_1)^2}=|\vec{u}||\vec{v}|\sin\theta.$$

■

This theorem is useful in 3D design for calculating areas and volumes and determining normals to surfaces, which are essential in computer graphics and 3D modeling.

■ **Example 15.10** Determine the equation of the plane passing through the points $A(1,0,2)$, $B(2,-1,3)$, and $C(0,1,5)$. ■

Demostración. First, calculate two vectors in the plane:

$$\vec{AB}=B-A=(2-1,\ -1-0,\ 3-2)=(1,\ -1,\ 1),$$

$$\vec{AC}=C-A=(0-1,\ 1-0,\ 5-2)=(-1,\ 1,\ 3).$$

Compute the cross product $\vec{n}=\vec{AB}\times\vec{AC}$:

$$\vec{n}=\begin{vmatrix}\vec{i} & \vec{j} & \vec{k}\\ 1 & -1 & 1\\ -1 & 1 & 3\end{vmatrix}=\vec{i}((-1)(3)-(1)(1))-\vec{j}((1)(3)-(1)(-1))+\vec{k}((1)(1)-(-1)(-1)).$$

Calculating each component:

$$n_x=(-1)(3)-(1)(1)=-3-1=-4,$$

$$n_y=-((1)(3)-(1)(-1))=-(3-(-1))=-(3+1)=-4,$$

$$n_z = (1)(1) - (-1)(-1) = 1 - 1 = 0.$$

Thus, the normal vector is $\vec{n} = (-4, -4, 0)$.
The equation of the plane is:

$$n_x(x - x_0) + n_y(y - y_0) + n_z(z - z_0) = 0,$$

where (x_0, y_0, z_0) is a point on the plane, e.g., $A(1,0,2)$. Substituting:

$$-4(x-1) - 4(y-0) + 0(z-2) = 0.$$

Simplifying:

$$-4(x-1) - 4y = 0 \implies -4x + 4 - 4y = 0 \implies 4x + 4y = 4 \implies x + y = 1.$$

■

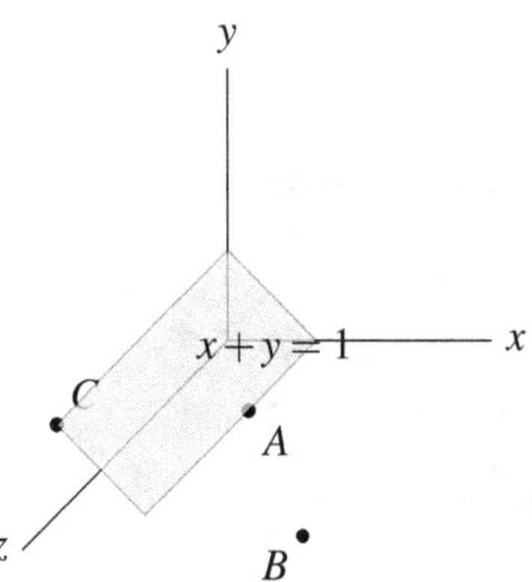

Figura 15.3.1: *Plane $x+y=1$ passing through points A, B, and C.*

■

In this example, we determined the equation of the plane passing through three non-collinear points in space using the cross product to find the plane's normal vector.

Exercise 15.9 Find the equation of the line passing through the point $P(1,2,3)$ and perpendicular to the plane $2x - y + 2z - 5 = 0$.

■

Exercise 15.10 Calculate the angle between the lines r_1 and r_2 defined by their direction vectors $\vec{v}_1 = (1, 2, -1)$ and $\vec{v}_2 = (2, -1, 2)$.

■

of Exercise 1. The normal vector to the plane is $\vec{n} = (2, -1, 2)$.
The line sought is perpendicular to the plane, so its direction vector is parallel to $\vec{n}$.
The parametric equation of the line is:

$$x = x_0 + at, \quad y = y_0 + bt, \quad z = z_0 + ct,$$

where $(x_0, y_0, z_0) = (1, 2, 3)$ and $\vec{d} = (a, b, c) = \vec{n} = (2, -1, 2)$.
Thus, the equations of the line are:

$$x = 1 + 2t, \quad y = 2 - t, \quad z = 3 + 2t.$$

■

In 3D design, working with planes and lines in space is common, and understanding the relationships between them is essential for modeling complex surfaces and structures.

The ability to calculate angles between lines and planes is fundamental for ensuring that elements in a 3D design fit correctly and behave as expected.

Theorem 15.3.4 The **angle** θ between two lines in space with direction vectors $\vec{u}$ and $\vec{v}$ is calculated using:

$$\cos\theta = \frac{\vec{u}\cdot\vec{v}}{|\vec{u}||\vec{v}|}.$$

Demostración. The angle between two vectors in $\mathbb{R}^3$ is given by the dot product formula:

$$\vec{u}\cdot\vec{v} = |\vec{u}||\vec{v}|\cos\theta.$$

Rearranging for $\cos\theta$ gives the theorem's expression. ■

■ **Example 15.11** Calculate the angle between the lines r_1 and r_2 with direction vectors $\vec{u} = (1,0,1)$ and $\vec{v} = (0,1,1)$. ■

Demostración. First, compute the dot product:

$$\vec{u}\cdot\vec{v} = (1)(0)+(0)(1)+(1)(1) = 0+0+1 = 1.$$

Calculate the magnitudes:

$$|\vec{u}| = \sqrt{1^2+0^2+1^2} = \sqrt{1+0+1} = \sqrt{2},$$

$$|\vec{v}| = \sqrt{0^2+1^2+1^2} = \sqrt{0+1+1} = \sqrt{2}.$$

Thus:

$$\cos\theta = \frac{1}{\sqrt{2}\times\sqrt{2}} = \frac{1}{2}.$$

Therefore, $\theta = \arccos\left(\frac{1}{2}\right) = 60^\circ$. ■

This example shows how to calculate the angle between two lines in space using their direction vectors.

In 3D designs, calculating angles between elements is crucial for ensuring structural integrity and the aesthetic quality of the model.

In conclusion, advanced mathematics provides essential tools for three-dimensional design. A deep understanding of vectors, planes, lines, and their interactions allows designers and engineers to create precise and efficient models, bringing ideas from concept to virtual or physical reality.

15.3.2 Solving Spatial Problems

.

Solving problems in three-dimensional space is an essential skill in advanced mathematics, particularly in analytic geometry and vector algebra. This section focuses on providing the tools and methods needed to address problems involving points, lines, and planes in space, as well as their interactions and relationships.

Definition 15.3.3 A **vector in three-dimensional space** is an element of $\mathbb{R}^3$, represented by an ordered triplet of real numbers (x,y,z). Geometrically, a vector can be interpreted as a directed segment with origin at the point $(0,0,0)$ and endpoint at the point (x,y,z).

Vectors are fundamental for describing positions and directions in space and allow the formulation of equations and the systematic solution of spatial problems.

Definition 15.3.4 The **parametric equation** of a line in space passing through a point $P_0(x_0,y_0,z_0)$ and having a direction vector $\vec{v}=(a,b,c)$ is:

$$x = x_0 + at, \quad y = y_0 + bt, \quad z = z_0 + ct, \quad t \in \mathbb{R}.$$

This parametric representation is useful for describing lines in space and facilitates the analysis of their behavior and relationships with other geometric objects.

Proposition 15.3.5 Two lines in space are **parallel** if and only if their direction vectors are proportional; that is, there exist real numbers $\lambda \neq 0$ such that:

$$\vec{v}_1 = \lambda \vec{v}_2.$$

Demostración. If $\vec{v}_1$ and $\vec{v}_2$ are proportional, then the lines have the same direction and, therefore, are parallel. Conversely, if the lines are parallel, their direction vectors must be proportional. ■

This property is fundamental when solving problems involving parallelism in three-dimensional space.

Theorem 15.3.6 The **minimum distance** d from a point $P(x_0,y_0,z_0)$ to a plane π given by the equation $Ax+By+Cz+D=0$ is:

$$d = \left| \frac{Ax_0 + By_0 + Cz_0 + D}{\sqrt{A^2+B^2+C^2}} \right|.$$

Demostración. The distance from a point to the plane is the orthogonal projection of the vector from any point on the plane to the given point onto the plane's normal vector. By normalizing the normal vector, we obtain the stated formula. ■

This formula is essential for solving problems involving relative positions of points and planes.

■ **Example 15.12** Calculate the distance from the point $P(3,-2,5)$ to the plane $\pi: 2x-y+2z-5=0$.

■

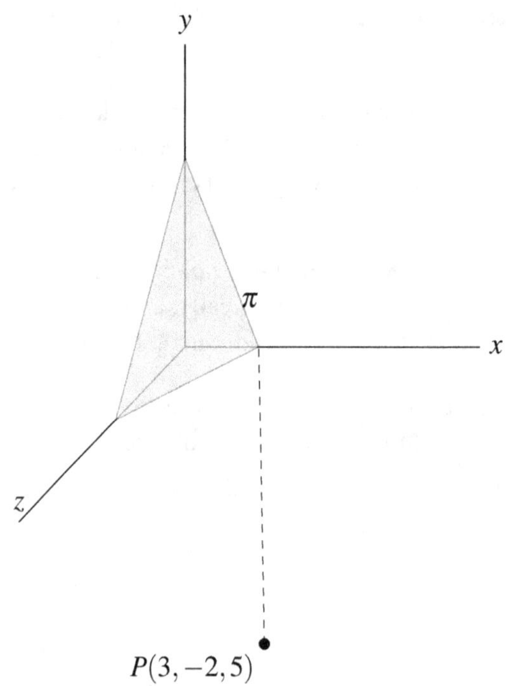

Figura 15.3.2: *Distance from point P to plane π.*

Demostración. Using the point-to-plane distance formula:

$$d = \left| \frac{2(3)-(-2)+2(5)-5}{\sqrt{2^2+(-1)^2+2^2}} \right| = \left| \frac{6+2+10-5}{\sqrt{4+1+4}} \right| = \left| \frac{13}{\sqrt{9}} \right| = \left| \frac{13}{3} \right| \approx 4{,}33.$$

Therefore, the distance is approximately 4,33 units. ■

This example demonstrates how to apply the formula to determine distances in three-dimensional space.

Lema 15.3.1 The **intersection** of a line and a plane in space can be found by solving the system of equations formed by the line's parametric equations and the plane's equation.

Demostración. By substituting the parametric expressions for x, y, and z from the line into the plane's equation, we obtain an equation in terms of the parameter t. Solving this equation yields the value of t corresponding to the intersection point. ■

This technique is essential for solving problems involving points of intersection in space.

■ **Example 15.13** Find the intersection point between the line r given by:

$$x = 1+2t, \quad y = -1+t, \quad z = 3-t,$$

and the plane $\pi : x+y+z = 6$.

■

Demostración. We substitute the equations of the line into the equation of the plane:

$$(1+2t)+(-1+t)+(3-t) = 6 \implies 1+2t-1+t+3-t = 6 \implies (2t+t-t)+(1-1+3) = 6.$$

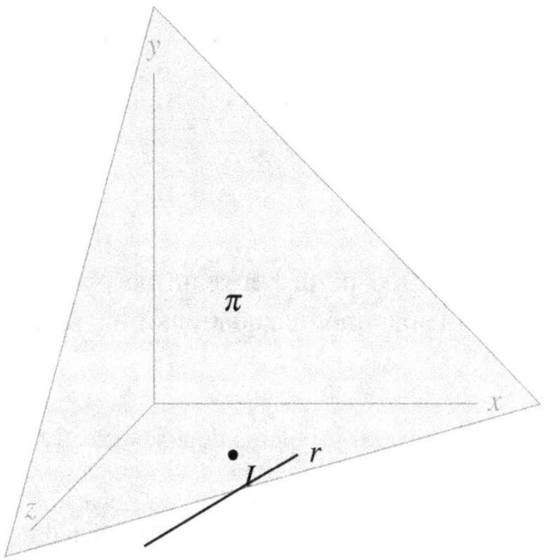

Figura 15.3.3: *Intersection of line r with plane π at point I.*

Simplifying:

$$2t+3=6 \implies 2t=3 \implies t=\frac{3}{2}.$$

Substituting $t=\frac{3}{2}$ into the parametric equations:

$$x=1+2\left(\frac{3}{2}\right)=1+3=4,$$

$$y=-1+\frac{3}{2}=-1+1{,}5=0{,}5,$$

$$z=3-\frac{3}{2}=3-1{,}5=1{,}5.$$

Thus, the intersection point is $I(4,0{,}5,1{,}5)$. ■

This example demonstrates how to determine the intersection point between a line and a plane in space.

Theorem 15.3.7 The **minimum distance** d between two skew lines in space, with direction vectors $\vec{u}$ and $\vec{v}$, is given by:

$$d=\frac{|(\vec{P_2}-\vec{P_1})\cdot(\vec{u}\times\vec{v})|}{|\vec{u}\times\vec{v}|},$$

where $\vec{P_1}$ and $\vec{P_2}$ are points on the respective lines.

Demostración. The minimum distance between two skew lines is the length of the common perpendicular segment to both lines. The numerator represents the volume of the parallelepiped defined by the vectors $\vec{P_2}-\vec{P_1}$, $\vec{u}$, and $\vec{v}$, and the denominator is the area of the base of the parallelepiped. Dividing these gives the height, i.e., the minimum distance. ■

This formula is crucial for solving problems requiring the calculation of distances between elements in three-dimensional space.

Exercise 15.11 Calculate the minimum distance between the lines r_1 and r_2 given by:

$$r_1 : \frac{x-1}{2} = \frac{y+1}{-1} = \frac{z}{3}, \quad r_2 : \frac{x}{1} = \frac{y-2}{2} = \frac{z-1}{-2}.$$

■

Exercise 15.12 Find the intersection point between the planes $\pi_1 : x+y+z=6$ and $\pi_2 : 2x-y+3z=4$, and determine the parametric equation of the resulting line of intersection. ■

(R) The ability to solve complex spatial problems depends greatly on a solid understanding of fundamental concepts such as vectors, lines, and planes, and their relationships. The techniques and theorems presented in this section are powerful tools for addressing and solving a wide range of problems in spatial geometry and related fields.

of Exercise 2. To find the intersection of planes π_1 and π_2, we look for a line that satisfies both equations.
We can parameterize one of the variables, for example, let $z=t$.
From the equation of π_1:

$$x+y+t=6 \implies x=6-y-t.$$

From the equation of π_2:

$$2x-y+3t=4.$$

Substituting x:

$$2(6-y-t)-y+3t=4 \implies 12-2y-2t-y+3t=4.$$

Simplifying:

$$12-3y+t=4 \implies -3y+t=-8 \implies 3y=t+8 \implies y=\frac{t+8}{3}.$$

Now, substituting y into x:

$$x=6-\frac{t+8}{3}-t=6-\frac{t+8+3t}{3}=6-\frac{4t+8}{3}=\frac{18-4t-8}{3}=\frac{10-4t}{3}.$$

Thus, the parametric equation of the line of intersection is:

$$x=\frac{10-4t}{3}, \quad y=\frac{t+8}{3}, \quad z=t.$$

■

In this exercise, we found the parametric equation of the line of intersection between two planes, a common problem in spatial geometry.
In conclusion, solving spatial problems requires a combination of theoretical understanding and practical skills in analytic geometry and vector algebra. The methods and examples presented provide a solid foundation for tackling more complex challenges in mathematics and related disciplines.

15.4 Solved Exercises

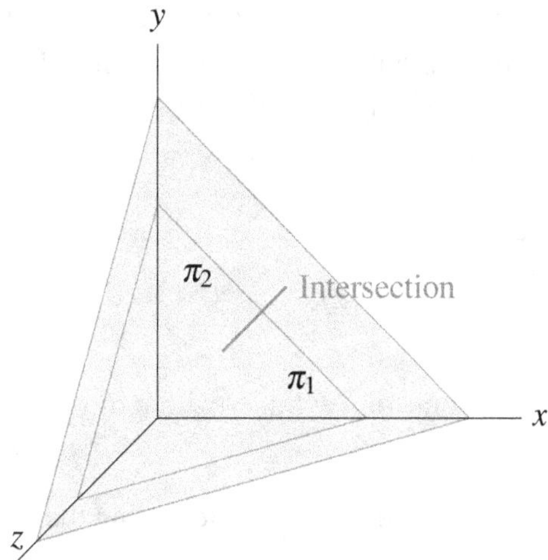

Figura 15.3.4: *Intersection of planes π_1 and π_2 resulting in a line.*

Exercise 15.13 Prove that the area of an equilateral triangle with side l is $\frac{\sqrt{3}}{4}l^2$. ■

Demostración. An equilateral triangle can be divided into two right triangles. The height h of the triangle can be found using the Pythagorean Theorem in one of these right triangles, where one leg is $\frac{l}{2}$ and the hypotenuse is l:

$$h = \sqrt{l^2 - \left(\frac{l}{2}\right)^2} = \sqrt{l^2 - \frac{l^2}{4}} = \sqrt{\frac{3l^2}{4}} = \frac{\sqrt{3}}{2}l.$$

The area of the equilateral triangle is then:

$$A = \frac{1}{2} \times l \times h = \frac{1}{2} \times l \times \frac{\sqrt{3}}{2}l = \frac{\sqrt{3}}{4}l^2.$$

■

Exercise 15.14 Calculate the area of a regular pentagon with side $l = 5$ cm and apothem $a = 6{,}9$ cm. ■

Demostración. The area A of a regular polygon is calculated using the formula:

$$A = \frac{n \cdot l \cdot a}{2},$$

where n is the number of sides, l is the side length, and a is the apothem. For a regular pentagon, $n = 5$. Substituting the values:

$$A = \frac{5 \cdot 5 \cdot 6{,}9}{2} = \frac{172{,}5}{2} = 86{,}25\,\text{cm}^2.$$

Therefore, the area of the pentagon is $86{,}25\,\text{cm}^2$. ■

Exercise 15.15 Prove that the distance between the points $A(x_1, y_1)$ and $B(x_2, y_2)$ in the Cartesian plane is $d = \sqrt{(x_2 - x_1)^2 + (y_2 - y_1)^2}$. ■

Demostración. The distance between two points in the Cartesian plane can be calculated using the Pythagorean Theorem. Considering the points $A(x_1, y_1)$ and $B(x_2, y_2)$, a right triangle can be formed with one leg of length $|x_2 - x_1|$ and the other of length $|y_2 - y_1|$. The hypotenuse of this triangle is the distance between the points A and B, so:

$$d = \sqrt{(x_2 - x_1)^2 + (y_2 - y_1)^2}.$$

■

Exercise 15.16 Calculate the volume of a cylinder with radius $r = 4$ cm and height $h = 10$ cm. ■

Demostración. The volume V of a cylinder is calculated using the formula:

$$V = \pi r^2 h.$$

Substituting the given values:

$$V = \pi(4)^2(10) = \pi \cdot 16 \cdot 10 = 160\pi\,\text{cm}^3.$$

Therefore, the volume of the cylinder is $160\pi\,\text{cm}^3$, or approximately $502{,}65\,\text{cm}^3$. ■

Exercise 15.17 Prove that the angle between two lines with slopes m_1 and m_2 is:

$$\theta = \arctan\left(\frac{|m_2 - m_1|}{1 + m_1 m_2}\right).$$

■

Demostración. Let α be the angle that the first line, with slope m_1, makes with the x-axis, and let β be the angle that the second line, with slope m_2, makes with the x-axis. Then, the slope of each line is $m_1 = \tan\alpha$ and $m_2 = \tan\beta$. The angle θ between the two lines is the difference $|\alpha - \beta|$. Using the tangent difference identity:

$$\tan(\alpha - \beta) = \frac{\tan\alpha - \tan\beta}{1 + \tan\alpha\tan\beta} = \frac{m_1 - m_2}{1 + m_1 m_2}.$$

Thus, the angle θ between the two lines is:

$$\theta = \arctan\left(\frac{|m_2 - m_1|}{1 + m_1 m_2}\right).$$

■

15.5 Proposed Exercises

15.5.1 Perimeters and Areas of Plane Figures: Circles, Triangles, and Polygons

Exercise 15.18 Calculate the perimeter and area of a circle with radius $r = 7$ cm. ■

Exercise 15.19 Find the area of an equilateral triangle with side $l = 10$ cm. ■

Exercise 15.20 Calculate the area and perimeter of a square with side length 12 cm. ■

Exercise 15.21 Determine the area of a regular pentagon inscribed in a circle with radius 5 cm. ■

Exercise 15.22 Find the area of a regular hexagon with side length 8 cm. ■

15.5.2 Coordinate System: Point Locations, Distances, and Slopes

Exercise 15.23 Calculate the distance between the points $A(3,4)$ and $B(-1,-2)$ in the Cartesian plane. ■

Exercise 15.24 Determine the slope of the line passing through the points $C(2,3)$ and $D(6,7)$. ■

Exercise 15.25 Find the coordinates of the midpoint of the segment connecting the points $E(1,-1)$ and $F(4,5)$. ■

Exercise 15.26 If the slope of a line is $\frac{3}{4}$, what is the angle the line makes with the x-axis? ■

Exercise 15.27 Calculate the area of a triangle in the Cartesian plane formed by the points $G(0,0)$, $H(3,0)$, and $I(0,4)$. ■

15.5.3 Lines and Planes in Space: Parallelism and Perpendicularity Relations

Exercise 15.28 Determine whether the lines $L_1 : \frac{x-1}{2} = \frac{y+2}{-1} = \frac{z}{3}$ and $L_2 : \frac{x}{1} = \frac{y-1}{2} = \frac{z+1}{-3}$ are parallel, perpendicular, or neither. ■

Exercise 15.29 Calculate the angle between the plane $\pi_1 : 2x - y + 2z = 5$ and the plane $\pi_2 : x + y + z = 3$. ■

Exercise 15.30 Find the equation of the plane passing through the points $P(1,2,3)$, $Q(4,5,6)$, and $R(7,8,9)$. ■

Exercise 15.31 Calculate the distance from the point $S(3,-2,5)$ to the plane $\pi : x - 2y + z = 4$. ■

16. Geometric Solids

16.1 Geometric Bodies: Prisms, Cylinders, Cones, and Spheres.

16.1.1 Volume of Prisms and Cylinders.

In the study of geometric solids, prisms and cylinders are fundamental figures that help understand more advanced concepts in spatial geometry and integral calculus. Analyzing their properties and methods for calculating their volumes is essential for applications in engineering, physics, and advanced mathematics.

Definition 16.1.1 A **prism** is a polyhedron formed by two congruent and parallel bases, which are polygons, and lateral faces that are parallelograms. The lateral edges are parallel to each other and perpendicular to the bases in a right prism.

Definition 16.1.2 A **cylinder** is a geometric solid generated by a line (generator) that moves parallel to itself while always remaining in contact with a closed planar curve (directrix), typically a circle in the case of a circular cylinder.

The main characteristic shared by prisms and cylinders is that their volume can be calculated by multiplying the area of the base by the height of the solid.

Theorem 16.1.1 The **volume** V of a prism or cylinder is:

$$V = A_b \times h,$$

where A_b is the area of the base and h is the height of the solid.

Demostración. The volume of a prism or cylinder is derived by integrating the base area along the height. For a prism or cylinder with a constant base, the volume is simply the product of the base area and the height, as the cross-sectional area remains constant throughout the solid's height. ■

This direct relationship between the base area and volume simplifies calculations and is applicable to any prism or cylinder, regardless of the shape of its base.

■ **Example 16.1** Calculate the volume of a rectangular prism with dimensions: length $l = 8$ cm, width $w = 5$ cm, and height $h = 10$ cm.

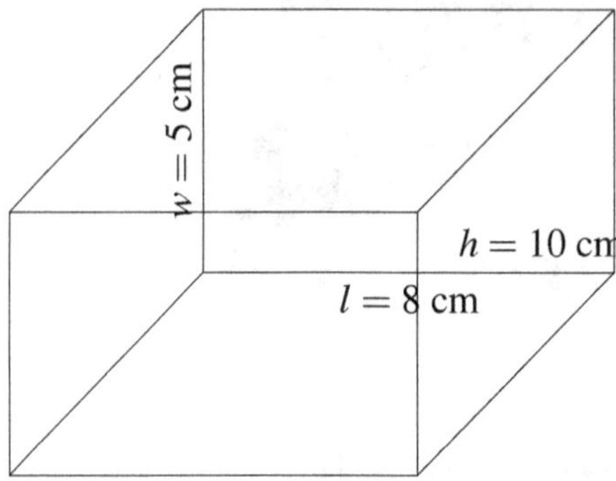

Figura 16.1.1: *Rectangular prism with given dimensions.*

■

Demostración. The area of the base is the area of the rectangle:

$$A_b = l \times w = 8\,\text{cm} \times 5\,\text{cm} = 40\,\text{cm}^2.$$

The volume is:

$$V = A_b \times h = 40\,\text{cm}^2 \times 10\,\text{cm} = 400\,\text{cm}^3.$$

■

This example shows the direct application of the volume formula for a rectangular prism, simplifying the calculation through the multiplication of its dimensions.

Definition 16.1.3 A **right circular cylinder** is a cylinder whose base is a circle and whose axis is perpendicular to the plane of the base.

Theorem 16.1.2 The **volume** V of a right circular cylinder is:

$$V = \pi r^2 h,$$

where r is the radius of the base and h is the height of the cylinder.

Demostración. The base area of a circular cylinder is $A_b = \pi r^2$. Applying the general volume formula for prisms and cylinders:

$$V = A_b \times h = \pi r^2 h.$$

■

This formula is fundamental in various fields, including engineering and physics, where cylinders are common shapes in structures and components.

■ **Example 16.2** Calculate the volume of a right circular cylinder with radius $r = 3$ m and height $h = 5$ m.

■

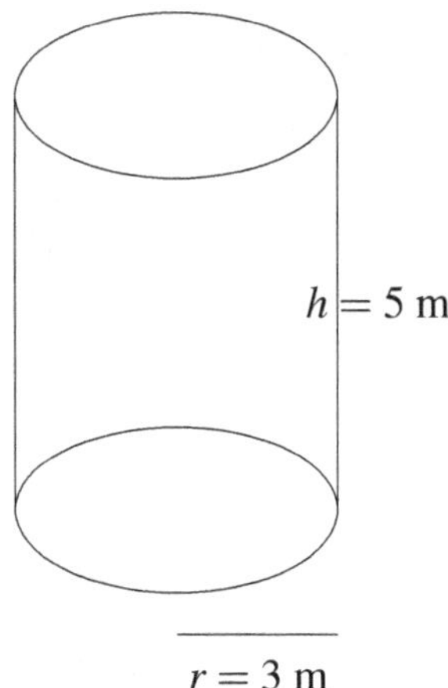

Figura 16.1.2: *Right circular cylinder with given radius and height.*

Demostración. Calculate the base area:

$$A_b = \pi r^2 = \pi(3\,\text{m})^2 = 9\pi\,\text{m}^2.$$

The volume is:

$$V = A_b \times h = 9\pi\,\text{m}^2 \times 5\,\text{m} = 45\pi\,\text{m}^3 \approx 141{,}37\,\text{m}^3.$$

■

This example illustrates how to use the volume formula of a right circular cylinder, obtaining an exact result in terms of π and a decimal approximation.

Lema 16.1.1 The volume of a prism or cylinder remains constant if the base area and height are preserved, regardless of the specific shape of the base.

Demostración. Given that $V = A_b \times h$, as long as the product of the base area and height remains constant, the volume does not change. The shape of the base may vary, but if its total area is unchanged, the solid's volume remains the same. ■

This result is useful in optimization and design, where maximizing or minimizing certain dimensions while maintaining constant volume is often required.

Corollary 16.1.3 Two prisms or cylinders with equal heights and equal base areas have equal volumes.

Demostración. Directly from the previous lemma, if $A_{b1} = A_{b2}$ and $h_1 = h_2$, then:

$$V_1 = A_{b1} \times h_1 = A_{b2} \times h_2 = V_2.$$

■

Exercise 16.1 Calculate the volume of a right triangular prism whose base is an equilateral triangle with side $l = 6$ cm and prism height $h = 12$ cm.

First, calculate the base area using the formula for the area of an equilateral triangle:

$$A_b = \frac{\sqrt{3}}{4} l^2.$$

Then, multiply the base area by the prism height to obtain the volume.

Exercise 16.2 A cylinder has a volume of 200π cm^3 and a height of $h = 8$ cm. Determine the radius r of the cylinder's base. ■

Use the formula for the cylinder's volume $V = \pi r^2 h$ and solve for r:

$$r = \sqrt{\frac{V}{\pi h}}.$$

(R) In advanced calculations, it is common to use **integral calculus** to determine the volume of solids of revolution, which can be considered generalizations of cylinders. This approach allows the calculation of volumes for solids with variable bases and heights.

To deepen the study of volumes, it is useful to explore how integrals enable the calculation of volumes for more complex solids, such as those obtained by rotating a function around an axis.

Proposition 16.1.4 The volume V of a solid of revolution generated by rotating a continuous, non-negative function $f(x)$ over the interval $[a,b]$ around the x-axis is:

$$V = \pi \int_a^b [f(x)]^2 dx.$$

Demostración. By rotating the function $f(x)$ around the x-axis, a solid is generated whose cross-sections perpendicular to the x-axis are circles with radius $f(x)$. The area of each cross-section is $A(x) = \pi[f(x)]^2$. Integrating these areas over $[a,b]$ gives the total volume of the solid. ■

This method is essential for calculating volumes of solids with shapes more complex than conventional prisms and cylinders.

■ **Example 16.3** Calculate the volume of the solid generated by rotating the function $f(x) = \sqrt{x}$ around the x-axis from $x = 0$ to $x = 4$.

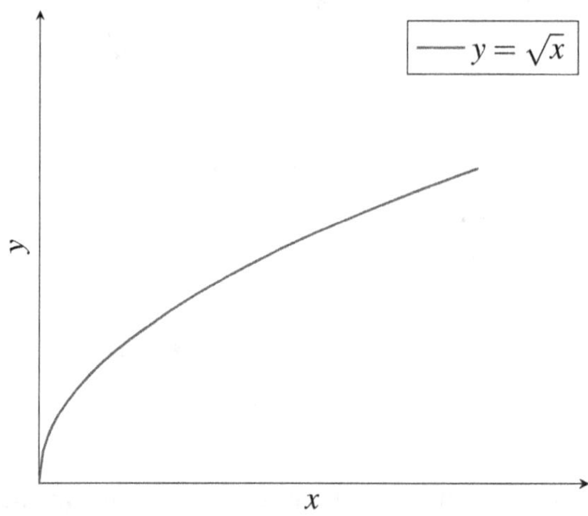

Figura 16.1.3: *Graph of* $y = \sqrt{x}$ *from* $x = 0$ *to* $x = 4$.

■

Demostración. We apply the formula for the volume of a solid of revolution:

$$V = \pi \int_0^4 [\sqrt{x}]^2 dx = \pi \int_0^4 x\,dx = \pi \left[\frac{x^2}{2}\right]_0^4 = \pi \left(\frac{16}{2} - 0\right) = 8\pi.$$

Thus, the volume of the solid is 8π cubic units. ■

This example demonstrates how to integrate concepts of integral calculus in volume calculation, extending the tools available for solving more complex problems.

R A deep understanding of the properties of prisms and cylinders, along with the ability to apply integral calculus techniques, is fundamental for tackling advanced problems in engineering and physics, where solids may have unconventional geometries.

In conclusion, the study of the volume of prisms and cylinders is an essential starting point for understanding more advanced concepts in spatial geometry and calculus, enabling the resolution of a wide variety of problems in mathematics and its applications.

16.1.2 Surface Areas of Cones and Spheres.

The study of surface areas of cones and spheres is fundamental in geometry, as it allows the calculation of physical quantities and the resolution of problems in engineering, physics, and applied mathematics. Below are key definitions and theorems, accompanied by examples and exercises to facilitate the understanding of these concepts.

Definition 16.1.4 A **cone** is a solid of revolution generated by rotating a right triangle about one of its legs. It consists of a circular base and a lateral surface that converges at a point called the **vertex** of the cone.

Definition 16.1.5 A **sphere** is the set of all points in space that are at a fixed distance r, called the **radius**, from a fixed point O, called the **center** of the sphere.

To calculate the surface areas of these solids, it is necessary to understand their geometric properties and apply integration techniques in advanced cases.

Theorem 16.1.5 The **surface area** A of a right cone with height h and base radius r is:

$$A = \pi r(r+s),$$

where $s = \sqrt{r^2+h^2}$ is the **slant height** of the cone.

Demostración. The surface area of a cone consists of two parts: the base area and the lateral area.
Base area:

$$A_{\text{base}} = \pi r^2.$$

Lateral area: The lateral surface can be ünrolledïnto a circular sector with radius s and arc length $l = 2\pi r$. The area of the sector is:

$$A_{\text{lateral}} = \pi r s.$$

Total area:

$$A = A_{\text{base}} + A_{\text{lateral}} = \pi r^2 + \pi r s = \pi r(r+s).$$

■

R The slant height s represents the distance from the vertex of the cone to any point on the circumference of the base. It is the hypotenuse of the right triangle formed by the radius r, the height h, and the slant height s.

Example 16.4 Calculate the surface area of a right cone with base radius $r = 4$ cm and height $h = 3$ cm.

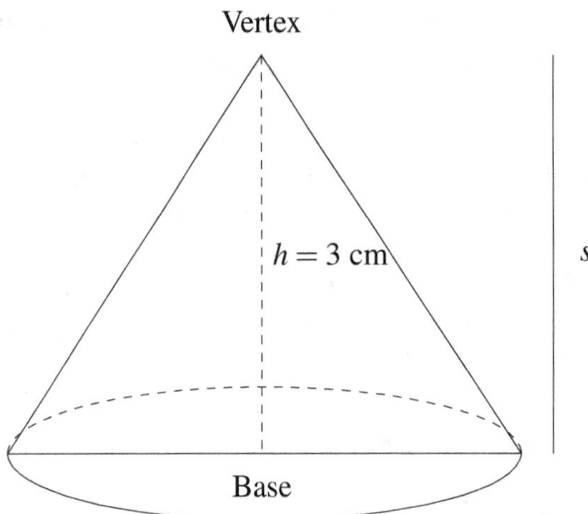

Figura 16.1.4: *Right cone with radius $r = 4$ cm and height $h = 3$ cm.*

■

Demostración. First, calculate the slant height s:

$$s = \sqrt{r^2 + h^2} = \sqrt{4^2 + 3^2} = \sqrt{16 + 9} = \sqrt{25} = 5\,\text{cm}.$$

Now, calculate the surface area:

$$A = \pi r(r + s) = \pi(4)(4 + 5) = \pi(4)(9) = 36\pi\,\text{cm}^2.$$

■

This example demonstrates the direct application of the surface area formula for a right cone. Similarly, we can analyze the sphere and its surface area.

Theorem 16.1.6 The **surface area** A of a sphere with radius r is:

$$A = 4\pi r^2.$$

Demostración. The surface area of a sphere can be derived using integration or by recognizing that it is four times the area of a great circle (maximum circle) of the sphere. Mathematically:

$$A = 4 \times (\text{Area of the great circle}) = 4\pi r^2.$$

Alternatively, using integral calculus, the area is obtained by integrating the infinitesimal rings that form the sphere when a semicircle is rotated around the x-axis:

$$A = 2\pi \int_{-r}^{r} \sqrt{r^2 - x^2}\, dx = 4\pi r^2.$$

■

R The formula $A = 4\pi r^2$ is fundamental in physics, especially in topics related to radiation and flux, where the surface area of a sphere plays a key role in the equations.

Example 16.5 Calculate the surface area of a sphere with radius $r = 7$ m.

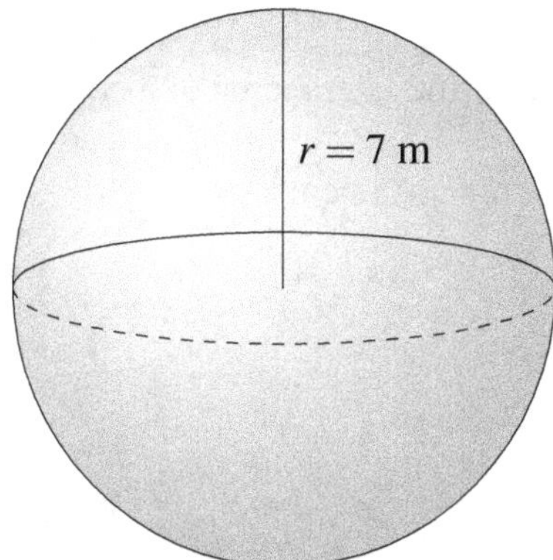

Figura 16.1.5: *Sphere with radius $r = 7$ m.*

■

Demostración. We apply the formula for the surface area of a sphere:

$$A = 4\pi r^2 = 4\pi(7\,\text{m})^2 = 4\pi(49\,\text{m}^2) = 196\pi\,\text{m}^2.$$

■

This example demonstrates how to calculate the surface area of a sphere using the established formula.
It is interesting to note how the surface area formulas for cones and spheres are related to the geometric properties of the solids and how they can be derived through analytical methods.

Lema 16.1.2 The lateral area of a cone can be obtained as the area of a circular sector whose central angle θ is given by:

$$\theta = 2\pi\left(\frac{r}{s}\right),$$

where r is the base radius and s is the slant height.

Demostración. When the lateral surface of the cone is unfolded, it forms a circular sector with radius s and arc length $l = 2\pi r$. The central angle is:

$$\theta = \frac{l}{s} = \frac{2\pi r}{s}.$$

The area of the sector is:

$$A_{\text{lateral}} = \frac{\theta}{2\pi} \times \pi s^2 = \frac{2\pi r}{s} \times \frac{s^2}{2\pi} = \pi r s.$$

■

This lemma provides a geometric interpretation of the lateral area of the cone, relating it to a circular sector.

Exercise 16.3 Determine the lateral area of a right cone with radius $r = 5$ cm and slant height $s = 13$ cm. ■

Exercise 16.4 Calculate the surface area of a sphere with a diameter $d = 10$ cm. ■

of Exercise 1. We use the formula for the lateral area of a cone:

$$A_{\text{lateral}} = \pi r s = \pi(5\,\text{cm})(13\,\text{cm}) = 65\pi\,\text{cm}^2.$$

■

of Exercise 2. First, find the radius:

$$r = \frac{d}{2} = \frac{10\,\text{cm}}{2} = 5\,\text{cm}.$$

Then, calculate the surface area of the sphere:

$$A = 4\pi r^2 = 4\pi(5\,\text{cm})^2 = 4\pi(25\,\text{cm}^2) = 100\pi\,\text{cm}^2.$$

■

(R) In design and construction problems, it is common to require the calculation of surface areas to determine the amount of material needed to cover a solid. The presented formulas are essential for precise estimations.

In conclusion, the study of the surface areas of cones and spheres is crucial in geometry and its applications. The derived formulas and calculation methods allow for the resolution of practical and theoretical problems and establish important connections between different areas of mathematics.

16.2 Calculation of Volumes and Surface Areas: Formulas and Applications.

16.2.1 Applications in Engineering Problems.

The calculation of volumes and surface areas is fundamental in engineering for designing and analyzing structures, tanks, mechanical components, and systems. Precision in these calculations ensures efficiency, safety, and resource optimization in engineering projects.

Definition 16.2.1 A **horizontal cylindrical tank** is a container shaped like a cylinder lying on its longitudinal axis. It is commonly used for storing liquids and gases in chemical, petroleum, and water storage industries.

Calculating the liquid volume in a horizontal cylindrical tank is essential for determining storage capacity and monitoring fill levels.

Theorem 16.2.1 The **liquid volume** V in a horizontal cylindrical tank of radius r, length L, and liquid height h is given by:

$$V = L\left(r^2\cos^{-1}\left(\frac{r-h}{r}\right) - (r-h)\sqrt{2rh-h^2}\right).$$

Demostración. The cross-section of the tank is a circle of radius r. The area A of the circular segment filled with liquid is:

$$A = r^2 \cos^{-1}\left(\frac{r-h}{r}\right) - (r-h)\sqrt{2rh-h^2}.$$

Multiplying the area A by the length L gives the volume V of the liquid in the tank. ■

■ **Example 16.6** Calculate the volume of oil in a horizontal cylindrical tank with radius $r = 3\,\text{m}$, length $L = 10\,\text{m}$, and filled to a height of $h = 4\,\text{m}$.

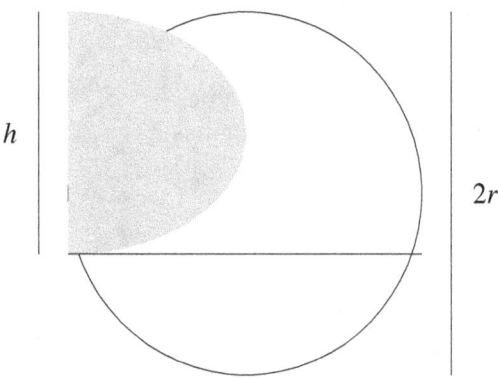

Tank cross-section

Figura 16.2.1: *Cross-section of a horizontal cylindrical tank with liquid height h.*

■

First, calculate $\cos^{-1}\left(\frac{r-h}{r}\right)$:

$$\cos^{-1}\left(\frac{3-4}{3}\right) = \cos^{-1}\left(-\frac{1}{3}\right) \approx 1{,}9106\,\text{rad}.$$

Next, calculate $\sqrt{2rh-h^2}$:

$$\sqrt{2\times 3\times 4 - 4^2} = \sqrt{24-16} = \sqrt{8} \approx 2{,}8284.$$

Now, the area A is:

$$A = 3^2 \times 1{,}9106 - (-1)\times 2{,}8284 = 9\times 1{,}9106 + 2{,}8284 \approx 17{,}1954 + 2{,}8284 = 20{,}0238\,\text{m}^2.$$

The volume V is:

$$V = 10 \times 20{,}0238 = 200{,}238\,\text{m}^3.$$

This calculation is essential for engineers needing to determine the amount of liquid stored in tanks for industrial processes.

Definition 16.2.2 A **helical slide** is a spiral-shaped structure used in water parks and amusement parks. Its design requires precise calculation of areas and volumes to ensure safety and functionality.

The design of a helical slide involves understanding the geometry of a helix and its development in space.

Theorem 16.2.2 The **length of a helix** with radius r, height h, and number of turns n is given by:

$$L = n\sqrt{(2\pi r)^2 + \left(\frac{h}{n}\right)^2}.$$

Demostración. The length of a helix is the length of a curve that advances in space, making turns around an axis while ascending in height. For a complete turn, the length L_1 is:

$$L_1 = \sqrt{(2\pi r)^2 + \left(\frac{h}{n}\right)^2}.$$

Multiplying by the number of turns n, the total length is:

$$L = nL_1 = n\sqrt{(2\pi r)^2 + \left(\frac{h}{n}\right)^2}.$$

■

■ **Example 16.7** Design a helical slide with radius $r = 5\,\text{m}$, total height $h = 15\,\text{m}$, and $n = 3$ turns. Calculate the length of the slide. ■

Calculate the length L:

$$L = 3\sqrt{(2\pi \times 5)^2 + \left(\frac{15}{3}\right)^2} = 3\sqrt{(10\pi)^2 + 5^2} = 3\sqrt{(100\pi^2) + 25}.$$

Simplify:

$$L = 3\sqrt{100\pi^2 + 25} = 3\sqrt{25(4\pi^2 + 1)} = 3 \times 5\sqrt{4\pi^2 + 1} = 15\sqrt{4\pi^2 + 1}.$$

Numerically calculate:

$$L \approx 15\sqrt{4 \times 9{,}8696 + 1} = 15\sqrt{39{,}4784 + 1} = 15\sqrt{40{,}4784} \approx 15 \times 6{,}3640 = 95{,}46\,\text{m}.$$

This result is crucial for determining the amount of material needed and estimating the user's travel time on the slide.

Exercise 16.5 A civil engineer must design a dam in the shape of a rectangular parallelepiped to retain a water volume of $V = 10{,}000\,\text{m}^3$. If the dam's base is a rectangle with dimensions $b = 50\,\text{m}$ and $a = 20\,\text{m}$, determine the required height h. ■

Exercise 16.6 In a factory, a metallic sphere with radius $r = 2\,\text{m}$ needs to be fully covered with a thermal insulation layer of thickness $e = 0{,}1\,\text{m}$. Calculate the volume of insulation material required. ■

(R) The applications of volume and surface area calculations in engineering are vast, spanning structural design, manufacturing processes, and resource optimization. Mathematical precision is essential for innovation and efficiency in engineering solutions.

Understanding and applying these advanced mathematical concepts enables engineers to solve complex problems and develop projects that meet technical and economic requirements.

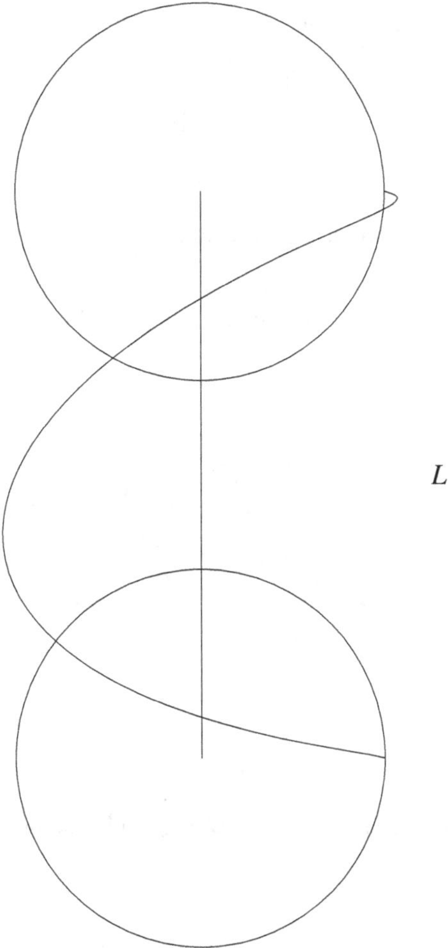

Figura 16.2.2: *Schematic representation of the helical slide.*

16.2.2 Volume Calculation for Irregular Objects.

The calculation of volumes for irregular objects is a fundamental topic in advanced mathematics with direct applications in engineering, physics, and other applied sciences. Unlike solids with simple geometric shapes, irregular objects require more sophisticated techniques, such as integral calculus and numerical methods, to accurately determine their volume.

Definition 16.2.3 An **irregular object** is a solid whose shape cannot be easily described using elementary geometric figures such as prisms, cylinders, cones, or spheres. Its volume cannot be calculated using simple formulas and requires advanced methods like integration or approximation techniques.

To address the volume calculation for irregular objects, it is essential to understand and apply concepts from integral calculus and differential geometry.

Theorem 16.2.3 — **Cavalieri's Principle.** If two solids in space have the same height and all cross-sections at the same height have equal areas, then the solids have equal volumes.

Demostración. Cavalieri's Principle states that the volume of a solid can be determined by considering the areas of its horizontal cross-sections. If for every height h, the cross-sectional areas $A(h)$

of two solids are equal, then integrating these areas over the common height yields equal volumes:

$$V = \int_0^H A(h)\,dh.$$

■

This principle is fundamental for calculating the volumes of irregular objects by comparing them to solids with known volumes.

■ **Example 16.8** Calculate the volume of the solid bounded by the surface $z = x^2 + y^2$ and the plane $z = 4$.

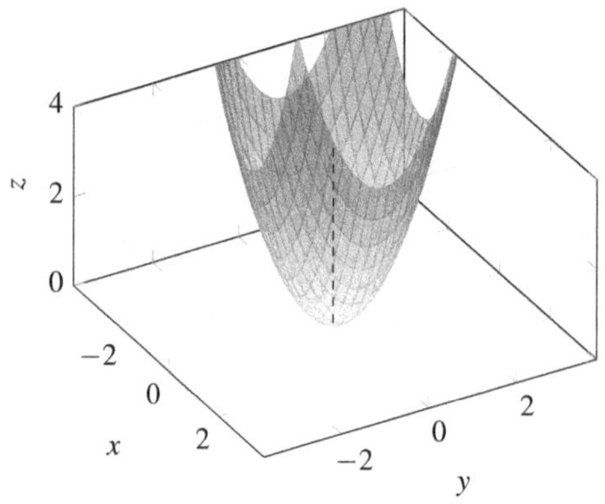

Figura 16.2.3: *Solid bounded by $z = x^2 + y^2$ and $z = 4$.*

■

Demostración. The solid is an upward circular paraboloid truncated at $z = 4$. To calculate the volume, we use cylindrical coordinates. Here, $x^2 + y^2 = r^2$, and the upper bound for z is $z = 4$. The limits of integration are:

$$0 \leq r \leq 2, \quad 0 \leq \theta \leq 2\pi, \quad z = r^2 \leq z \leq 4.$$

The volume V is:

$$V = \int_0^{2\pi} \int_0^2 \int_{r^2}^4 r\,dz\,dr\,d\theta.$$

Integrating with respect to z:

$$V = \int_0^{2\pi} \int_0^2 [z]_{r^2}^4\, r\,dr\,d\theta = \int_0^{2\pi} \int_0^2 (4 - r^2)\,r\,dr\,d\theta.$$

Simplify the integral:

$$V = \int_0^{2\pi} \int_0^2 (4r - r^3)\,dr\,d\theta = \int_0^{2\pi} \left[2r^2 - \frac{r^4}{4}\right]_0^2 d\theta = \int_0^{2\pi} (8-4)\,d\theta = \int_0^{2\pi} 4\,d\theta.$$

Finally, integrate with respect to θ:

$$V = 4\theta\Big|_0^{2\pi} = 4(2\pi - 0) = 8\pi.$$

Thus, the volume of the solid is 8π cubic units. ■

In this example, we used cylindrical coordinates and triple integral calculus to determine the volume of a solid defined by a curved surface and a plane.

Lema 16.2.1 The volume integral in cylindrical coordinates for a solid of revolution around the z-axis is given by:

$$V = \int_{0}^{2\pi} \int_{r_{\min}}^{r_{\max}} \int_{z_{\min}(r)}^{z_{\max}(r)} r\,dz\,dr\,d\theta.$$

Demostración. In cylindrical coordinates, a differential volume element is $dV = r\,dz\,dr\,d\theta$. Integrating over the appropriate limits for r, z, and θ yields the total volume of the solid. ■

This lemma is fundamental for calculating the volumes of solids with rotational symmetry or those that adapt well to cylindrical coordinates.

Theorem 16.2.4 **Pappus's Theorem** states that the volume V of a solid of revolution generated by rotating a plane figure of area A around an axis external to the figure and coplanar with it is:

$$V = A \times d,$$

where d is the distance traveled by the centroid of the figure during the rotation, i.e., $d = 2\pi r$, with r being the distance from the centroid to the axis of rotation.

Demostración. Pappus's Theorem is proven by considering that the volume of the solid of revolution is equivalent to the area of the plane figure multiplied by the circular trajectory traveled by its centroid when rotated 360 degrees about the given axis. ■

This theorem is particularly useful for calculating the volumes of irregular solids generated by rotation, where the area and centroid of the plane figure are known or can be easily determined.

■ **Example 16.9** Calculate the volume of the solid generated by rotating the region under the curve $y = \sqrt{x}$ from $x = 0$ to $x = 4$ around the x-axis.

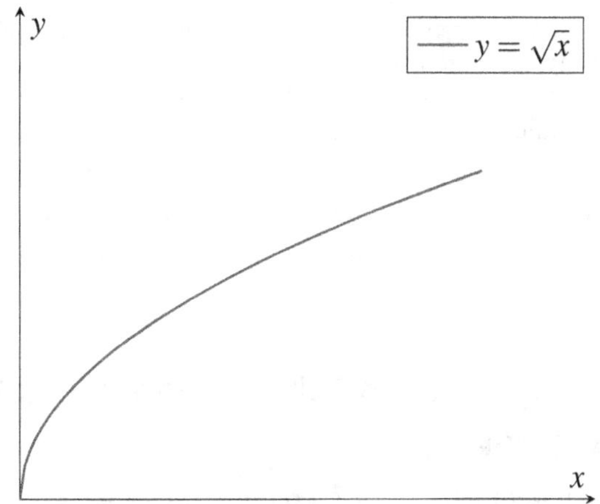

Figura 16.2.4: *Region under the curve $y = \sqrt{x}$ from $x = 0$ to $x = 4$.*

■

Demostración. First, calculate the area A under the curve:

$$A = \int_{0}^{4} \sqrt{x}\,dx = \frac{2}{3}x^{3/2}\Big|_{0}^{4} = \frac{2}{3}(4)^{3/2} - 0 = \frac{2}{3}(8) = \frac{16}{3}.$$

The centroid $(\bar{x},\bar{y})$ of the region under the curve $y=\sqrt{x}$ is:

$$\bar{x}=\frac{\int_a^b xf(x)\,dx}{A},\quad \bar{y}=\frac{\int_a^b \frac{1}{2}[f(x)]^2\,dx}{A}.$$

Calculate $\bar{x}$:

$$\bar{x}=\frac{\int_0^4 x\sqrt{x}\,dx}{\frac{16}{3}}=\frac{\int_0^4 x^{3/2}\,dx}{\frac{16}{3}}=\frac{\frac{2}{5}x^{5/2}\Big|_0^4}{\frac{16}{3}}=\frac{\frac{2}{5}(4)^{5/2}}{\frac{16}{3}}=\frac{\frac{2}{5}(32)}{\frac{16}{3}}=\frac{\frac{64}{5}}{\frac{16}{3}}=\frac{64\times 3}{5\times 16}=\frac{192}{80}=\frac{12}{5}=2{,}4.$$

The distance traveled by the centroid during the rotation about the x-axis is $d=2\pi\bar{y}$. However, as the area lies under the curve and is rotated about the x-axis, the centroid moves along a straight line without traversing a distance. Therefore, this method is not applicable here. Instead, we use the disk method to calculate the volume.
The volume V is:

$$V=\pi\int_0^4[\sqrt{x}]^2\,dx=\pi\int_0^4 x\,dx=\pi\left.\frac{x^2}{2}\right|_0^4=\pi\left(\frac{16}{2}-0\right)=8\pi.$$

Thus, the volume of the solid is 8π cubic units. ■

This example demonstrates that, in certain cases, it is more appropriate to use direct integration methods rather than Pappus's Theorem.

Exercise 16.7 Calculate the volume of the solid obtained by rotating around the y-axis the region bounded by $x=y^2$ and $y=x-2$. ■

Exercise 16.8 Determine the volume of the solid formed by rotating the region enclosed by the curves $y=\ln x$, $y=0$, $x=1$, and $x=e$ around the x-axis. ■

(R) The calculation of volumes for irregular objects is essential in fields such as civil engineering, mechanical engineering, and architecture, where structures and components often have complex shapes. The integral techniques and theorems presented allow these problems to be addressed with mathematical rigor.

A deep understanding of these methods and the ability to apply them in diverse situations are key skills in advanced mathematical reasoning, crucial for students and professionals in mathematics and related fields.

16.3 Practical Problems: Using Solids to Solve Everyday Challenges

16.3.1 Applications in Packing Problems.

Packing is a field of study in geometry focused on determining the best way to arrange objects within a given space to maximize space utilization and minimize waste. This topic is fundamental in logistics, material design, communications, and other areas where spatial optimization is crucial.

Definition 16.3.1 A **geometric packing** is an arrangement of geometric figures within a defined space such that they do not overlap and optimize a given criterion, such as density (the proportion of occupied space) or the number of placed objects.

A classic example is the packing of circles in a plane or spheres in three-dimensional space.

Theorem 16.3.1 The **maximum density** of a packing of equal circles in a plane is $\frac{\pi}{2\sqrt{3}} \approx$ 0,9069, meaning approximately 90,69 % of the plane can be covered by circles in an optimal packing.

Demostración. The densest packing of equal circles in a plane is the *hexagonal packing*, where each circle is surrounded by six circles in a honeycomb-like arrangement. The density δ of this packing is calculated as:

$$\delta = \frac{\text{Area of a circle}}{\text{Area of a hexagonal cell}} = \frac{\pi r^2}{2\sqrt{3}r^2} = \frac{\pi}{2\sqrt{3}} \approx 0{,}9069.$$

■

This result is fundamental in packing theory and has practical applications in space and material optimization.

■ **Example 16.10** Determine how many cylindrical cans with a diameter $d = 6$ cm and height $h = 12$ cm can be packed into a rectangular box with dimensions $L = 30$ cm, $W = 24$ cm, and $H = 12$ cm using **hexagonal packing**.

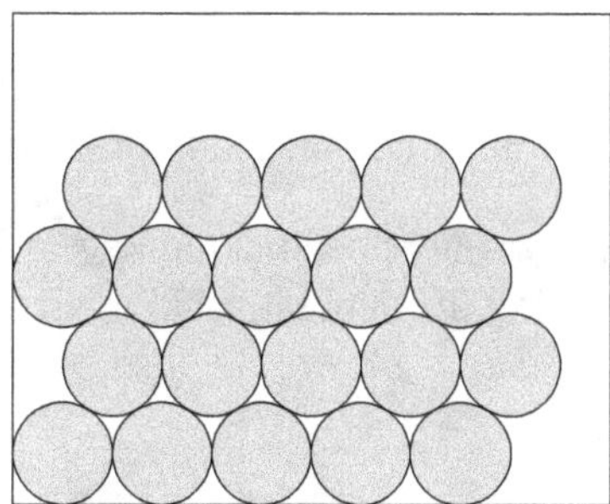

Figura 16.3.1: *Hexagonal packing of cylinders in a rectangular box.*

■

Step 1: Calculate the number of cylinders in the horizontal direction.
The distance between the centers of the cylinders horizontally is equal to the diameter $d = 6$ cm.
Maximum number of cylinders horizontally:

$$n_x = \left\lfloor \frac{L}{d} \right\rfloor = \left\lfloor \frac{30}{6} \right\rfloor = 5.$$

Step 2: Calculate the vertical distance between rows.
In hexagonal packing, the vertical distance between rows is:

$$\Delta y = d \times \frac{\sqrt{3}}{2} = 6 \times \frac{\sqrt{3}}{2} \approx 5{,}196 \text{ cm}.$$

Step 3: Calculate the number of rows that fit vertically.
Maximum number of rows:

$$n_y = \left\lfloor \frac{W - \frac{d}{2}}{\Delta y} \right\rfloor + 1 = \left\lfloor \frac{24-3}{5{,}196} \right\rfloor + 1 = \left\lfloor \frac{21}{5{,}196} \right\rfloor + 1 = 5.$$

Step 4: Calculate the total number of cylinders.

Odd rows contain $n_x = 5$ cylinders, and even rows contain $n_x - 1 = 4$ cylinders.
Total number of cylinders:

$$\text{Total} = \left\lfloor \frac{n_y}{2} \right\rfloor (5+4) + (n_y \mod 2) \times 5 = 2 \times 9 + 1 \times 5 = 18 + 5 = 23.$$

Thus, up to **23 cans** can be packed into the box using hexagonal packing.
This example illustrates how to apply hexagonal packing to maximize the number of objects in a limited space.

Lema 16.3.1 The **maximum density** of a packing of equal spheres in three-dimensional space is $\frac{\pi}{3\sqrt{2}} \approx 0{,}7405$, meaning approximately 74,05 % of the space can be occupied by spheres in an optimal packing.

Demostración. The densest packing of equal spheres in three-dimensional space is the *close-packed arrangement*, which can be face-centered cubic (FCC) or hexagonal close-packed (HCP). The density δ of this packing is calculated as:

$$\delta = \frac{\pi}{3\sqrt{2}} \approx 0{,}7405.$$

■

This result is significant in crystallography and the study of solid material structures.

Exercise 16.9 Calculate how many spheres with a radius $r = 3$ cm can be packed into a cubic box with side length $L = 18$ cm using the densest packing arrangement.

Exercise 16.10 Determine the maximum number of cylindrical cans with a diameter $d = 5$ cm and height $h = 12$ cm that can be packed into a cylindrical container with diameter $D = 30$ cm and height $H = 24$ cm.

R Packing problems are critical in logistics, packaging design, and storage, where optimizing space usage is essential to reduce costs and improve efficiency.

In conclusion, the study of packing problems combines concepts from geometry, optimization, and number theory and has practical applications in various fields of science and engineering. Understanding these principles allows for the design of efficient solutions for optimal space utilization.

16.3.2 The Use of Solids in Object Design.

Object design is a field that combines aesthetics with functionality, where the geometry of solids plays a fundamental role. Understanding the mathematical properties of solids allows designers to create efficient, attractive, and functional objects. This section explores how geometric solids are applied in object design, incorporating advanced mathematical concepts to optimize and enhance designs.

Definition 16.3.2 A **geometric model** is a mathematical representation of a physical object using basic geometric shapes, such as solids, surfaces, and curves, to describe its structure and properties.

Geometric models are essential in object design, as they enable the analysis and modification of an object's characteristics before production, facilitating optimization and problem detection.

Theorem 16.3.2 Any three-dimensional object can be approximated through the combination and transformation of basic geometric solids using operations of **union**, **intersection**, and **difference**, known as Boolean operations.

Demostración. In computational geometry and computer-aided design (CAD), complex objects are constructed from primitive solids (such as prisms, cylinders, cones, and spheres) by applying Boolean operations. Through union, intersection, and difference, it is possible to combine these solids to approximate any three-dimensional shape with the desired precision. ■

This theorem is fundamental in 3D modeling, enabling the creation of complex designs from simple shapes.

■ **Example 16.11** Design a vase as a solid of revolution generated by rotating the curve defined by $y = x^3 - 3x$ about the y-axis between $x = -2$ and $x = 2$.

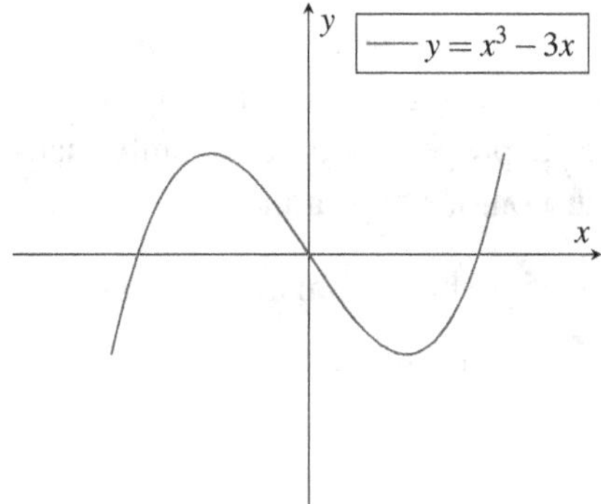

Figura 16.3.2: *Curve to be rotated to generate the vase.*

■

Demostración. To design the vase, we rotate the curve $y = x^3 - 3x$ about the y-axis. This generates a solid of revolution whose surface and volume can be calculated using integrals.
The volume V of the vase is:

$$V = 2\pi \int_{y_{\min}}^{y_{\max}} x(y)A(y)\,dy,$$

where $x(y)$ is the inverse of the function $y = x^3 - 3x$, and $A(y)$ is the differential area at height y. However, due to the complexity of the inverse function, it is more practical to parametrize the curve. Parametrize x in terms of t:

$$x(t) = t, \quad y(t) = t^3 - 3t, \quad t \in [-2, 2].$$

The volume is calculated using:

$$V = 2\pi \int_{-2}^{2} x(t) \left| \frac{dy}{dt} \right| dt = 2\pi \int_{-2}^{2} t|3t^2 - 3|dt.$$

Calculate $\frac{dy}{dt}$:

$$\frac{dy}{dt} = 3t^2 - 3.$$

Take the absolute value because the volume must be positive.
Finally, evaluate the integral:

$$V = 2\pi \int_{-2}^{2} t|3t^2 - 3|dt.$$

This integral can be solved piecewise considering the intervals where $3t^2 - 3$ changes sign. ■

This example demonstrates how to use solids of revolution and integral calculus to design aesthetically and functionally pleasing objects.

Proposition 16.3.3 The use of **symmetry** in object design reduces modeling complexity and can improve the structural strength and aesthetics of the object.

Demostración. Symmetry allows a part of the object to represent the whole through operations of reflection, rotation, or translation. Mathematically, if an object is symmetric with respect to a plane or axis, its structural and geometric properties are uniformly distributed, which can lead to better stress distribution and a more harmonious appearance. ■

Symmetry is a powerful tool in design, both from aesthetic and functional perspectives.

Lema 16.3.2 Using the **golden ratio** in the dimensions of an object can positively influence its aesthetic perception.

Demostración. The golden ratio $\phi = \frac{1+\sqrt{5}}{2} \approx 1{,}618$ has been historically used in art and architecture for its pleasing aesthetic effect. In object design, adjusting key dimensions to be in the golden ratio can enhance the visual appeal of the object. Mathematically, if the ratio between two dimensions a and b is $\frac{a}{b} = \phi$, they are said to be in the golden ratio. ■

16.3.3 The Use of Solids in Object Design.

Object design combines aesthetics with functionality, where the geometry of solids plays a crucial role. Understanding the mathematical properties of solids enables designers to create efficient, attractive, and functional objects. This section demonstrates how geometric solids are applied in object design, using advanced mathematical concepts to optimize and enhance designs.

■ **Example 16.12** Design a rectangular screen where the ratio between the length L and the width W is the golden ratio. If the width is $W = 24$ cm, determine the length L.

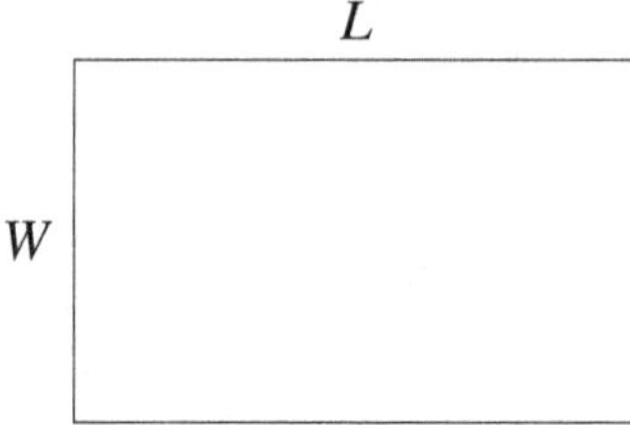

Figura 16.3.3: *Screen with the golden ratio.*

■

Demostración. We apply the golden ratio:

$$\frac{L}{W} = \phi \implies L = \phi W = 1{,}618 \times 24\,\text{cm} \approx 38{,}83\,\text{cm}.$$

Thus, the length of the screen should be approximately 38,83 cm to achieve the golden ratio. ■

This example illustrates how mathematical relationships can be applied in the design of everyday objects to enhance their aesthetics.

Theorem 16.3.4 Material optimization in object design can be achieved by minimizing the **surface-to-volume ratio** (SVR) of the object.

Demostración. The surface-to-volume ratio is given by $\text{SVR} = \frac{S}{V}$, where S is the surface area, and V is the volume of the object. Minimizing the SVR reduces the amount of material needed to cover or construct the object relative to the space it occupies. Mathematically, for a given object, finding the shape that minimizes S for a fixed V is a problem of calculus of variations and optimization. For example, the sphere minimizes the SVR for a given volume, as demonstrated by the principle that among all solids of equal volume, the sphere has the smallest surface area. ■

This theorem is fundamental in designing containers and structures where efficient material usage is a priority.

Exercise 16.11 Design a closed cylindrical tank to hold a volume of $V = 1000$ liters ($1\ \text{m}^3$). Determine the dimensions (radius r and height h) that minimize the total surface area of the cylinder. ■

Formulate the optimization problem by minimizing the surface area $S = 2\pi rh + 2\pi r^2$ subject to the volume constraint $V = \pi r^2 h = 1\ \text{m}^3$.
Use the method of Lagrange multipliers or express h in terms of r from the volume constraint and minimize $S(r)$.

of Exercise. First, express h in terms of r:

$$V = \pi r^2 h \implies h = \frac{V}{\pi r^2} = \frac{1}{\pi r^2}.$$

Now, express the surface area S in terms of r:

$$S(r) = 2\pi rh + 2\pi r^2 = 2\pi r\left(\frac{1}{\pi r^2}\right) + 2\pi r^2 = \frac{2}{r} + 2\pi r^2.$$

To minimize $S(r)$, differentiate with respect to r and set the derivative equal to zero:

$$S'(r) = -\frac{2}{r^2} + 4\pi r = 0.$$

Multiply through by r^2:

$$-2 + 4\pi r^3 = 0 \implies 4\pi r^3 = 2 \implies r^3 = \frac{1}{2\pi} \implies r = \left(\frac{1}{2\pi}\right)^{1/3}.$$

Calculate r:

$$r \approx \left(\frac{1}{2\pi}\right)^{1/3} \approx 0{,}535\,\text{m}.$$

Now, calculate h:

$$h = \frac{1}{\pi r^2} = \frac{1}{\pi(0{,}535)^2} \approx \frac{1}{\pi \times 0{,}286} \approx 1{,}114\,\text{m}.$$

Thus, the dimensions that minimize the surface area are approximately $r = 0{,}535$ m and $h = 1{,}114$ m. ■

This exercise demonstrates how to apply mathematical optimization techniques in object design to minimize material usage.

(R) Object design requires the integration of advanced mathematical concepts, including geometry, calculus, and optimization. The ability to mathematically model objects and analyze their properties enables the creation of more efficient, aesthetic, and functional designs.

In conclusion, the use of solids in object design is a direct application of mathematics to the real world. Through understanding and applying geometric and analytical concepts, it is possible to innovate and improve designs, optimizing resources and creating objects that meet both functional and aesthetic needs.

16.4 Solved Exercises

Exercise 16.12 Calculate the volume of a right triangular prism whose base is an equilateral triangle with side $l = 6$ cm and height of the prism $h = 10$ cm. ■

Demostración. First, calculate the area of the base of the equilateral triangle. The formula for the area A of an equilateral triangle with side l is:

$$A = \frac{\sqrt{3}}{4} l^2.$$

Substituting $l = 6$ cm:

$$A = \frac{\sqrt{3}}{4} \times (6)^2 = \frac{\sqrt{3}}{4} \times 36 = 9\sqrt{3}\,\text{cm}^2.$$

The volume V of the prism is the area of the base multiplied by the height:

$$V = A \times h = 9\sqrt{3} \times 10 = 90\sqrt{3}\,\text{cm}^3.$$

■

Exercise 16.13 Calculate the surface area of a right cone with a base radius $r = 4$ cm and height $h = 3$ cm. ■

Demostración. First, calculate the slant height s of the cone using the Pythagorean theorem:

$$s = \sqrt{r^2+h^2} = \sqrt{4^2+3^2} = \sqrt{16+9} = \sqrt{25} = 5\,\text{cm}.$$

The total surface area A of the cone includes the base area and the lateral area:

$$A = \pi r^2 + \pi r s = \pi(4)^2 + \pi(4)(5) = 16\pi + 20\pi = 36\pi\,\text{cm}^2.$$

■

Exercise 16.14 Determine the volume of a sphere with radius $r = 7$ m. ■

Demostración. The formula for the volume V of a sphere is:

$$V = \frac{4}{3}\pi r^3.$$

Substituting $r = 7$ m:

$$V = \frac{4}{3}\pi(7)^3 = \frac{4}{3}\pi \times 343 = \frac{1372}{3}\pi \approx 457{,}33\pi\,\text{m}^3.$$

■

Exercise 16.15 Calculate the lateral area of a cylinder with radius $r = 5$ cm and height $h = 12$ cm. ■

Demostración. The lateral area A_{lateral} of a cylinder is calculated using the formula:

$$A_{\text{lateral}} = 2\pi r h.$$

Substituting $r = 5$ cm and $h = 12$ cm:

$$A_{\text{lateral}} = 2\pi(5)(12) = 120\pi\,\text{cm}^2.$$

■

Exercise 16.16 Calculate the volume of a cylinder with radius $r = 4$ cm and height $h = 10$ cm. ■

Demostración. The formula for the volume V of a cylinder is:

$$V = \pi r^2 h.$$

Substituting $r = 4$ cm and $h = 10$ cm:

$$V = \pi(4)^2 \times 10 = \pi \times 16 \times 10 = 160\pi\,\text{cm}^3.$$

■

16.5 Proposed Exercises

16.5.1 Geometric Solids: Prisms, Cylinders, Cones, and Spheres

Exercise 16.17 Calculate the volume of a quadrangular prism whose base is a square with side $l = 8$ cm and height $h = 15$ cm. ■

Exercise 16.18 Determine the surface area of a cylinder with radius $r = 5$ cm and height $h = 12$ cm. ■

Exercise 16.19 Calculate the volume of a right cone with radius $r = 6$ cm and height $h = 9$ cm. ■

Exercise 16.20 A sphere has a diameter of 10 cm. Calculate its surface area and volume. ■

Exercise 16.21 Determine the volume of a triangular prism whose base is an equilateral triangle with side $l = 7$ cm and the prism's height is $h = 10$ cm. ■

16.5.2 Volume and Surface Area Calculations: Formulas and Applications

Exercise 16.22 Calculate the volume of a cylinder with radius $r = 3$ m and height $h = 5$ m, and express the result in terms of π. ■

Exercise 16.23 Determine the lateral area of a right cone whose base has a radius of 4 cm and height of 9 cm. ■

Exercise 16.24 Calculate the volume of a sphere with radius $r = 6$ cm and round the result to two decimal places. ■

Exercise 16.25 A cylinder has a lateral area of 150π cm^2 and a height of $h = 15$ cm. Determine the radius of the base of the cylinder. ■

Exercise 16.26 Calculate the volume of a pentagonal prism whose height is $h = 8$ cm, and each side of the pentagonal base measures 5 cm. (Consider the base as a regular pentagon). ■

16.5.3 Practical Problems: Using Solids to Solve Everyday Problems

Exercise 16.27 A company wants to construct a cylindrical storage tank with a radius of 2 m and a height of 6 m. Calculate the volume of the tank in cubic meters. ■

Exercise 16.28 Determine how many juice cans, each shaped like a cylinder with a diameter of 5 cm and height of 12 cm, can fit in a rectangular box measuring 30 cm in length, 20 cm in width, and 15 cm in height. ■

Exercise 16.29 A metal sphere has a radius of 10 cm. If it is melted to create small cones, each with a base radius of 2 cm and a height of 4 cm, how many cones can be formed? ■

Exercise 16.30 A swimming pool is cylindrical with a radius of 4 m and a depth of 1.5 m. Calculate how many liters of water are needed to fill it completely. (Note: 1 m^3 = 1000 liters). ■

Exercise 16.31 Calculate the surface area of a grain silo with a cylindrical shape, having a radius of 3 m and a height of 12 m. Include the surface area of the top and base. ■

Índice alfabético

Symbols

A

B

C

D

E

F

G

H

I

L

T

U

V

W

X

Y

LOGICAL MATHEMATICAL REASONING

Helbert Justo Luque Zevallos

YEAR 2024

First Edition
ISBN:9798303552661

1 Series: Bachelor's Degree in Mathematics

- Basic Mathematics
- Logical Mathematical Reasoning
- Mathematical Analysis I
- Algebra
- Statistics and Probability
- Mathematical Analysis II
- Linear Algebra I
- Statistical Inference
- Real Analysis I
- Numerical Analysis
- Linear Algebra II
- Algebraic Structures
- Topology
- Real Analysis II
- Ordinary Differential Equations
- Linear Optimization
- Partial Differential Equations
- Introduction to Hyperbolic Geometry
- Galois Theory
- Numerical Methods for Solving Differential Equations
- Measure and Integration
- Nonlinear Optimization
- Qualitative Theory
- Functional Analysis
- Differential Geometry I
- Introduction to Algebraic Topology
- Differentiable Manifolds
- Introduction to Variational Methods for Differential Equations
- Introduction to Differential Topology
- Minimal Surfaces I
- Differential Geometry II
- Introduction to the Finite Element Method
- Introduction to the Geometry of Differential Forms

www.ingramcontent.com/pod-product-compliance
Lightning Source LLC
Chambersburg PA
CBHW082243220526
45469CB00009B/2860

9798303552661